Troubleshooting & Repairing Diesel Engines

Troubleshooting & Repairing Diesel Engines

3rd Edition

Paul Dempsey

TAB Books
Division of McGraw-Hill

New York San Francisco Washington, D.C. Auckland Bogotá
Caracas Lisbon London Madrid Mexico City Milan
Montreal New Delhi San Juan Singapore
Sydney Tokyo Toronto

©1995 by **Paul Dempsey**.
Published by TAB Books, a division of McGraw-Hill, Inc.

pbk   5 6 7 8 9 10 FGR/FGR 9 9 8

Library of Congress Cataloging-in-Publication Data
Dempsey, Paul.
    Troubleshooting and repairing diesel engines / Paul K. Dempsey. —
3rd ed.
        p.   cm.
    ISBN 0-07-016348-0
    1. Diesel motor—Maintenance and repair.   I. Title.
TJ799.D44 1994   94-37456
621.43'6'0288—dc20   CIP

Acquisitions editor: April Nolan
Editorial team: David M. McCandless, Managing Editor
               Melanie Holscher, Book Editor
               Joanne M. Slike, Executive Editor
Production team: Katherine G. Brown, Director
               Susan E. Hansford, Coding
               Ollie Harmon, Coding
               Brenda Wilhide, Computer Artist
               Rose McFarland, Desktop Operator
               Lorie L. White, Proofreading
               Joann Woy, Indexer   0163480
Designer: Jaclyn J. Boone   EL1

Contents

Preface

Attached to diesel engines is a certain mystique that makes owners and mechanics alike call for professional help at the first sign of trouble. There is, in fact, something intimidating about an engine that has no visible means of ignition, the torque characteristics of a bull ox, and fuel-system tolerances expressed as wavelengths of light.

Yet, working on diesels is no more difficult than servicing the current crop of gasoline engines. In some ways, the diesel is an easier nut to crack—symptoms of failure are less ambiguous, specifications are more complete, and the quality of design and materials is generally superior.

The purpose of this book is to ease the transition from gasoline to diesel engines for mechanics and those owners who want to take a hand in keeping their engines running. Nothing in human experience quite compares to the frustration created by an engine that refuses to start. By the same token, no music is sweeter than the sound of an engine that you have just repaired.

Emphasis is on the uniquely diesel aspects of the technology—diagnostics, fuel systems, turbo charging, and the kind of major engine work not often languished on gasoline engines. The book also discusses fairly esoteric subjects, such as injection-pump rebuilding and dynamic balancing. You might not actually perform this kind of specialized work, but you might be asked to pay for it, so you should know how to evaluate its quality.

I have searched hard for a substitute for understanding but have yet to find one—hence, the emphasis here on how things work. What is not understood cannot be fixed, except by accident or through an enormous expenditure of time and parts. Consequently, I have combined "how-to" information with theory. The recipes change with engine make and model; however, the theory has currency for all.

Fleet operators and supervisors will find useful information on the new technology of computerized maintenance management. The effectiveness of these programs is well documented. One program I am acquainted with reduced maintenance costs by 70% and added 10,000 hours to the working hours at the EMD engine. In one typical application, computer-orchestrated maintenance reduces costs by 70% and added 10,000 hours to the working lives of EMD engines.

The computer revolution has also impacted truck, bus, some marine, and many stationary engines. First encounters with computer-controlled engines can send technicians into culture shock. None of the old rules apply, or that is the way it seems. Actually these "green" engines are still diesels and subject to all the ills that compression ignition is heir to. But the control hardware is electronic, and that is where new skills and new ways of thinking are required. Chapter 13 is devoted to making this transition easier.

1
CHAPTER

Rudolf Diesel

Rudolf Diesel was born of German parentage in Paris in 1858. His father was a self-employed leather worker who, by all accounts, managed to provide only a meager income for his wife and three children. Their stay in the City of Light was punctuated by frequent moves from one shabby flat to another. Upon the outbreak of the Franco-Prussian War in 1870, the family became political undesirables and was forced to emigrate to England. Work was almost impossible to find, and in desperation, Rudolf's parents sent the boy to Augsburg to live with an uncle. There he was enrolled in school.

Diesel's natural bent was for mathematics and mechanics. He graduated at the head of his class, and on the basis of his teachers' recommendations and a personal interview by the Bavarian director of education, he received a scholarship to the prestigious Polytechnikum in Munich.

His professor of theoretical engineering was the renowned Carl von Linde, who invented the ammonia refrigeration machine and devised the first practical method of liquefying air. Linde was an authority on thermodynamics and high-compression phenomena. During one of his lectures he remarked that the steam engine had a thermal efficiency of 6–10%; that is, one-tenth or less of the heat energy of its fuel was used to turn the crankshaft, and the rest was wasted. Diesel made special note of this fact. In 1879 he asked himself whether heat could not be directly converted into mechanical energy instead of first passing through a working fluid such as steam.

On the final examination at the Polytechnikum, Diesel achieved the highest honors yet attained at the school. Professor Linde arranged a position for the young diploma engineer in Paris, where, in few months, he was promoted to general manager of the city's first ice-making plant. Soon he took charge of distribution of Linde machines over southern Europe.

By the time he was thirty, Diesel had married, fathered three children, and was recognized throughout the European scientific community as one of the most gifted engineers of the period. He presented a paper at the Universal Exposition held in Paris in 1889—the only German so honored. When he received the first of several citations of merit from a German university, he announced wryly in his acceptance speech: "I am an iceman . . ."

The basis of this acclaim was his preeminence in the new technology of refrigeration, his several patents, and a certain indefinable air about the young man that marked him as extraordinary. He had a shy, self-deprecating humor and an absolute passion for factuality. Diesel could be abrupt when faced with incompetence and was described by relatives as "proud." At the same time he was sympathetic to his workers and made friends among them. It was not unusual for Diesel to wear the blue cotton twill that was the symbol of manual labor in the machine trades.

He had been granted several patents for a method of producing clear ice, which, because it looked like natural ice, was much in demand by the upper classes. Professor Linde did not approve of such frivolity, and Diesel turned to more serious concerns. He spent several years in Paris, working on an ammonia engine, but in the end was defeated by the corrosive nature of this gas at pressure and high temperatures.

The theoretical basis of this research was a paper published by N.L.S. Carnot in 1824. Carnot set himself to the problem of determining how much work could be accomplished by a heat engine employing repeatable cycles. He conceived the engine drawn in Fig. 1-1. Body 1 supplies the heat; it can be a boiler or other heat exchanger. The piston is at position C in the drawing. As the air is heated, it expands in correspondence to Boyle's law. If we assume a frictionless engine, its temperature will not rise. Instead, expansion will take place, driving the piston to D. Then A is removed, and the piston continues to lift to E. At this point the temperature of the air falls until it exactly matches cold surface 2 (which can be a radiator or cooling tank). The air column is now placed in contact with 2, and the piston falls because the air is compressed. Note, however, that the temperature of the air does not change. At B cold body 2 is removed, and the piston falls to A. During this phase the air gains temperature until it is equal to 2. The piston climbs back into the cylinder.

The temperature of the air, and consequently the pressure, is higher during expansion than during compression. Because the pressure is greater during expansion, the power produced by the expansion is greater than that consumed by the compression. The net result is a power output that is available for driving other machinery.

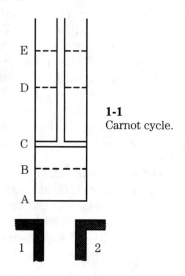

1-1
Carnot cycle.

Of course this is an "ideal" cycle. It does not take into account mechanical friction nor transfer of heat from the air to the piston and cylinder walls. The infinitesimal difference of heat between 1 and 2 is sufficient to establish a gradient and drive the engine. It would be completely efficient.

In 1892 and 1893 Diesel obtained patent specifications from the German government covering his concept for a new type of *Verbrennungskraftmaschinen*, or heat engine. The next step was to build one. At the insistence of his wife, he published his ideas in a pamphlet and was able to interest the leading Augsburg engine builder in the idea. A few weeks later the giant Krupp concern opened negotiations. With typical internationalism he signed another contract with the Sulzer Brothers of Switzerland.

The engine envisioned in the pamphlet and protected by the patent specifications had these characteristics:

- *Compression of air prior to fuel delivery.* The compression was to be adiabatic; that is, no heat would be lost to the piston crown or cylinder head during this process.
- *Metered delivery of fuel so compression pressures would not be raised by combustion temperatures.* The engine would operate on a *constant-pressure* cycle; expanding gases would keep precisely in step with the falling piston. This is a salient characteristic of Carnot's ideal gas cycle, and stands in contrast to the Otto cycle, in which combustion pressures rise so quickly upon spark ignition that we describe it as a *constant-volume* engine.
- *Adiabatic expansion.*
- *Instantaneous exhaust at constant volume.*

It is obvious that Diesel did not expect a working engine to attain these specifications. Adiabatic compression and exhaust phases are, by definition, impossible unless the engine metal is at combustion temperature. Likewise, fuel metering cannot be so precise as to limit combustion pressures to compression levels. Nor can a cylinder be vented instantaneously. But these specifications are significant in that they demonstrate an approach to invention. The rationale of the diesel engine was to save fuel by as close an approximation to the Carnot cycle as materials would allow. The steam, or *Rankine cycle*, engine was abysmal in this regard; and the *Otto* four-stroke-cycle spark or hot-tube-ignition engine was only marginally better.

This approach, from the mathematically ideal to materially practical, is exactly the reverse of the one favored by inventors of the Edison, Westinghouse, and Kettering school. When Diesel visited America in 1912, Thomas A. Edison explained to the young inventor that these men worked *inductively*, from the existing technology, and not *deductively*, from some ideal or model. Diesel felt that such procedure was at best haphazard, even though the results of Edison and other inventors of the inductive school were obviously among the most important. Diesel believed that productivity should be measured by some absolute scientific standard.

The first Diesel engine was a single-cylinder four-cycle design, operated by gasoline vapor. The vapor was sprayed into the cylinder near top dead center by means of an air compressor. The engine was in operation in July of 1893. However, it was discovered that a misreading of the blueprints had caused an increase in the size of the chamber. This was corrected with a new piston, and the engine was connected

to a pressure gauge. The gauge showed approximately 80 atmospheres before it shattered, spraying the room with brass and glass fragments. The best output of what Diesel called his "black mistress" was slightly more than 2 hp—not enough power to overcome friction and compression losses. Consequently, the engine was redesigned.

The second model was tested at the end of 1894. It featured a variable-displacement fuel pump to match engine speed with load. In February of the next year, the mechanic Linder noted a remarkable development. The engine had been sputtering along, driven by a belt from the shop power plant, but Linder noticed that the driving side of the belt was slack, indicating that the engine was putting power into the system. For the first time the Diesel engine ran on its own.

Careful tests—and Diesel was nothing if not careful and methodical—showed that combustion was irregular. The next few months were devoted to redesigning the nozzle and delivery system. This did not help, and in what might have been a fit of desperation, Diesel called upon Robert Bosch for an ignition magneto. Bosch personally fitted one of his low-tension devices to the engine, but it had little effect on the combustion problem. Progress came about by varying the amount of air injected with the fuel, which, at this time, was limited to kerosene or gasoline.

A third engine was built with a smaller stroke/bore ratio and fitted with two injectors. One delivered liquid fuel, the other a mixture of fuel and air. This was quite successful, producing 25 hp at 200 rpm. It was several times as efficient as the first model. Further modifications of the injector, piston, and lubrication system ensued, and the engine was deemed ready for series production at the end of 1896.

Diesel turned his attention to his family, music, and photography. Money began to pour in from the patent licensees and newly organized consortiums wanting to build engines in France, England, and Russia. The American brewer Adolphus Busch purchased the first commercial engine, similar to the one on display at the Budweiser plant in St. Louis today. He acquired the American patent rights for one million marks, which at the current exchange rate amounted to a quarter of a million dollars—more than Diesel had hoped for.

The next stage of development centered around various fuels. Diesel was already an expert on petroleum, having researched the subject thoroughly in Paris in an attempt to refine it by extreme cold. It soon became apparent that the engine could be adapted to run on almost any hydrocarbon from gasoline to peanut oil. Scottish and French engines routinely ran on shale oil, while those sold to the Nobel combine in Russia operated well on refinery tailings. In a search for the ultimate fuel, Diesel attempted to utilize coal dust. As dangerous as this fuel is in storage, he was able to use it in a test engine.

These experiments were cut short by production problems. Not all the licensees had the same success with the engine. In at least one instance, a whole production run had to be recalled. The difficulty was further complicated by a shortage of trained technicians. A small malfunction could keep the engine idle for weeks, until the customer lost patience and sent it back to the factory. With these embarrassments came the question of whether the engine had been oversold. Some believed that it needed much more development before being put on the market. Diesel was confident that his creation was practical—if built and serviced to specifications. But

he encouraged future development by inserting a clause in the contracts that called for pooled research: the licensees were to share the results of their research on Diesel engines.

Diesel's success was marred in two ways. For one, he suffered exhausting patent suits. The Diesel engine was not the first to employ the principle of compression ignition; Akroyd Stuart had patented a superficially similar design in 1890. Also, Diesel had a weakness for speculative investments. This weakness, along with a tendency to maintain a high level of personal consumption, cost Rudolf Diesel millions. His American biographers, W. Robert Nitske and Charles Morrow Wilson, estimate that the mansion in Munich cost a million marks to construct at the turn of the century.

The inventor eventually found himself in the uncomfortable position of living on his capital. His problem was analogous to that of an author who is praised by the critics but who cannot seem to sell his books. Diesel engines were making headway in stationary and marine applications, but they were expensive to build and required special service techniques. True mass production was out of the question. At the same time, the inventor had become an international celebrity, acclaimed on three continents.

Diesel returned to work. After mulling a series of projects, some of them decidedly futuristic, he settled on an automobile engine. Two such engines were built. The smaller, 5-hp model was put into production, but sales were disappointing. The engine is, by nature of its compression ratio, heavy and, in the smaller sizes, difficult to start. (The latter phenomenon is due to the unfavorable surface/volume ratio of the chamber as piston size is reduced. Heat generated by compression tends to bleed off into the surrounding metal.) A further complication was the need for compressed air to deliver the fuel into the chamber. Add to these problems precision machine work, and the diesel auto engine seemed impractical. Mercedes-Benz offered a diesel-powered passenger car in 1936. It was followed by the Austin taxi (remembered with mixed feelings by travelers to postwar London), by the Land Rover, and more recently, by the Peugeot. However desirable diesel cars are from the point of view of fuel economy and longevity, they are still not competitive with gasoline-powered cars.

Diesel worked for several months on a locomotive engine built by the Suizer Brothers in Switzerland. First tests were disappointing, but by 1914 the Prussian and Saxon State Railways had a diesel in everyday service. Of course, most of the world's locomotives are diesel-powered today.

Maritime applications came as early as 1902. Nobel converted some of his tanker fleet to diesel power, and by 1905 the French navy was relying on these engines for their submarines. Seven years later, almost 400 boats and ships were propelled solely or in part by compression engines. The chief attraction was the space saved, which increased the cargo capacity or range.

In his frequent lectures Diesel summed up the advantages of his invention. The first was efficiency, which was beneficial to the owner and, by extension, to all of society. In immediate terms, efficiency meant cost savings. In the long run, it meant conserving world resources. Another advantage was that compression engines could be built on any scale from the fractional horsepower to the 2400-hp Italian Tosi of 1912. Compared to steam engines, the diesel was compact and clean. Rudolf Diesel was very much concerned with the question of air pollution, and mentioned it often.

But the quintessential characteristic, and the one that might explain his devotion to his "black mistress," was her quality. Diesel admitted that the engines were expensive, but his goal was to build the best, not the cheapest.

During this period Diesel turned his attention to what his contemporaries called "the social question." He had been poor and had seen the effects of industrialization firsthand in France, England, and Germany. Obviously machines were not freeing men, or at least not the masses of men and women who had to regulate their lives by the factory system. This paradox of greater output of goods and intensified physical and spiritual poverty had been seized on by Karl Marx as the key "contradiction" of the capitalistic system. Diesel instinctively distrusted Marx because he distrusted the violence that was implicit in "scientific socialism." Nor could he take seriously a theory of history whose exponent claimed it was based on absolute principles of mathematical integrity.

He published his thoughts on the matter under the title *Solidarismus* in 1903. The book was not taken seriously by either the public or politicians. The basic concept was that nations were more alike than different. The divisions that characterize modern society are artificial to the extent that they do not have an economic rationale. To find solidarity, the mass of humanity must become part owners in the sources of production. His formula was for every worker to save a penny a day. Eventually these pennies would add up to shares or part shares in business enterprises. Redistributed wealth and, more important, the sense of controlling one's destiny would be achieved without violence or rancor through the effects of the accumulated capital of the workers.

Diesel wrote another book that was better received. Entitled *Die Enstehung des Dieselmotors*, it recounted the history of his invention and was published in the last year of his life.

For years Diesel had suffered migraine headaches, and in his last decade, he developed gout, which at the end forced him to wear a special oversized slipper. Combined with this was a feeling of fatigue, a sense that his work was both done and undone, and that there was no one to continue. Neither of his two sons showed any interest in the engine, and he himself seemed to have lost the iron concentration of earlier years when he had thought nothing of a 20-hour workday. It is probable that technicians in the various plants knew more about the current state of diesel development than he did.

And the bills mounted. A consultant's position, one that he would have coveted in his youth, could only postpone the inevitable; a certain level of indebtedness makes a salary superfluous. Whether he was serious in his acceptance of the English-offered consultant position is unknown. He left his wife in Frankfort in apparent good spirits and gave her a present. It was an overnight valise, and she was instructed not to open it for a week. When she did, she found it contained 20,000 marks. This was, it is believed, the last of his liquid reserves. At Antwerp he boarded the ferry to Warwick in the company of three friends. They had a convivial supper on board and retired to their staterooms. The next morning Rudolf Diesel could not be found. One of the crew discovered his coat, neatly folded under a deck rail. The captain stopped the ship's progress, but there was no sign of the body. A few days later a pilot boat sighted a body floating in the channel, removed a coin purse and spectacle case from the pockets, and set the corpse adrift. The action was not unusual or callous; seamen had, and still do have, a horror of retrieving bodies from the sea. These items were considered by the family to be positive identification. They accepted the death as suicide, although the English newspapers suggested foul play at the hands of foreign agents who did not want Diesel's engines in British submarines.

2
CHAPTER

Diesel basics

At first glance, a diesel engine looks like a sturdy gasoline engine, minus the ignition system (Fig. 2-1). Major mechanical parts are similar and often interchangeable, as when a manufacturer builds gasoline versions of its diesel engines.

But there are differences in the way compression ignition (CI, or Diesel cycle) engines and spark ignition (SI, or Otto cycle) engines work that cut deeper than the mode of igniting the fuel.

Compression ratio

The compression ratio, or the amount of "squeeze" applied to the fuel and air prior to combustion, is a fundamental parameter of all engines, regardless of type. Figure 2-2 illustrates the concept, expressed as:

Compression ratio = Swept volume + Clearance volume/Swept volume

where

Swept volume = Cylinder volume traversed by the piston as it completes a full stroke from top dead center (tdc) to bottom dead center (bdc).

Clearance volume = Combustion chamber volume.

Diesel engines employ high compression ratios to generate the heat necessary for ignition. In theory, diesel compression ratios could be much higher than they commonly are, for some marginal increases in power. However, several considerations work against ratios higher than about 23:1. The first consideration is the high energy input necessary to overcome compression during cranking. One tries to put reasonable limits on the size and energy draw of the starter. Once the engine starts, the limiting factor becomes the strength of materials; the energy released by ultrahigh compression ratios pushes metallurgy. And finally, at some point, the tiny combustion chambers necessary to support extremely high compression mask gas flow around the valves, costing power.

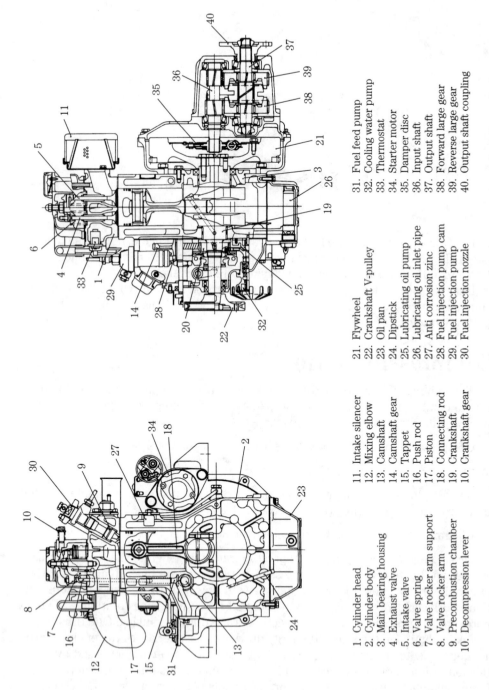

1. Cylinder head
2. Cylinder body
3. Main bearing housing
4. Exhaust valve
5. Intake valve
6. Valve spring
7. Valve rocker arm support
8. Valve rocker arm
9. Precombustion chamber
10. Decompression lever

11. Intake silencer
12. Mixing elbow
13. Camshaft
14. Camshaft gear
15. Tappet
16. Push rod
17. Piston
18. Connecting rod
19. Crankshaft
10. Crankshaft gear

21. Flywheel
22. Crankshaft V-pulley
23. Oil pan
24. Dipstick
25. Lubricating oil pump
26. Lubricating oil inlet pipe
27. Anti corrosion zinc
28. Fuel injection pump cam
29. Fuel injection pump
30. Fuel injection nozzle

31. Fuel feed pump
32. Cooling water pump
33. Thermostat
34. Starter motor
35. Damper disc
36. Input shaft
37. Output shaft
38. Forward large gear
39. Reverse large gear
40. Output shaft coupling

2-1 Yanmar 1GM10, shown with marine transmission, provides auxiliary power for small sailboats. The 19.4-cu in. unit develops 9 hp and forms the basic module for twin-and three-cylinder versions.

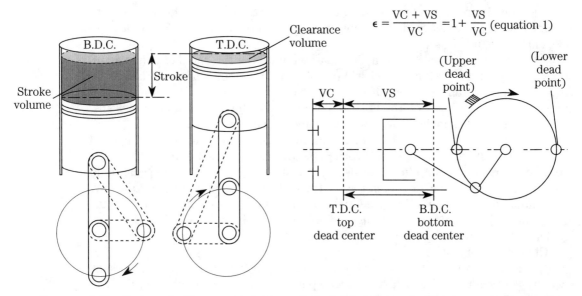

$$\epsilon = \frac{VC + VS}{VC} = 1 + \frac{VS}{VC} \text{ (equation 1)}$$

. Compression ratio is a simple concept, but one that mathematics and pictures express better than ·ds. R = Compression ratio VC = Combustion chamber volume (clearance volume) VS = Swept volume ;placement).

In some cases, older diesels built with c/r's of 16 or 17:1 can benefit from a point or two of higher compression. Starting becomes easier or at least faster, and the modified engine might produce less exhaust smoke.

Compression ratio has profoundly different implications for SI engines. Industrial versions of these engines are usually limited to c/r's in the 8 or 9:1 range. Going much higher crosses the threshold of detonation with disastrous results for engine longevity. Detonation, or spark knock, is a maverick kind of combustion that occurs after normal ignition has commenced. The tag-ends of the air/fuel mixture spontaneously ignite and create sharp, rabbit punchlike peaks in cylinder pressure that literally knock holes in pistons. The surest cure is to reduce c/r.

Induction

An SI engine uses a low-pressure fuel injector or some form of carburetor to mix air and fuel in the intake manifold. A throttle valve, mounted upstream of the fuel delivery point, regulates the amount of air delivered to the combustion chambers, which is only slightly in excess of the amount needed to ensure complete combustion. Fuel delivery changes with the throttle-plate angle, so that available air and fuel vapor remain in a proportion that, for gasoline, is roughly 15 parts air to 1 part vaporized fuel. For a given load, engine speed is directly proportional to throttle angle.

The air/fuel mix is drawn into the cylinder during the intake stroke, when cylinder pressures are quite low. Once admitted, the intake valve or port closes and the rising piston compressed the mixture. That the mixture must be compressed before ignition was the great discovery of Nicklaus Otto, which made SI engines practical and attached his name to the operating cycle. Only four men in the history of technology have been so honored. Note, however, that compression in an SI, or Otto cycle, engine occurs after induction.

In a CI engine, air and fuel remain separated until the threshold of combustion. Fuel is injected late in the compression stroke, when cylinder pressure is already high. In other words, the work of compression precedes fuel induction. Injection pressures range from about 1600 psi to more than 25,000 psi for modern designs. The amount of fuel supplied varies with load and speed requirements.

Air enters through the intake manifold, in a manner roughly analogous to SI practice. But the air supply is unthrottled, so that the CI engine draws the same volume of air per revolution, regardless of its speed. At idle, the open manifold supplies a large surplus air, which passes through the engine without taking part in combustion. Typically idle speed air consumption averages about 100 pounds per pound of fuel consumed; at high speed or under heavy load, the additional fuel supplied drops the ratio to roughly 20:1. That is still more air than is needed for CI combustion.

Without a throttle valve, a diesel breathes easily at low speeds, which explains why truck drivers can idle their rigs for long periods without consuming appreciable fuel. (An SI engine requires a fuel-rich mixture at idle to generate power necessary to overcome the throttle restriction.)

The absence of an air restriction and an ignition system that operates as a function of engine architecture can conspire to take control of the engine away from the operator. All that's needed is for lube oil to enter the combustion chamber where it can be burned as fuel. Oil might slop past worn piston rings or be drawn in through the intake manifold from a leaking turbocharger seal. A runaway engine generally accelerates itself to perdition because few operators have the presence of mind (or courage) to stuff a rag into the air intake.

Ignition & combustion

SI engines are fired by an electrical spark timed to occur just before the piston reaches the top of the stroke. Because the full charge of fuel and air are present, combustion proceeds rapidly, as a kind of controlled explosion. Cylinder pressure rises sharply, peaking during the span of a few crankshaft degrees of piston motion. Thus, the volume of the cylinder above the piston undergoes little change between the moment of ignition and the point of peak pressure. Engineers, exaggerating a bit, describe SI engines as "constant volume" engines (Fig. 2-3).

Compared to spark ignition, the onset of compression ignition is a leisurely process (Fig. 2-4). Some time is required for the fuel spray to vaporize and more time is required for the spray to reach ignition temperature. Fuel continues to be injected during the delay period.

Once ignited, the accumulated fuel burns rapidly with correspondingly rapid increases in cylinder temperature and pressure. The injector continues to deliver fuel through the period of rapid combustion and into the period of controlled combustion that follows. Injection stops and combustion enters what is known as the afterburn period.

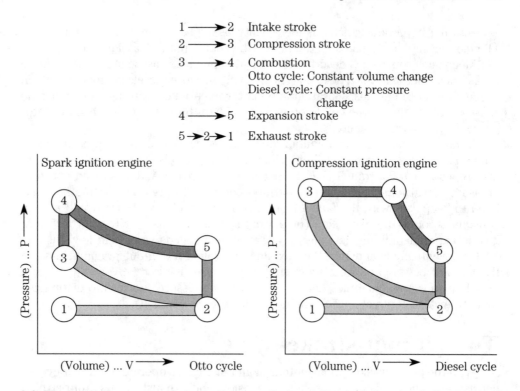

1 ——→ 2   Intake stroke
2 ——→ 3   Compression stroke
3 ——→ 4   Combustion
                 Otto cycle: Constant volume change
                 Diesel cycle: Constant pressure
                            change
4 ——→ 5   Expansion stroke
5 →→ 2 →→ 1   Exhaust stroke

2-3 These cylinder pressure/volume diagrams distort reality somewhat, but indicate why SI engines are described as "constant volume" and CI engines as "constant pressure" devices.

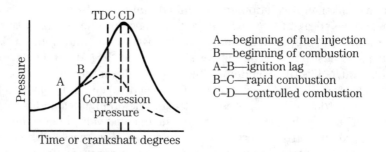

A—beginning of fuel injection
B—beginning of combustion
A–B—ignition lag
B–C—rapid combustion
C–D—controlled combustion

2-4 Diesel combustion process plotted against time (crankshaft rotation) and cylinder pressure.

    The delay between the onset of fuel delivery and ignition (A–B in the drawing) should be as brief as possible to minimize the amount of unbent fuel accumulated in the cylinder. The greater the ignition lag, the more violent the subsequent combustion.

    Ignition lag is always worst upon starting cold, when engine metal acts as a heat sink. Clatter, white exhaust smoke, and rough combustion should disappear when the engine reaches operating temperature. It also helps to time injection late in the compression stroke, when air temperature is high.

Another important variable is fuel quality, as expressed by the cetane number. The more easily the fuel ignites, the higher the cetane number assigned to it.

Mechanics sometimes describe the rough combustion associated with ignition lag as "diesel detonation." The term is misleading, because diesels do not detonate in the manner of SI engines. Vibration and clatter experienced during startup are no cause for alarm unless the condition persists. When this is the case, check injector performance and fuel quality.

In normal operation, with ignition delay under control, cylinder pressures and temperatures rise more slowly (but to higher average levels) than for SI engines. In his proposal of 1893, Rudolf Diesel went one step further and visualized constant pressure expansion. That is, fuel input and combustion pressures would be regulated to keep step with the falling piston, so that cylinder pressure would remain constant through the expansion (or power) stroke. He was able to approach that goal in experimental engines, but only if rotational speeds were held low. His colleagues eventually abandoned the idea and controlled fuel input pragmatically, on the basis of power produced. Even so, the pressure rise is relatively smooth and diesel engines are sometimes known as "constant pressure" devices to distinguish them from "constant volume" SI engines (refer back to Fig. 2-3).

Two- & four-stroke-cycle

With the previous exceptions noted, CI and SI engines operate on similar cycles, consisting of the intake, compression, expansion, and exhaust events. Four-stroke-cycle engines of either type allocate one up or down stroke of the piston for each of the four events. Two-stroke-cycle engines telescope events in two strokes of the piston, or one crankshaft revolution. In the United States, the term *stroke*, is generally dropped and we speak of two-or four-cycle engines; in other parts of the English-speaking world, the preferred nomenclature is "two stroke" and "four stroke."

Four-cycle diesel engines operate as shown in Fig. 2-5. Air, entering around the open intake valve, fills the cylinders as the piston falls on the intake stroke. The intake valve closes as the piston rounds bdc (bottom dead center). Injection begins at some point near tdc on the compression stroke. The fuel ignites and drives the piston down the bore on the expansion stroke. The exhaust valve opens and the piston rises on the exhaust stroke, purging the cylinder of spent gases. When the piston again reaches tdc, the four-stroke cycle is complete, two crankshaft revolutions from its beginning.

Figure 2-6 illustrates the operation of Detroit Diesel two-cycle engines, which employ blower-assisted scavenging. As shown in the upper left drawing, pressurized air enters the bore through radial ports and forces the exhaust gas out through the open exhaust valve. In the case of the Detroit, the blower purges, or scavenges, the cylinder without raising its pressure much above atmospheric. The exhaust valve remains open until the ports are closed to eliminate a supercharge effect.

The exhaust valve closes and the piston continues to rise, compressing the air charge ahead of it. Near tdc, the injector fires, combustion begins, and a few crankshaft degrees later, the piston drives down under the influence of the expansion event. The exhaust valve opens just before the scavenge ports are uncovered to give

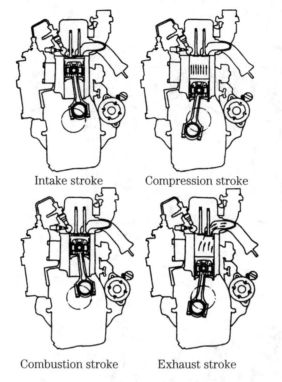

Intake stroke   Compression stroke

Combustion stroke   Exhaust stroke

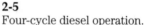

2-5
Four-cycle diesel operation.

the spent gases opportunity to blow down. These four events (intake, compression, expansion, and exhaust) occur in two piston strokes.

The engine illustrated employs unidirectional scavenging. Purge air enters from the bottom of the cylinder and moves upward, driving the exhaust gases ahead of it. The layout can also be reversed, so that purge air enters from above and exhaust exits through ports near the bottom of the cylinder.

Not all two-cycle diesel engines have valves. Combining scavenge air with combustion air eliminates the intake valve, and a port located above the air inlet port replaces the exhaust valve. Such engines employ cross-flow or loop scavenging (Fig. 2-7, A & B) to purge upper reaches of the cylinder and to minimize the loss of scavenge air to the exhaust. In the cross-flow scheme, a deflector cast into the piston top directs the incoming air stream into the stagnant region above the piston. Loop scavenging employs angled inlet air ports to the same end.

It is also possible to eliminate the external air pump by using the crankcase as part of the air inlet tract. Piston movement provides the necessary compression to pump the air, via a transfer port, into the cylinder. Not many crankcase-scavenged diesel engines are seen in this country, but the German manufacturer Fichtel & Sachs has built thousands.

Because it fires every revolution, a two-cycle engine should be twice as powerful as an equivalent four-cycle engine. Such is not the case, principally because of the difficulties associated with scavenging. Four-cycle engines mechanically purge the exhaust gases, typically through some 440 degrees of crankshaft rotation. (The ex-

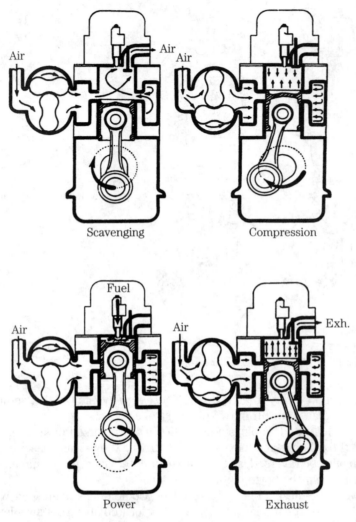

Air   Air Air

Scavenging   Compression

Fuel

Air   Air   Exh.

Power   Exhaust

2-6 Two-cycle diesel operation. The Detroit Diesel engine depicted employs a Roots-type positive-displacement blower for scavenging.

haust valve opens early and closes after the intake valve opens.) Two-cycles scavenge in a less positive way during an abbreviated interval of about 130 degrees. Consequently, some exhaust gas remains in the cylinder to blanket combustion.

Initial cost

The fuel system accounts for much of the cost differential between CI and SI engines. This system consists of a low-pressure pump to transfer fuel from the tank to the engine bay, a high-pressure pump to generate the force needed to overcome engine compression, and one injector per cylinder, which function as fuel entry points and as check valves against ingress of cylinder pressure into the fuel system.

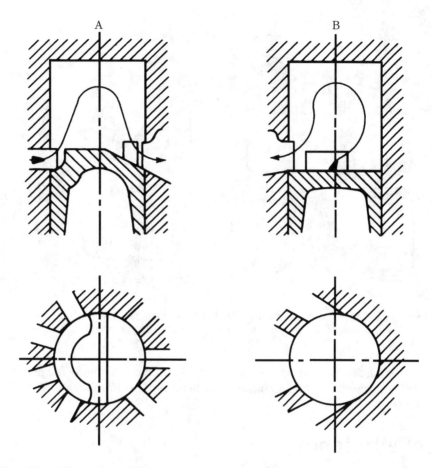

2-7 Cross-flow (A) and loop (B) scavenging.

In theory, a diesel engine requires 10,000 volumes of air per volume of fuel oil. In practice, air volume is increased by one-quarter to reduce smoke. Thus, the high-pressure fuel pump has only $\frac{1}{12,500}$ the capacity of the cylinder it serves. The scale effect is formidable and made more so by the need to inject a finely divided fuel spray with enough force to penetrate deeply against cylinder compression. Injection pressures of 25,000 psi are not uncommon in contemporary, emissions-rated systems; older engines made do with a tenth or less of that pressure.

Fuel metering parts are manufactured with extreme precision and lapped to tolerances expressed as wavelengths of light. Not only are pump plungers non-interchangeable, plungers that have been exposed to direct sunlight cannot be fitted to their barrels without first equalizing the temperatures.

Diesel engines must be built sturdier than their SI counterparts and assembled with higher grade components. Anodized pistons, forged crankshafts, sophisticated oil systems, all drive up the price.

In an attempt to control costs and reduce inventory, most diesel manufacturers rely heavily on modular construction (Fig. 2-8). Cylinder liners, pistons, rods, valves,

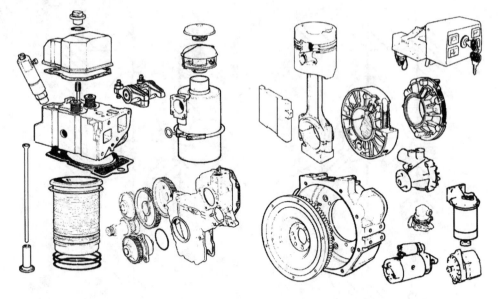

2-8 Two-, three-, four-, and six-cylinder VM HR 94 Industrial engines share these parts in common. Modular construction reduces costs and simplifies maintenance.

injectors, and machine tool center distances are common to engine families, so that building a 12-cylinder engine amounts to little more than 12 replications of the single-cylinder model.

Fuel efficiency

High compression ratios (or more exactly, large ratios of expansion) give CI engines superior thermal efficiency. Under optimum conditions, a well designed SI engine utilizes about 30% of the heat liberated from the fuel to turn the crankshaft. The rest goes out the exhaust and into the cooling system and lubricating oil. Thermal efficiencies for diesel engines reach 40% or better. By this measure—which might become quite critical if suspicions about global warming turn out to be well founded—diesel engines are the most efficient engines known. (Gas turbines do better but only at a constant speed.)

Excellent thermal efficiency, plus good volumetric efficiency and the capability to recycle some exhaust heat by turbocharging, translate as fuel economy. It is not unreasonable to expect a specific fuel consumption of 0.35 pound per brake horsepower hour from one of these engines, operating near its torque peak. A gasoline engine can burn 0.5 lb of fuel per hp-hour under the same conditions.

The weight differential between diesel fuel (7.6 lb/US gal for No. 2-D) and gasoline (about 6.1 lb/US gal) gives the diesel even greater advantage when consumption is figured in gallons per hour or mile. Diesel passenger cars and light trucks deliver 30% to 50% better mileage than the same vehicles with gasoline engines, although at some tradeoff in acceleration and top speed.

Fuel

Conventional diesel fuel is a middle distillate, slightly heavier than kerosene or jet fuel. The composition varies with the source crude, the refining processes used, the additive mix, and the regulatory climate. Fuel sold in this country must conform to standards set by the Environmental Protection Agency or to the more rigorous version of these standards prevailing in California.

Performance parameters

From the operator's point of view, cetane number is the most critical fuel variable. High-cetane numbers accord with good ignition quality, which means that the fuel ignites at a low temperature and burns quickly for minimum ignition lag. Engines start easily on high-cetane fuels and, once started, tend to run smoothly.

Interestingly enough, high-octane fuels, appropriate for SI engines, have low cetane numbers. Thus, aviation gasoline is a poor diesel fuel, usable only if the compression ratio is raised to generate the necessary ignition temperature. On the other hand, ether or amyl nitrate, both of which would detonate badly in SI engines, are excellent starting fuels for diesels.

Fuel sold in this country must, by law, have a cetane number of at least 40. Heavier crudes tend to have higher numbers; lighter, more aromatic crudes can be boosted to this level during the refining process.

Viscosity, or the "pourability" of the fuel, also affects performance. Lower, more gasoline-like, viscosity fuels tend to atomize better and (probably for that reason) have been shown to generate less exhaust smoke. A pronounced change in viscosity will affect the volume of fuel delivered for a given rack setting. Low viscosity fuels tend to leak past the pump plungers, so that less is available for injection. Measured at room temperature, the viscosities of light diesel fuels sold in this country vary from about two centistrokes to more than four.

Fuel quality can also affect engine and component life. The abrasive ash left after combustion collects on injector nozzles where it disrupts spray patterns and can measurably increase upper cylinder and ring wear. At present, U.S. fuels contain 0.01% ash by weight, which appears to be the practical minimum. Water, present in trace amounts in all fuels, plays havoc with bright parts, and when present in sufficient quantities, promotes bacterial growth.

Red, white, and blue

As of Jan. 1, 1994, light diesel fuel sold in this country comes in red, blue, and clear varieties. Some history is necessary to understand the rationale for three versions of what was formerly known as No. 2-D.

The 1990 amendments to the Clean Air Act called for a reformulated diesel fuel, with significantly less sulfur than the 0.5% norm then in effect. Sulfur and its compounds are major pollutants that cannot be effectively addressed by vehicle emissions systems.

The new fuel, which contains only 0.05% sulfur by weight, was phased in during late 1993 for use in commercial trucks and diesel passenger cars operating on the public roads. Small refiners, who have difficulties meeting the sulfur limit, were allowed to market fuels with slightly higher sulfur contents as an interim measure. But, by early 1995, all highway diesel sold in this country should conform to the 0.05% limit.

For reasons of political expediency, exhaust emissions from off-road diesels received little attention during the period when the three-tier fuels strategy was developed. It was decided to continue to provide high-sulfur fuel for agricultural and construction equipment, railroads, commercial vessels, and diesel-fired heating systems. Users of this fuel do not pay the 24.3-cent-a-gallon federal highway tax.

To avoid confusion and loss of tax revenue, the Environmental Protection Agency (EPA) decreed that low-sulfur fuel would be colorless and that the high-sulfur variant would be tinted blue. As things turned out, the blue dye (1.2 dialkylamino-anthraquinone) turned the sulfurous base stock a bilious green, but no matter.

Red diesel came about because of the anxiety of the Internal Revenue Service. Vehicles operated on public highways by government agencies and the American Red Cross (but not other charitable organizations) are exempted from the highway tax. At the same time, the fuel for these vehicles falls under the low-sulfur rule. To keep the books straight, the IRS insisted that this untaxed fuel be stained red.

Congressional action and bureaucratic rules decide who burns what. We have already mentioned the tax exemption given to the Red Cross; if truckers can find a place to buy blue fuel, they can use it to power their engine-driven refrigeration machinery; homeowners can heat with blue or red, but will fall afoul of the law if they mix the two fuels to obtain purple.

Table 2-1 lists applications for the various tints of diesel. But it is by no means complete. For a complete list, contact the Petroleum Marketers Association to obtain a three-hour video tape on the subject.

Table 2-1. Red, white, and blue

White (clear) stands at the top of the hierarchy, blue at the bottom. Users legally able to use blue can upgrade with red or clear. But woe betide the hapless operator who gets caught using a lower grade of fuel than decreed. Penalties for misfueling range up to $25,000 per day, per offense.

Color of fuel	User
Blue (high-sulfur, untaxed)	Off-road equipment
	Commercial vessels
	Heating plants
	Truck-borne refrigeration plants
	Locomotives
Clear (low-sulfur, taxed)	Commercial and privately owned motor vehicles operating on public roads and highways
	Pleasure boats
Red	Government vehicles
	American Red Cross vehicles

There is yet another kind of diesel fuel, mandated in California for over-the-road vehicles. It conforms to federal sulfur regulations, but contains no more than 10% by volume of aromatics. Federal fuel can contain as much as 35%. Aromatics contribute to particulates—the carboniferous solids that blacken diesel exhaust—and, by general consensus, to a public health hazard.

In actual fact, California Air Resources Board (CARB) tolerates fuels with more than 10% light hydrocarbons, if these fuels test as clean as the board's reference fuel. Chevron and other refiners have received approval for extremely low sulfur blends containing twice the mandated percentage of aromatics. But "alternative fuel," as it is sometimes called, costs six or seven cents more a gallon than federal white.

Thus, the fueling situation has become quite complicated for engines that can run on almost any liquid with BTU value. But most people in the industry believe that the alternative—abandoning diesel fuel for methanol or methane (natural gas)—would be even more disruptive.

Weight vs. horsepower

It is a truism that diesel engines are heavier and produce less power than equivalent SI engines. Attempts to quantify that statement by dividing the power outputs of various representative engines into their weights are less reliable than one might suppose. The resulting power-to-weight ratio can be expressed as pounds per horsepower or, in the European manner, as kilograms per kilowatt. Sometimes you will find the same data presented as Watts/kilograms. In any case, the computation can be no more meaningful than the data that comprises it.

One might think that the weight of an engine would be a straightforward concept, verifiable with a scale. Most manufacturers list fan-to-flywheel weights, less liquids and electrical gear. Omission of the weight of the battery is understandable, but the starter motor, lube oil, and coolant appear to be as much a part of the package as the exhaust manifold.

Power, or the rate of doing work, is even more slippery a concept because of its effect on sales. At some remove, brake horsepower derives from a test of the engine's ability to overcome the restraint imposed by a dynamometer. Air temperature, exhaust restrictions, engine accessory load, and operator intentions all influence the results. It is possible to obtain high flash readings on a dyno by momentarily drawing down the energy contained in the flywheel and other rotating masses.

It is true that protocols have been developed by various certification agencies to make dyno testing more consistent. The U.S. Society of Automotive Engineers, the Japanese Institute of Standards, the European DIN, and now the European Community have detailed how these tests will be run. But a worldwide standard horsepower does not exist.

For example, the VM "Sun" series 1105 engine develops 17 hp at 2600 rpm when tested in accordance with "A" DIN 6271 protocols. "B" DIN procedures are good from another two horsepower at the same rpm. "F" DIN 70020 acts like a phantom turbocharger to boost output to 21 hp. SAE protocols could give even more optimistic numbers.

Although you will not attain precision, you can make rough comparisons between representative types of diesel engines and their gasoline counterparts.

The Tecumseh XL OHV180 is a single-cylinder, air-cooled, four-cycle SI engine used in light industrial roles. According to its manufacturer, it develops 18 hp and weighs 124 lb. The AV 1105, previously mentioned, makes about the same power and tilts the scales at a massive 385 lb. Both engines have cast iron blocks and overhead valves.

Once you leave the domain of small stationary engines, you encounter a large number of four-and six-cylinder power plants intended for agricultural, construction, and marine applications. Outputs generally fall into the 60 to 150 hp range; most are naturally aspirated, although newer designs have at least the option of turbocharging. Power-to-weight ratios reflect some compromise between the durability expected by commercial users and light weight necessary for mobility.

The 5.9-liter, air-cooled Tatra 924 is an Eastern European example of the type once supplied to the Soviet military. It develops 120 hp from a platform that weighs nearly 1000 lb, or 8.25 lb/hp. The L439T Lugger marine engine could be expected to do better than the archaic Tatra, because it is a product of the international design community and turbocharged to boot. However, this 3.9-liter engine carries the same horsepower rating as the Tatra and weighs more for a calculated power-to-weight ratio of 9.6 lb/hp. According to the supplier, Seattle-based Alaska Diesel Electric, the Lugger series was deliberately de-rated to reduce operating stresses. Certainly it could make more than the 120 hp claimed for it.

By some ways of reckoning, the most advanced piston engines in the world power large (Class 8 and above) highway trucks. No other engine type exhibits the same combination of more-or-less antithetical characteristics: moderate weight, high power outputs, very low emissions, and extreme durability. A 500-hp version of the Caterpillar 3406E, scheduled for release in mid-1994, should come in at 5.7 lb/hp.

Diesels intended for passenger car service almost reach parity with SI performance. The 2500 cc VM/HR 4929, which powers an Alfa Romeo Sports tourer, carries only 3.8 16 for each of its 120 horses.

Another example is the new PSA Peugoet Citroen DJST, a replacement for the venerable XD3P that, in its day, powered a quarter of all European diesel passenger cars. At 409 lb (dry, without electrical system), the DJST is significantly lighter than its predecessor and develops 10% more power from roughly the same displacement. The manufacturer says that its 3.97 lb/hp power-to-weight ratio, coupled with low noise and a good emissions signature, makes the DJS series virtually indistinguishable from gasoline engines. Transparency is important for auto diesel makers, who can expand their market only by weaning customers away from the SI cycle.

Power was increased by better breathing, made possible by two intake valves per cylinder and the supercharging effect of a tuned intake manifold. A wastegate-controlled turbocharger develops 13.2 psi at wide-open throttle adds even more boost at the torque peaking speed. The company plans to phase in an aftercooler later in the production run, when additional power might be needed.

Following contemporary practice, the head is cast in aluminum and the block iron. Both castings are split for better dimensional control. The block, with only 9 mm separating adjacent cylinder bores, pushes the state of foundry art. Casting thickness is ultrasonically verified before and after machining operations.

Durability

One of the first Caterpillar 3176 truck engines was returned to the factory for teardown after logging more than 600,000 miles. The main and connecting rod bearings were the only major components to be replaced (at 450,000 and 225,000 miles respectively) during this long history. The original bearings were not available for inspection, but the replacements were said to be in good condition.

The crankshaft remained within new-parts tolerance, as did the rocker arms, camshaft journals, and lower block casting. Valves exhibited nominal wear and were judged reusable. Connecting rods could have gone another 400,000 miles, which would make million-mile components of them. Pistons and rings were good for another 200,000 estimated miles. Cylinder liners showed the most wear, but original honing marks were still visible. Cat engineers estimated that the liners would have run 100,000 additional miles without a perceptible increase in oil consumption.

Durability is not a Cat exclusive. Detroit Diesel warranties its coach engines for 100,000 miles in urban traffic; a fairly large number of Mercedes-Benz passenger car engines have passed the three-quarter million mile mark without major repairs. Stationary engines, built when Rudolf Diesel held the patents, are still working.

Most diesels come out of a very conservative design tradition, appropriate for high-pressure devices sold to commercial and industrial users. In this market, high initial cost, weight, and moderate levels of performance are acceptable tradeoff against early failure.

The classic diesel engine is founded on heavy, fine-grained iron castings, liberally reinforced with webbing and aged, by one process or another, to relieve foundry stresses before machining. One of the most remarkable aspects of the Cat engine mentioned above is that main-bearing saddles remained in alignment after the thermal cycling received during 600,000 miles of use.

Buttressed main-bearing caps, pressed into the block and, in some cases, cross-drilled for additional rigidity, support the crankshaft. Pistons generally run against replaceable liners, whose metallurgy can be precisely controlled. Some manufacturers use gear-driven accessories and many dispense with the timing chain.

There are also less obvious features in the form of detailed design changes incorporated during the long production runs most of these engines enjoy. For example, the Cat 3176B, successor to the 3176, includes redesigned pistons, rings, connecting rods, head gasket, rocker shafts, injectors, and water pump. The crankshaft was modified and new tooling developed to improve finish on the liners.

None of this is to say that diesel engines are zero-defect products designed by corporations who always put customer satisfaction above immediate profit. Some engines are dogs. The worst are cloned SI engines, such as the now-defunct 315 Oldsmobile, which broke head bolts, crankshafts, and almost everything in between. This engine helped to set back the cause of diesel passenger cars in this country for decades. The Volkswagen passenger car diesel, since evolved into a reasonably good industrial engine, was not much better.

There are quality variations among industrial/commercial engines as well. One truck engine maker has been bedeviled by porous castings that leak coolant into the oil sump; a well-known tractor manufacturer has trouble grinding straight crankshafts. No engine enjoys immunity from stress-related cracks that, in many instances, are nonrepairable.

But, these difficulties are the kind of things that mechanics are overly sensitive to. In all, the diesel engine has been one of the most successful technologies of the twentieth century.

3
CHAPTER

Planned maintenance

In the old days, mechanics worked on the edge of crisis because engines and related equipment seemed to break down at random. When that happened, the whole shop would be mobilized to make whatever makeshift repairs were necessary to get the equipment back on line. Every emergency repair further compromised reliability and set the stage for the next breakdown. This kind of maintenance, variously called *corrective*, *reactive*, or *unscheduled*, is extremely inefficient, whether measured by dollar cost or mean time between failures.

Even so, the better mechanics developed a sense of the engines in their charge and seemed to know what part would fail next. But this kind of empathy took years of close observation to develop and was not the sort of thing that could be easily communicated. Some had it and others didn't. Those who did had an almost total recall of past failures, the extent of repairs, and the wear rates of critical parts. In other words, they kept equipment history files in their heads. It would have been much better had these files been accessible to everyone involved.

Inventory control was an annual affair, and it was not unusual for mechanics to build their own private stores of parts. That contributed to the confusion on the job and helped fuel a cottage industry in after-hours diesel repair.

The advent of computerized maintenance management systems (CMMS) has changed all that, at least as far as the more progressive operations are concerned. A CMMS makes it possible to orchestrate the maintenance effort, create a detailed and accessible history for each piece of equipment, and control inventory on a real-time basis. That sounds very abstract, but the results can be dramatic. One Houston offshore operator was spending three-quarters of a million dollars a year for upkeep on each of the ten vessels in its fleet. After a CMMS was installed and running four years, unit maintenance costs dropped 70% to $241,571 per vessel. Time between overhauls on main engines was stretched by more than a third without compromising reliability.

What is CMMS?

CMMS programs print out work order (WO) cards that initiate maintenance actions for specific engines and other pieces of equipment (Fig. 3-1). Technicians check off the tasks completed on the cards and note any delayed for lack of parts or other reasons.

MAJESTIC TIDE	001	EXHAUST SILENCER MAIN ENGINE			175
FACILITY NAME	FAC NO	EQUIPMENT NAME			MACH I D NO
08/28/89	MAIN ENGINES		ENGR	W	4015.02
DUE DATE		LOCATION	SKILL	FREQ	PMWO NO

PREVENTATIVE MAINTENANCE WORK ORDER

```
01   TEST START THE ENGINE___/
02   ON THE TURBOCHARGER, CLEAN THE COMPRESSOR AND TURBINE BY WATER
     INJECTION___/
03   CHECK AND RECORD READINGS OF ALL TEMPERATURE GAUGES, PRESSURE GAUGES,
     LOAD INDICATORS, ETC.___/
04   CHECK THE PRESSURE DROP INDICATORS ON ALL FILTERS___/
05   CHECK OIL LEVEL AND CONDITION OF OIL IN THE SUMP, GOVERNOR, AND THE
     TURBOCHARGERS___/
06   CHECK EXPANSION TANK LEVEL OF THE FUEL NOZZLE TEMPERATURE CONTROL
     SYSTEM___/
```

3-1 A work order card describes equipment, its location, and the required maintenance action, together with the frequency of that action and the skill level necessary to perform it. When the work is finished, the mechanic signs the card and returns it to management.
Engineering Management Consultants, Inc., Houston

The status of maintenance actions—completed or overdue—is inputted into the computer memory bank, together with the technician's notes on equipment condition. This material provides clues for troubleshooting and, analyzed intelligently, has predictive value for similar engines operating in similar environments. Thus, if a DD governor bushing fails at 300,000 miles on one engine, you can anticipate—but not predict with certainty—failure of the same bushing on other high-mileage Detroits.

In addition, computerized maintenance management programs track spare parts inventories. A competent CMMS lists parts by manufacturer's nomenclature and number, and includes ordering information. Alternative vendors can also be listed, together with cost data.

TQM

The benefits described in the preceding section do not accrue automatically. Probably two out of three maintenance management programs perform less well than promised and some fail spectacularly. A nationally known manufacturer of floor coverings spent $1.4 million developing an in-house CMMS that has never worked. A major oil field service company had a similar experience with a purchased program.

What went wrong? Every story is different, but the fundamental failure was conceptual or, if you will, philosophical. Computers and software are merely tools, and no more capable of defining the maintenance effort than wrenches or screwdrivers.

An inappropriate and haphazardly implemented CMMS merely superimposes a layer of computerized chaos on a situation that was already confused.

The late C. Edwards Deming spoke of Total Quality Management, or TQM. This concept, used to good effect by the Japanese auto industry and since adopted by Ford Motor and elements of the U.S. Navy, can be distilled into five broad principles:

- Identify job requirements.
- Devise an appropriate strategy to meet job requirements.
- Select a team capable of implementing that strategy.
- Provide the team with appropriate tools.
- Fine tune the operation with inputs from team members.

Although promulgated as a philosophy of manufacturing, TQM provides valuable guidelines for companies making the transition from purely corrective maintenance to computerized maintenance. Material that follows was developed by Engineering Management Consultants, Inc., a Houston-based engineering and software company.

Identify job requirements

Begin by establishing the scope of responsibility of the maintenance department. Think about what you have been doing, its importance to the rest of the organization, and list all equipment for which you or your group have responsibility. Tally the resources—human, budgetary, and institutional—under your control.

Devise an appropriate strategy

Select the type of maintenance appropriate for the situation and give careful thought to the content of the work orders, which are program outputs and the primary interface with the technicians.

Basically, there are three types of maintenance activity:

Planned (also called Preventive)

Time-line maintenance is triggered by date or equipment running hours. Past performance, manufacturer's recommendations, and employee feedback determine scheduling. In general, planned maintenance is the most cost-efficient and least technologically risky approach.

Predictive

In its purest form, predictive maintenance is a reactive activity, triggered by empirical data about equipment condition. Data comes by way of sensors and monitoring devices that identify abnormal temperatures, pressures, vibrational patterns, and so on.

Our experience with predictive maintenance has not been entirely encouraging. State-of-the-art sensing equipment is expensive and temperamental. Base-line data requires months or even years to accumulate and might not, finally, have predictive reliability. For example, oil analysis might show a high silver content, when actually the bearings are still good. Close analysis of vibration signatures borders on tea-leaf reading, at least, in our view. Technicians who rely solely on sensor readings find themselves replacing more parts than they should.

The relative unreliability of sensed data, plus the fact that parts wear at different rates, can reproduce the crisis mentality that we had hoped the computer would eliminate. Man-loading and downtime planning is difficult at best, impossible at worst.

Combination

This approach, used effectively by operators of large vehicle and naval vessel fleets, combines elements of predictive and planned maintenance. It works to the degree that emphasis is put on the planned aspects of the program and that low-tech sensing has priority.

CMMS-generated work orders drive the maintenance effort by identifying the tasks to be performed. As mentioned previously, WOs should be distilled from manufacturer's service recommendations, technician's input, and equipment history.

It is enough that WOs name the task and any special tools that might be required. Avoid the "how-to" format, which requires endless work to compile and might, in spite of best efforts, omit critical information. In no way should technicians be "given an out" through legalistic interpretations of what was ordered done.

Select the team

The maintenance manager should be an expert technician with a deep understanding of computer technology. As the individual in charge, he or she will be expected to initiate or, at least, remain current on changes in the program. A CMMS requires almost continual tinkering.

Maintenance personnel should be familiar with the equipment and knowledgeable enough about computers to input data. Some companies try to upgrade worker's skills by writing super-detailed WOs. This tactic never works and, in the long run, costs more than hiring competent people to begin with.

Provide the tools

A personal computer, or several personal computers linked in a network, should be sufficient for all but the largest operations.

Much second-guessing can be avoided if the technicians who will use the program have a real say in its selection. Management should evaluate several programs and give at least as much weight to the technicians' views as to their own. Hopefully, an open selection process will create a reservoir of goodwill that will make the transition easier for all concerned.

The selected program should be user-friendly, dynamic, and include a full spectrum of useful features.

User-friendly

A user-friendly program explains itself as it unfolds. Menus, such as the one shown in Fig. 3-2, define available options. The operator selects the desired program function (to accept data inputs, to report, print WO cards, and so on) and the software does the rest.

```
MENU

TO SELECT THE PORTION OF THE PROGRAM YOU WISH TO WORK WITH,
TYPE THE NUMBER CORRESPONDING TO YOUR SELECTION AND PRESS F5

*********************************************************************

   1. FACILITY INFORMATION      2. EQUIPMENT INFORMATION

   3. EDIT TEXT LIBRARY          4. ADD OR UPDATE HISTORY

   5. TEXT LIBRARY INFORMATION   6. REVIEW HISTORY

   7. EQUIPMENT LISTS            8. COPY EQUIPMENT

   9. SPECIFIC EQUIPMENT TEXT   10. MENU CONTINUATION


*********************************************************************
===>                    F1 - EXIT COMPUMAINT
```

Engineering Management Consultants, Inc., Houston

3-2 CompuMaint main menu screen lists selections of primary interest to the line mechanic. A second menu screen is devoted to report-generating and other managerial functions.

Instructions appear on the screen at appropriate junctures, so that there is never a need to go thumbing through a program manual. Nor should comprehensive, multilayered "Help" screens be necessary.

User-friendly software is forgiving, in the sense that it is virtually impossible for the operator to crash the program. An operator mistake triggers an error message and on-screen help messages. An automatic save feature, invoked in the event of power failure, should be part of the package.

Dynamic

A dynamic program exhibits the ability to grow and change with the maintenance effort. Structurally, the database must be large enough to accommodate future maintenance needs, which are nearly always greater than anticipated. The better programs enable individual pieces of equipment or whole facilities (repair stations, vehicle fleets, ships) to be "cloned," or replicated.

Truly dynamic programs impact the thrust and content of the maintenance effort in ways that cannot be entirely foreseen. The report function, which organizes and retrieves stored data, helps fuel this dynamism by posing leading questions. For example, a technician should be able to generate a report that correlates cold starts with bearing life or compares the durability of v-belts purchased from different suppliers. Such reports should be easy to craft, which means that English, and not some programming language, should be the medium. The fourth-generation database language Focus offers this capability.

Full-spectrum

A full-spectrum program encompasses the whole range of maintenance responsibilities. These programs accommodate capital equipment records, work scheduling, corrective maintenance, equipment history, cost analysis, running-hour tracking, parts management, and data base administrative functions.

Listen to the team

The last of Deming's principles requires a certain humility on the part of management. Adjust maintenance procedures, frequency, and methodology to reflect the views of the men and women who do the work. They are, after all, the experts.

4
CHAPTER

Diagnostics

This chapter describes and illustrates methods used to isolate and identify the most commonly encountered malfunctions. Because an engine is a collection of mutually dependent systems, the first step is to identify, at least provisionally, the system at fault. Once a likely candidate is selected, turn to the appropriate chapter for an overview of system theory and repair procedures. No discussion of diagnostics can be of much use without a theoretical understanding of the various engine systems and their components.

Most malfunctions are straightforward, involving a single failure in a single system. Complex malfunctions, linked in unexpected ways to partial failures in several systems, sometimes happen, but not nearly so often as one might suspect. Simple hypotheses are nearly always the best, but there is no high road to diagnostics—one must understand the systems involved.

This chapter deals with a hypothetical engine that either will not start or balks when it does. This means that you must have a theoretical and practical acquaintance with the starting system, the engine itself, and especially with the fuel system, the bugaboo of diesel engines. The best way to learn significant things about the engine is to study the lube oil circuitry, preferably in the hardware. (Factory manuals usually skip over this with a hard-to-follow diagram.) As for the fuel system, you will need to study the factory literature and, in addition, come to terms with the system as it actually is. For example, Series 400, 500, 700, and 900 Navistar trucks have a small filter under the fuel inlet banjo fitting on the outboard side of the pump that is not documented by the factory.

Background

Gather as much information as you can, beginning with what the operator knows. When was the problem first noticed? Was it slow to develop or did it arise suddenly, without prior warning? Exactly how does the engine behave—cold or hot, at what speeds and loads? Are there any secondary problems? Was the engine under load when it malfunctioned? Factual questions like these can elicit useful information without drawing the operator into speculation about the cause of failure.

When available, check the maintenance log, looking for patterns of failure and for chronic problems that have never been adequately addressed. Note the level and intensity of routine maintenance—especially the frequency of lube oil and filter changes. Pay particular attention to any recent changes in oil and fuel consumption figures.

A computerized PM system, documenting failures in similar engines and the actions taken to correct these failures, is always useful. At any rate, make note of this failure for future reference, because even the most thorough repair merely resets the clock. Nearly all failures originate on the drawing board. It is also useful to have a file of factory service bulletins, which update the manual with lessons learned in the field.

In the absence of documentation, a clean engine can usually be taken as evidence that some effort was made toward proper maintenance. Other telltales are the condition of the drive belts and battery terminals. Sloppy maintenance practices do not always mean that the malfunction is serious—a neglected engine can, for a time, remain healthy—but don't bet on it.

Basic techniques

Good mechanics go into action quickly, making simple observations and tests that set limits to the problem.

- Use the dipstick and the oil pressure gauge for insight into engine internals. Fuel contamination can be smelled and in severe cases will flood the sump. Water in the oil breaks into steam when the dipstick is held against a hot exhaust manifold. Ethylene glycol reacts with lube oil to produce a sticky brown varnish, which will be visible on the dipstick. Oil pressure should come up quickly and, if crank bearings are good, will hold steady under load.
- Starting system malfunctions can be quickly diagnosed by measuring battery terminal voltage during cranking. If voltage drops 25% or more, make a hydrometer check, looking for a shorted cell. Check the starter motor and cables by monitoring current draw: higher-than-normal current means a shorted or binding motor; low current means resistance in the supply circuit.
- Try to distinguish between conditions that affect all cylinders and those that affect just one or two. Too much advance on the pump timing will send all cylinders into detonation; the clatter caused by a faulty injector will be limited to the associated cylinder. In like manner, single-cylinder malfunctions can generate puffs of smoke; general malfunctions discolor the whole of the exhaust stream.
- Treat the injectors as you would spark plugs, disconnecting the high-pressure lines one at a time to "short" the associated cylinders (Fig. 4-1). Unit injectors are neutralized by depressing the cam followers with a pry bar. Disabling a healthy cylinder will produce a marked increase in engine vibration and a decrease in rpm. If there is no change, the cylinder was already misfiring, because of either lack of compression or lack of fuel. If a knock goes away or sharply diminishes when a particular cylinder is disabled, the cylinder was detonating or has a bearing failure.

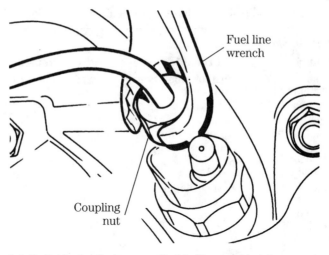

Fuel line
wrench

Coupling
nut

4-1 Individual cylinders are disabled by cracking the associated high pressure fuel lines.

- Crank the engine with the suspect injector removed and connected to its fuel pipe or mounted in a test fixture. Observe the spray pattern, keeping well clear of the fuel stream. This establishes whether fuel gets through the injector and should indicate any gross anomalies in the spray pattern. When in doubt, substitute a known-good injector.
- Many conditions, loss of power, smoke generation, poor idle, roughness, and hard starting, can be engine or fuel related. If substitution of known-good injectors does not solve the difficulty, make the compression and cylinder leakdown tests described below. Engine tests are simpler than fuel system tests and can be done on premises.
- Locate bearing noises with the help of a stethoscope or, better, a length of broomstick held next to the ear for bone conduction. Probe the castings, and note the oil pan and other stampings that pick up vibration from remote sources. If noise from belt-driven accessories is suspected, temporarily remove the belts.

  Note that a faulty injector can mimic bearing noise, particularly on two-cycle engines.
- When stymied, buy time and possibly gain insight by checking the obvious and accessible, i.e., fuel and air filter elements, water traps, and injector timing.

Symptoms

Table 4-1 details the general significance of exhaust smoke. White smoke means one or more fueled, but misfiring cylinders. The condition usually accompanies cold starts and, when present in a warm engine, often indicates low compression. Reports have it that white smoke and "wet stacking" (i.e., raw fuel in the exhaust) are en-

Table 4-1. Exhaust smoke diagnosis

Black or dark gray smoke

Sympton	Probable cause	Remedial action
Smokes under load, especially at high and medium speed. Engine quieter than normal.	Injector pump timing retarded.	Set timing.
Smokes under load, especially at low and medium speed. Engine noisier than normal.	Injector pump timing advanced.	Set timing.
Smokes under load at all speeds, but most apparent at low and medium speeds. Engine may be difficult to start.	Weak cylinder compression.	Repair engine.
Smokes under load, especially at high speed.	Restricted air cleaner.	Clean/replace air filter element.
Smokes under load, noticeable loss of power.	Turbocharger malfunction.	Check boost pressure.
Smokes under load, especially at high and medium speeds. Power may be down.	Dirty injector, nozzle(s).	Clean/replace injectors.
Smokes under load, especially at low and medium speeds. Power may be down.	Clogged/restricted fuel lines.	Clean/replace fuel lines.
Puffs of black smoke, sometimes with blue or white component. Engine may knock.	Sticking injectors.	Repair/replace injectors.

Blue, blue-gray or gray-white smoke

Symptom	Probable cause	Remedial action
Whitish or blue smoke at high speed and light load, especially when engine is cold. As temperature rises, smoke color changes to black. Power loss across the rpm band, especially at full throttle.	Injector pump timing retarded.	Set timing.
Whitish or blue smoke under light load after engine reaches operating temperature. Knocking may be present.	Leaking injector(s).	Repair/replace injector(s).

Blue, blue-gray or gray-white smoke

Symptom	Probable cause	Remedial action
Blue smoke under acceleration after prolonged period at idle. Smoke may disappear under steady throttle.	Leaking valve seals.	Replace seals, check valve guides/stems.
Persistent blue smoke at all speeds, loads and operating temperatures.	Worn rings/ cylinders.	Overhaul/rebuild engine.
Light blue or whitish smoke at high speed under light load. Pungent odor.	Over-cooling.	Replace thermostat.

demic in low-compression (16.5:1) Cummings engines. Black smoke is the sooty residue of partially burned fuel, normally present during hard acceleration. Blue, blue-white, or gray-white smoke results from lube-oil combustion.

Table 4-2 amounts to a short course in diesel diagnostics, which if it does not isolate the problem, should at least point in the right direction.

Table 4-2. Common malfunctions and probable causes

Engine cranks slowly, does not start

Possible cause	Remedial action
Cranking system malfunction	Check batteries, cable terminals, starter motor.
Crankshaft bound	Bar engine over by hand, if it appears tight, check oil
viscosity	and for ethylene glycol leaks into sump.
Faulty injectors	Check injector opening pressure.

Cranking speed erratic, engine does not start

Possible cause	Remedial action
Valves out of time	Check valve timing.

Engine cranks normally, does not start

Possible cause	Remedial action
Glow plugs inoperative	Check glow plugs, control circuitry.
Incorrect or contaminated fuel	Check fuel supply.
No fuel to cylinders	Trace the supply forward from tank to transfer pump, transfer pump to high pressure pump (which may be a single unit), high pressure pump to injectors, injectors to nozzles.

Table 4-2. Continued.

Possible cause	Remedial action
Incorrect pump timing	Check timing.
Air intake restriction	Check filter element.
Exhaust system restriction	Check system.
Low compression	Test cylinder compression.

Engine does not idle

Possible cause	Remedial action
Incorrect idle adjustment	Correct adjustment.
Air in fuel system	Bleed system, tighten connections.
Clogged fuel filter(s)	Replace filter(s).
Fuel return line restricted	Disconnect line to verify and remove obstruction.
Air inlet restriction	Replace filter element.
Exhaust system restriction	Repair system.

Engine runs erratically at low speed

Probable cause	Remedial action
Any of the malfunctions listed above that affect idle	Check as indicated.
Loss of compression in one or more cylinders	Make a cylinder compression test.

Engine does not develop normal power

Probable cause	Remedial action
Insufficient fuel supply	Check filters, transfer pump, cap vent and for possible air leaks into system.
Incorrect or contaminated fuel	Check fuel supply.
Injection timing error	Check timing.
Air inlet system restriction	Check filter element.
Loss of turbocharger boost connections	Check boost pressure, exhaust upstream of unit and turbocharger.
High pressure fuel system malfunction	Check high pressure system, beginning with leaks and concluding with pump calibration.
Loss of engine compression	Make a cylinder compression check.

Engine knocks excessively

Probable cause	Remedial action
Defective injector(s)	Test by substitution.
Restricted fuel return line	Open line to verify, remove restriction.
Injection pump malfunction	Test pump.
Bearing knock	Verify and repair engine.

Engine mechanical

Engine malfunctions can be grouped into four categories:
- Fluid leaks.
- Oil burning.
- Loss of compression.
- Bearing wear.

Fluid leak

We are concerned with three fluids—fuel, oil, and coolant—each of which can leak to the outside of the engine or internally into another system.

Fuel

External fuel leaks are discussed under the heading Fuel system. Fuel contamination of the lube oil (as from Roosa Master pump seal failure) can be detected by the loss of viscosity and telltale odor.

Lube oil leaks

Lube oil leaks rarely lead to engine failure (when they do, failure is catastrophic), but the mechanic should correct such leaks before releasing the engine for service. In rough order of frequency, the most common sites of oil leaks are:
- Valve cover gasket
- Oil pan gasket
- Timing case gasket
- External accessory mounting gaskets
- Front and rear crankshaft seals
- Camshaft and oil gallery plugs

Tracing the source of a small leak can be difficult, because oil moves down and back, toward the rear of the engine. A black light detector or aerosol powder (which leaves a tracing) helps, but the ultimate tactic is to pressurize the crankcase. Connect a low-pressure (5 psi maximum) air line at the oil filter or dipstick tube. Soap the engine down and look for bubbles. Light foaming at valve cover gaskets is permissible. Rope-type crankshaft seals might leak air and still function when the engine runs; lip seals might foam, but they should not hemorrhage.

Internal oil leaks show up as loss of oil pressure, a condition that should be verified with an accurate test gauge, connected to the main oil gallery (Fig. 4-2). Assuming that the engine develops some oil pressure, run it up to operating temperature and compare gauge/engine rpm readings with manufacturer's specifications.

Usual causes of no pressure include oil pump drive failure, and the pump relief valve stuck in the open position.

Low oil pressure might be caused by pump wear or worn crankshaft/camshaft bearings (usually accompanied by increased oil consumption).

Coolant

A hot pressure test, as described in chapter 12, will verify system integrity. External leaks are usually visible; identifying internal leak paths requires some detec-

4-2 When in doubt, verify oil pressure with a Bourdon-type test gauge. Engine manufacturers supply pressure specs at idle and at full-governed speed.

tive work. Massive leaks into the combustion chamber produce white exhaust smoke at normal engine temperatures and, on engines so equipped, can sometimes be seen as coolant flow through the exhaust gas recirculation (EGR) discharge port. Leaks into the oil sump can usually be inferred from an examination of the dipstick, as indicated above. Large amounts of water fill the sump with a mayonnaise-like emulsion. A cylinder leakdown test, described under the heading Compression, will also identify coolant leak paths.

Oil burning

Blue smoke is the primary indicator of oil burning, which might be accompanied by oily exhaust stacks. The question facing the mechanic is the source of the oil. Table 4-1 should give direction. Blue smoke at all speeds, hot or cold, points to cylinder/ring wear. This condition can be verified by compression and leakdown tests, described in the following section. Smoke during acceleration after periods of idle and immediately upon start-up indicates leaking valve seals. Oil in the intake manifold and/or air boxes originates from an upstream source, usually from leaking blower or turbocharger seals, although other possibilities should not be overlooked. In some cases, merely changing the air filter element will correct the problem.

Loss of compression

There are three gauges in common use that test the engine as an air pump.

Cylinder compression test

The cylinder compression gauge (Fig. 4-3) screws into the injector boss and normally reads to 1000 psi.

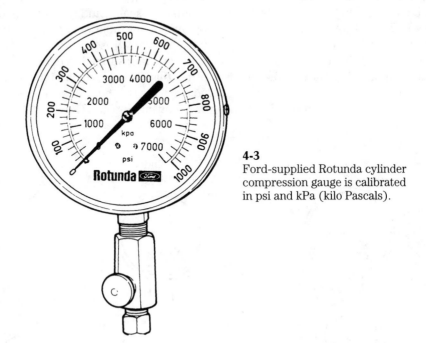

4-3
Ford-supplied Rotunda cylinder compression gauge is calibrated in psi and kPa (kilo Pascals).

WARNING: Do not attempt to substitute a gauge used for low compression SI engines. The gauge might explode.

Test procedures vary with make and model (e.g., some manufacturers specify a cold reading, while others require that the engine be at normal operating temperature). But the general procedure is as follows:

1. Disable all cylinders, except the one under test. The easiest way to do this is to remove the glow plugs, being careful to isolate the plugs from engine grounds.
2. Cut off the fuel supply. This will prevent the cylinder under test from starting (and possibly destroying the gauge) and will eliminate the fire hazard from fuel ejected through the glow plug bosses.
3. Test each cylinder in turn, allowing about six compression strokes per gauge reading. Record the readings.
4. Check the readings against published specifications for the engine and against each other. Variations of 20% or so between cylinders will result in noticeable roughness.

Weak compression in two adjacent cylinders points to a blown head gasket or possibly a cracked head. Do not squirt oil into a cylinder to determine if the compression loss is due to rings or valves. The cylinder might fire. Perform a leakdown test to locate the compression leak.

Cylinder leakdown test

Figure 4-4 illustrates a GM-supplied leakdown test kit with injector adapter. Unlike several competitive testers, the GM unit does not measure the percentage of leakdown. But it will indicate the source of compression leak. The engine should be warmed to operating temperature, with the crankcase sealed and radiator cap open.

CAUTION: Be extremely careful when opening a hot radiator cap. Mask the cap with a rag, turn it slowly to the release position, and allow pressure to vent before removal. Each cylinder in turn is brought up to tdc and pressurized to between 70 and 200 psi. Finding Fig. 4-4 tdc for cylinders other than No. 1 (which is marked) is a challenging proposition, but it can be achieved with the help of a degree wheel mounted in front of the harmonic balancer or, following Sun and GM practice, with a compression whistle.

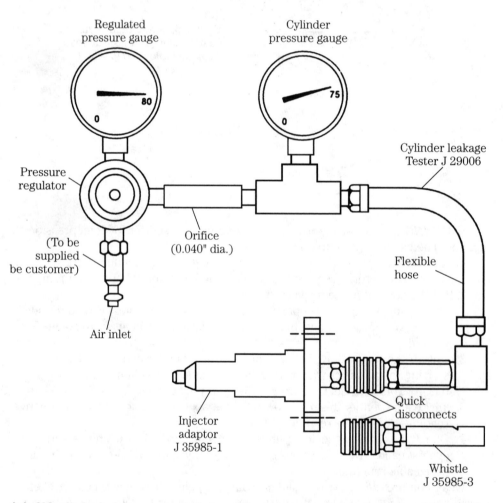

4-4 GM cylinder leakdown tester includes air line adapters, regulator, gauge, and tdc whistle. The whistle, which mounts to the adaptor, sounds as the piston nears top dead center on the compression stroke.

WARNING: If a piston is not at tdc, the engine will "motor," catching the unaware mechanic in the fan or belts.

Air escape paths are also escape paths for compression, as illustrated in Fig. 4-5.

- Leaks at the *air inlet* mean a bad intake valve.
- Leaks at the *dipstick* or *oil filler cap* mean worn cylinders or rings. Leakdown testers do not detect broken or stuck oil rings.
- Leaks at the *radiator header tank* mean a blown cylinder head gasket or a cracked head.
- Leaks at the *glow plug boss* of an adjacent cylinder mean a blown head gasket, or cracked head.
- Leaks at the *tailpipe* mean a bad exhaust valve.

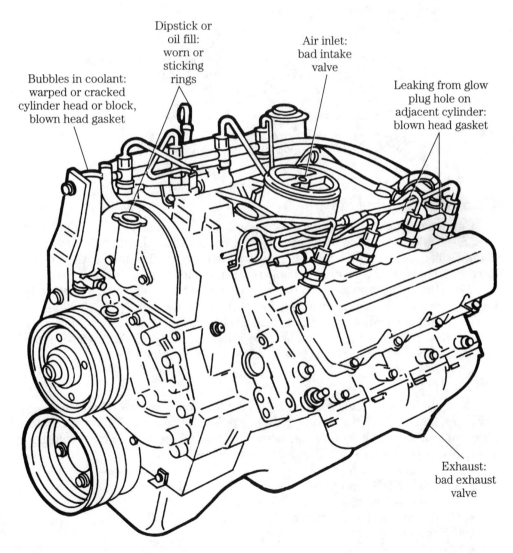

4-5 Leak points show where the compression escapes.

Crankcase pressure (blowby) test

The amount of blowby past the rings is a good, but limited, indication of cylinder bore/ring wear. Figure 4-6 illustrates the test apparatus, consisting of a pressure gauge and adapter, which replaces the oil filler cap. The limitation of this test is that not all manufacturers provide crankcase rpm/blowby pressure specifications. More often than not, the mechanic must establish the norm while the engine is still healthy.

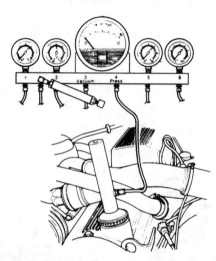

4-6
Ford supplies a crankcase pressure tester as part of the Rotunda gauge bar set (PN 0190002).

Crankshaft bearing noise

Crankshaft bearing knocks mean major and expensive repairs. By the time bearings reach the point of audible protest, the crankshaft journals will have been pounded flat and bearing fragments will have circulated throughout the engine. What could have been a relatively simple repair turns into a major project.

While this is slight consolation to a mechanic faced with a knocking engine, the technology exists to detect bearing failures early, while the damage is still local.

Frequent crankpin bearing clearance checks help (crankpin bearings wear about three times faster than main bearings), but short of periodically dismantling the engine, the best protection is spectroscopic analysis of the lube oil.

Oil analysis

Oil analysis has become routine, especially for larger engines. The technology, developed initially for diesel locomotives, was perfected by the U.S. Navy during the fifties and, in its present state, can predict engine failure with some exactitude. Trace materials found in lube oil and their sources are:

- Lead found in bearings.
- Silver found in bearings.
- Tin found in bearings.
- Aluminum found in bearings and pistons.
- Iron found in cylinder bores and rings.
- Chromium found in rings.

- Nickel found in bearings and valves.
- Copper found in bushings.

The presence of silicon and aluminum filings signals air cleaner failure. Other metallic elements, such as boron, calcium, and zinc, are present in the oil as additives, but in high concentrations can be considered contaminates.

Spectroscopic analysis involves vaporizing an oil sample in an electric arc. Each element has its own "signature" in the sense that it gives off light at a single frequency. Normally some 16 elements are tracked in concentrations of as little as one part per million. This information, together with engine history and use profile, gives a fairly definitive picture of engine condition.

The cost of the service is nominal, but if the transaction is handled by mail, the report can take a week or more to get back to the mechanic.

Fuel system

Fuel system diagnostics is simple enough to demonstrate on a particular engine; it is much more difficult to describe in the abstract. Some malfunctions can produce multiple effects, as outlined in Table 4-2. For example, a massive air leak on the low pressure side of the system can deny prime to the injection pump and prevent the engine from starting. Less serious leaks introduce bubbles into the fuel stream. Symptoms include hard starting, refusal to idle, and loss of power; leaks can also cause engine speed to rise and fall, or "hunt." A restricted fuel filter can affect idle quality, throttle engine output, and cause engine speed to surge.

However, the situation is not so ambivalent as it might appear. A mechanic can eliminate obvious possibilities early in the procedure by changing fuel filters and by verifying timing and fuel quality.

Fuel evaluation

The possibility of inappropriate or contaminated fuel should be looked into if the malfunction affects all cylinders, as when the engine refuses to start, misfires at random, or exhibits loss of power. Take a fuel sample from some convenient point upstream of the filter. Allow the fuel to settle for a few minutes in a glass container, and inspect for cloudiness (an indication of water), organic contamination (jellylike particles floating on the surface), and solids. Placing a few drops of fuel between two pieces of glass will make it easier to see impurities. Other ad hoc tests for evaluating fuel quality include:

- Fuel that burns cleanly in an oil lamp, emitting little smoke, will also burn cleanly in the engine.
- Water can be detected by wetting a strip of paper with fuel, setting it alight and listening for the crackle as the water bursts into steam.
- Mixing a small quantity of fuel with sulfuric acid will release carbon and resins, which appear as black spots. The fewer of these spots, the better.
- Litmus paper will show the presence of acids.

If fuel quality is unknown, it would be prudent to determine the cetane value. Figure 4-7 illustrates a fuel hydrometer used to make this determination.

4-7
Fuel quality is checked with a hydrometer calibrated in cetane numbers. The higher the cetane number, the less dense the fuel and the more easily it ignites.

Table 4-3 details the rather complex relationships between fuel properties and engine performance.

Fuel system tests

The fuel system consists of three circuits, each operating at a different pressure:

- The low-pressure circuit. This includes the tank strainer, water/fuel separator, filter(s), and lift pump. Pressures vary with application, but rarely exceed 75 psi. Failure in this circuit is most often associated with pressure-robbing restrictions, although air leaks and mechanical failures of the lift pump are also possible.

- The fuel-return circuit. Operating at almost zero pressure, the circuit consists of the return line, which conveys surplus fuel from the injector pump and injectors to the tank. Restrictions are the major problem.

- The high-pressure circuit. Most applications use a remote injection pump, connected to the injectors with high-pressure piping. (The high-pressure circuit is internal on engines with unit injectors, which are fed directly by the lift pump either through lines or, DDA-fashion, through a fuel manifold, integral with the cylinder head.) Fuel leaks, restrictions in the plumbing, and malfunctioning injectors are the major causes of failure. Unit injector pressure can exceed 20,000 psi on late model engines; distributor-type and in-line pump pressures typically range from about 1800 to 6000 psi; occasionally a very high-pressure in-line pump, such as the UTDS pump used on Mack trucks, might be encountered. Pressures achieved by that unit can exceed 15,000 psi.

Most of the tests described here follow industry practices. Specifications should be readily available. The exception is the Low-Pressure Circuit Test developed by Ford Motor Company for its diesel passenger cars and fight trucks. This test, or test series, locates malfunctions by measuring pressure drops across the individual components of the system *while the engine is running*. In other words, Ford uses the engine as a test bench. Most other manufacturers would have the mechanic disassemble the system and test each component separately. If you want to adapt to the

Table 4-3. Effects of fuel properties on engine performance*

Fuel property	Starting characteristic	Lubrication effectiveness	Smoke generation	Exhaust odor	Power output	Fuel consumption	Combustion chamber deposits
Firing cetane value	Directly related—starting improves as cetane value increases	Directly related—lubrication improves as cetane value rises	Closely related—smoke increases as cetane value decreases	Directly related—decreased by increasing cetane value	Irrelevant	Related	Related—decreased by reducing cetane value
Volatility 90% end point	No clear relationship	Related—becomes poor when volatility is poor	Directly related—increases as volatility decreases	No direct relationship	Irrelevant	Irrelevant	Related—increases as volatility decreases
Viscosity	No clear relationship	Some relationship—becomes poor when viscosity increases	Related—increases as viscosity increases	No independent relationship	Irrelevant	Irrelevant	Related—increases with viscosity
Specific gravity	Irrelevant	Irrelevant	Related—increases as specific gravity increases	No independent relationship	Directly related—associated with calorific value	Related—associated with calorific value	Related—depends on properties of engine
10% residual carbon	Irrelevant	Irrelevant	Related—improves as residual carbon decreases	No independent relationship	Irrelevant	Irrelevant	Related—decreases as residual carbon decreases
Sulfur				No independent relationship			
Flash point				No independent relationship			

*Courtesy Yanmar Diesel Engine Co., Ltd.

Ford approach to an engine built by another manufacturer, you should establish baseline pressure specifications while the system is still healthy and operating normally.

Low pressure circuit Most fuel-related malfunctions occur in the low-pressure circuit, which connects the fuel tank with the injection pump (or unit injectors). Before making any programmed tests, do the obvious—replace the fuel filter(s), drain the water separator, and prime the system.

Pressure drop test This test is made while the engine is running to locate excessive pressure drops, or restrictions, relative to the fuel filter.

Which gauge will be used depends on fuel filter location, relative to the transfer pump. If the filter is on the suction side of the pump, in a negative pressure area, a 0–30 in. Hg (inches of mercury) vacuum gauge will be required. If the filter is between the transfer pump and injection pump, it will be subject to positive pressure, and the test requires a 0–15 psi pressure gauge.

Follow this procedure:
1. Connect the appropriate (pressure or vacuum) gauge at the output side of the filter (Fig. 4-8).
2. Start the engine and note the gauge reading at the specified rpm.
3. Connect the appropriate gauge to the inlet side of the filter (Fig. 4-9).
4. Start the engine and note the gauge reading at the specified rpm.
5. Connect the gauge across the inlet side of the filter, and compare readings against factory specs or a previously established baseline.

Air leaks Large air leaks rob the injection pump (or, more rarely, the transfer pump) of prime. Smaller leaks can result in hard starting, sudden shutdowns, rough running, and loss of power.

Figure 4-10 illustrates a leak detector—a transparent section of fuel line spliced into the suction side of the injection pump. Some engines come from the factory with

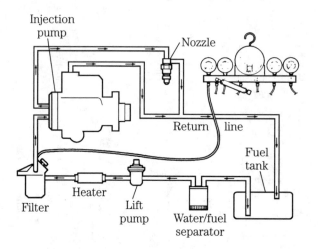

4-8 Filter location relative to the pump determines whether a pressure or vacuum gauge would be used. In this case, the filter is on the outlet side to the pump and sees positive pressure.

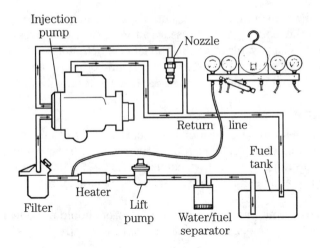

4-9 The next step is to install the gauge at the suction side of the filter. The difference between readings represents the pressure drop across the unit.

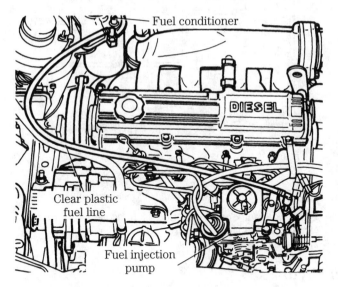

4-10 A section of clear plastic tubing installed between the lift and injection pumps will detect air leaks.

"peak-a-boo" fuel lines; those that don't probably should be modified, especially when responsibility for maintenance is split between several mechanics.

An alternate procedure is to disconnect the fuel return line at the tank. Submerge the free end of the line in a bucket of clean diesel fuel.

Start the engine and look for bubbles in the fuel stream. The usual sources of intrusion are finger-tight filter bleed screws or loose fittings. Tighten and retest. If the leak persists, it might be necessary to pressure-test the circuit. Break the connections at the tank and injection pump. Plug one end of the line and, using an adapter, make up a hand pump or other source of low-pressure air to the free end. Pressurize to no more than 15 psi. Apply soapy water or a commercial leak detector solution to line connection points, filter-canister parting surfaces, valves, and other potential leak sites.

WARNING: Do not attempt to pressurize the fuel tank—it might rupture.

Transfer pump

Figure 4-11 illustrates the academically correct way of checking pump output pressure. A container of diesel fuel supplies the pump, thus isolating it from possible suction-side restrictions. However, visual inspection of the tank screen and fuel line will usually suffice, and mechanics rarely go to the trouble of rigging an auxiliary fuel supply.

In addition to output pressure, the manufacturer should provide a pump volume specification, which is especially critical for diaphragm-type pumps.

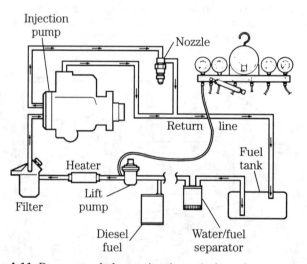

4-11 Recommended practice is to isolate the pump from suction-side pressure drops when making the pump output pressure test.

Fuel return circuit

The return line recirculates surplus fuel from the injection pump and injectors back to the tank. Restrictions in this circuit raise the internal pressure of the pump, opening the injectors early and producing symptoms such as abnormal combustion noise, hard starting, stumbling, and shutdown on deceleration.

Most manufacturers check the circuit by measuring the flow rate at a prescribed engine speed. The Ford approach is to measure pressure, either at the injection pump return-line outlet or at some convenient point in the main return line, downstream of the injector circuit tie-in.

Returned fuel must first pass through an orifice under the pump outlet fitting. Because the orifice is small and flow through it might be further constrained by the presence of a check valve, the orifice is a prime candidate for stoppages. The test just described will not detect a restriction across the orifice (unless the restriction is absolute and no fuel flows). When the engine shows symptoms of return line restrictions and a pressure test comes up negative, remove the orifice for inspection. Figure 4-12 illustrates two of the several possible orifice configurations.

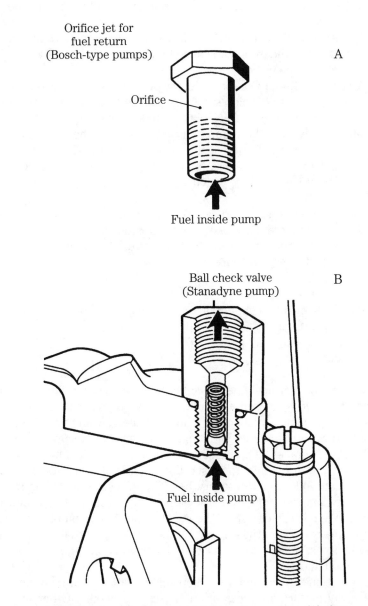

4-12 Returned fuel is metered through an orifice (A) that might incorporate a check valve (B).

High pressure fuel circuit

Malfunctions in this circuit can be categorized as leaks, injector failure, or pump failure.

Leaks

Most high pressure leaks develop at the fuel pipe unions on engines with pump-fed injectors. Loosening and retightening the union usually corrects the problem. See chapter 5 for additional information.

WARNING: Keep well clear of high-pressure fuel spray, which will penetrate the skin and should it strike the eyes, can cause blindness. Use a piece of cardboard to detect leaks in hard-to-see places.

UTDS, APD, 6BB, T, and Q series injector pumps, used on current Mack truck engines, have a tendency to develop small, hairline leaks at the delivery valves (just below the fuel pipe connections). Leak paths can develop across the O-ring seal, between the delivery valve and the pump body, and between the valve hold-down nut and pump body. Replace the O-ring seal and copper ring gasket. Carefully inspect the lapped surfaces for scoring. These faults can often be traced to overtightening the fuel pipe/delivery valve union, which twists the delivery valve and scores the mating surfaces.

The UTDS pumps in question develop more than 15,000 psi at engine torque peak; hard-used vehicles are more likely to develop leaks than those that are operated conservatively. Restrictions in the fuel line, overfueling, and clogged injectors will further increase delivery pressure.

Injectors

Like spark plugs, injectors should be presumed guilty until proven innocent. Injector failure has many symptoms, including:

- Black or gray smoke
- White smoke
- Detonation.

All these symptoms are cylinder specific. In some cases, a faulty pump-fed injector can be identified by merely feeling the pressure pulses in the individual fuel pipes. If that approach fails (long lines and multiple bends can mask the pulsations), disable the injectors, one at a time. With engine ticking over at slow idle, open the injector bleed screw or crack the hp fuel line fitting to reduce fuel pressure below the opening valve for pump-fed injectors (see Fig. 4-1). Unit injectors are disabled by depressing the cam followers with a pry bar. The effects of disabling a fully functional injector are unmistakable: the associated cylinder stops firing, and vibration and exhaust smoke levels increase. If the injector has failed completely, disabling it will have no further effect on the engine; if the injector is partially functional, combustion roughness associated with that cylinder will cease. When in doubt, substitute a known-good injector.

Most malfunctions are caused by carbon accumulation on the nozzle tips, which distorts the spray pattern and, less often, seizes the needle. Other faults are detected by test with the apparatus shown in Fig. 4-13. As noted earlier, keep well clear of the discharge when making these tests.

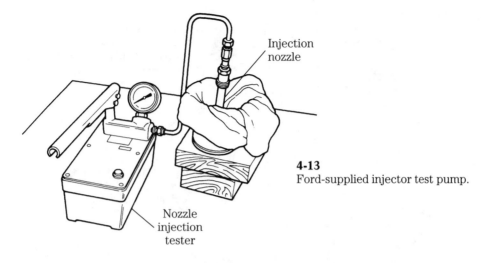

4-13
Ford-supplied injector test pump.

There are four standard tests—*spray pattern, opening pressure, sealing effectiveness*, and *chatter*—which apply to most injectors.

Spray pattern Connect the tester as shown, and operating the pump lever in short, fast strokes, observe the spray pattern. What constitutes an acceptable spray pattern depends, in part, upon injector type. As shown in Fig. 4-14, pintle and variations on pintle nozzles should produce a finely divided spray of uniform consistency and penetration. Spray pattern width varies with the application. Multihole nozzles produce a splayed pattern, typically as shown in Fig. 4-15. It is helpful to direct the discharge against a piece of paper; the pattern can then be compared to the pattern produced by a new nozzle.

Opening pressure Slowly depress the pump lever and note the peak reading, which corresponds to the injector opening pressure. Compare with manufacturer's specifications for the unit under test. Normally an injector that falls within the specification range is acceptable. However, some engines are very sensitive and do not idle properly unless injectors cluster at one or the other end of the specification range. Most injector springs can be stiffened by means of shims or a threaded spring seat.

Sealing Dribble standards vary with injector design and spring tension. The unit illustrated in Fig. 4-16 should remain fuel tight under 80% of opening pressure for at least 10 seconds without fuel droplets forming or running off the tip. Note any other sources of leakage and correct.

Chatter Injector opening is normally accompanied by a sharp "pop," which means that nozzle and spring assembly retract cleanly. Verify the action by listening to the chatter as the pump is cycled rapidly. Note, however, that leaking injectors can be made to chatter and that some perfectly good throttling pintle injectors pop open, but refuse to chatter, no matter how adroitly the pump handle is worked.

Injection pump

Replacing an injection pump is an expensive proposition, second only to rebuilding the engine. Before you decide to do that, make absolutely certain that the

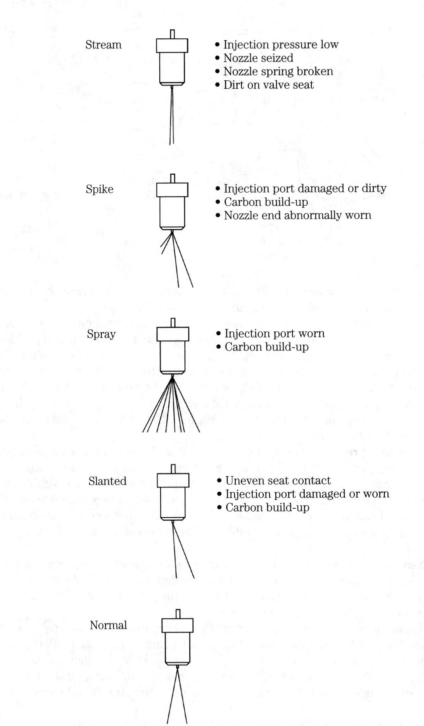

Stream
- Injection pressure low
- Nozzle seized
- Nozzle spring broken
- Dirt on valve seat

Spike
- Injection port damaged or dirty
- Carbon build-up
- Nozzle end abnormally worn

Spray
- Injection port worn
- Carbon build-up

Slanted
- Uneven seat contact
- Injection port damaged or worn
- Carbon build-up

Normal

4-14 Spray pattern diagnosis—modified pintle-type nozzle used on small, indirect injection engines.

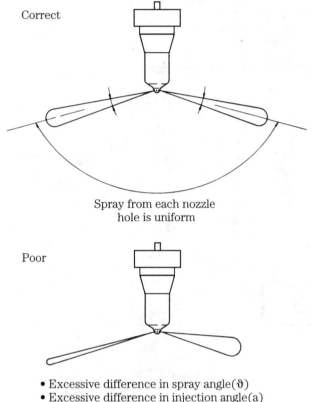

Correct

Spray from each nozzle
hole is uniform

Poor

- Excessive difference in spray angle(ϑ)
- Excessive difference in injection angle(a)
- Incomplete atomization
- Sluggish starting/stopping of injection

4-15 Spray pattern diagnosis—multihole nozzle used on
direct injection engines.

pump mechanicals are at fault. Many pumps include an accessory altitude-compensation mechanism that alters the air/fuel ratio and, in some examples, adjusts delivery timing (Fig. 4-17). A microscopic pinhole in the diaphragm (Caterpillar) or bellows can mimic symptoms of major pump failure.

Internal pressure test This test addresses itself to the central function of the pump—its ability to generate pressure at a proscribed engine rpm and load—and thus carries considerable weight. But do not take this or any other field test as definitive. If the pump should fail the test, recheck the low-pressure side of the system. If no faults are found there, remove the pump for bench testing before condemning it.

Test port location varies with make and model. Most pumps include test ports, as shown in Fig. 4-18; those that do not, develop full pressure below the return-line restrictor.

Controls Contemporary pumps might incorporate a fuel cutoff solenoid valve, as shown in Fig. 4-19. The valve remains closed until voltage is applied to the windings. For a quick field check, listen for the telltale click as the circuit is energized.

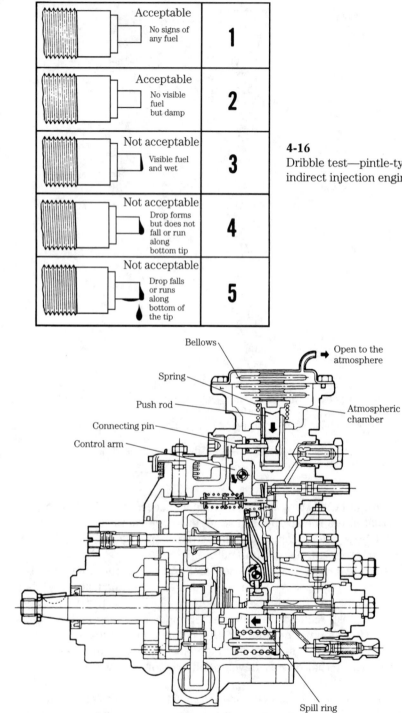

Acceptable No signs of any fuel	**1**	
Acceptable No visible fuel but damp	**2**	
Not acceptable Visible fuel and wet	**3**	
Not acceptable Drop forms but does not fall or run along bottom tip	**4**	
Not acceptable Drop falls or runs along bottom of the tip	**5**	

4-16
Dribble test—pintle-type nozzle,
indirect injection engines.

Bellows

Open to the
atmosphere

Spring

Push rod

Atmospheric
chamber

Connecting pin

Control arm

Spill ring

4-17 Injection pump for Mazda-built 2L Ford Escort/Lynx engine adjust
timing and fuel volume for reduced smoke at high altitude.

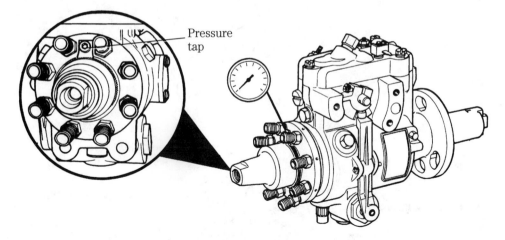

4-18 Internal pressure test port—Stanadyne pump.

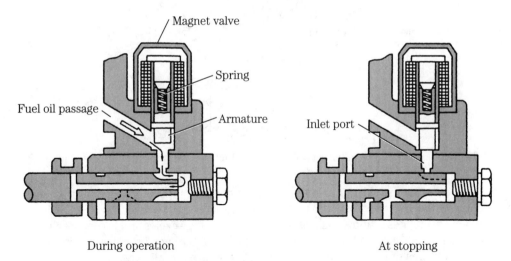

During operation

At stopping

4-19 A solenoid-operated fuel shutoff valve is a standard feature on modern pumps, some of which also incorporate a manual control.

The governor section of the pump can include an array of adjustment screws, most of which are sleeping dogs, better left undisturbed. The two that a mechanic needs to know about set the idle speed and maximum governed speed, usually by restricting the range of the throttle control lever. Figure 4-20 is typical, although more elaborate designs sometimes split the idle into two ranges.

Air intake system

A naturally aspirated four-cycle engine pumps about 3.0 cu. ft. of air per minute per developed horsepower. If that engine were to run 40 hours a week, it would ingest

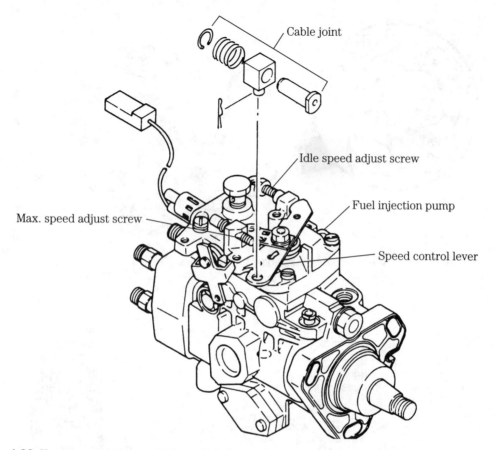

4-20 Yanmar pumps have a fairly typical idle and high speed limiter screw arrangement.

one-and-a-quarter million cubic feet of air a year. In a typical dust-laden urban environment, this translates as more than 175 lb. of solids, most of which are hard silicates. Two-cycle engines require half again as much air as four-cycles and turbocharging further increases the air volume. Obviously, diesel air filters need regular attention.

Symptoms of filter and intake tract restrictions generally are:

- Hard starting
- Black exhaust smoke
- Loss of power
- Failure to reach governed speed
- Erratic idle (naturally aspirated engines)
- Turbocharger speed surges and possible oil pullover.

Figure 4-21 illustrates a test hookup using a vacuum gauge. This drawing is a bit atypical, because the test tap is incorporated in the air cleaner cover; the usual practice is to locate the tap below the air cleaner assembly.

The test is made with the engine running at the speed that generates maximum air flow. Air flow peaks at fast idle for naturally aspirated (nonsupercharged) engines; turbocharged engines produce greatest air flow under full power.

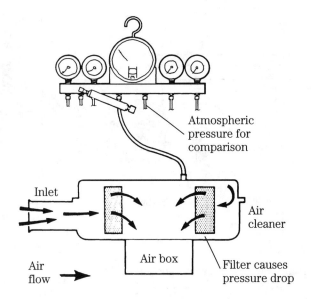

Inlet

Atmospheric
pressure for
comparison

Air
cleaner

Air box

Air
flow

Filter causes
pressure drop

4-21 Stationary and marine engineers use a manometer—a U-shaped tube, partially filled with water and calibrated in quarter-inch increments—to measure manifold pressure. No mechanical gauge can match the sensitivity of this device. Field mechanics prefer the convenience of a vacuum gauge. The Ford-supplied gauge shown has an expanded scale, calibrated in inches of water.

The usual culprit is the filter. Replace pleated paper elements, clean and reoil permanent types, and retest.

CAUTION: Do not operate the engine with the air cleaner removed. Unlike SI intake manifolds, which are obstructed by venturis and throttle plates, most diesel intake systems open directly (or via expensive compressor wheels) to the combustion chambers. A dropped wing nut or, in the case of turbocharged engines, careless fingers will have dramatic consequences.

Turbochargers

Turbocharger failure has a number of causes, most of them associated with the inflexible demands imposed on the lubrication system by these devices, which under peak load, rotate between 120,000 and 140,000 times a minute (Fig. 4-22). Turbocharger bearings require large quantities of clean motor oil for lubrication and cooling. Momentary interruptions in the oil supply, which might occur during initial start-up after an oil change or during sudden acceleration immediately after a cold start, will destroy the bearings. Engine oil change intervals must be religiously observed. If the engine is run hard and abruptly shut down, oil trapped in the turbocharger case absorbs heat from the turbine and carburizes into an abrasive.

Other problems associated with the device include carbon and scale accumulations on the wheels, which degrade efficiency and upset balance, exhaust-side leaks (most of them caused by thermal expansion of the piping), pressure-side leaks, blade-tip erosion, and bearing wear that, if severe, allows the rotating elements to rub against the sides of the case.

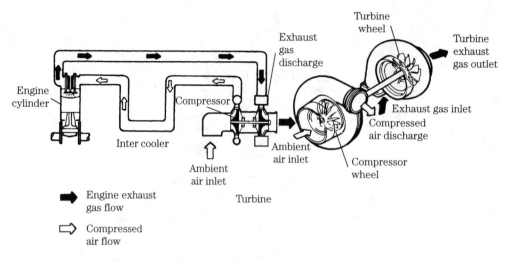

4-22 Gas flow through a Detroit Diesel turbocharger.

Some malfunctions—noise, excessive bearing clearances, cracked exhaust plumbing—obviously involve the turbocharger. Performance decline, such as loss of power, sooty exhaust, low or erratic turbo boost might have its causes outside the unit. Before assuming that the turbocharger is at fault, check the fuel system, particularly the injectors, and verify that the pressure drop across the air filter element is within specification. Carefully inspect turbocharger oil and vacuum lines.

Start the engine and listen to the sound the turbocharger makes. With a little practice, you will be able to distinguish the shrill sound of air escaping between the compressor and engine from exhaust leaks, which are pitched at a lower register. If the sound changes in intensity, check for a dogged filter, loose sound-deadening material in the intake duct, and dirt accumulations on the compressor wheel and housing. Determine the boost pressure at prescribed load and speed.

Loss of power/black smoke Assuming that the turbocharger is not throttled by upstream flow restrictions, check for loose bolts, blown gaskets, cracked pipes, and in short, any condition that would compromise gas flow to and from the unit.

Turn the impeller by hand—some seal drag is permissible, but light finger pressure should be sufficient (Fig. 4-23). Hard spots or binds mean that the bearings have gone and, quite possibly, taken the turbine housing with them. Installed bearing clearances vary between models. Most turbocharger bearings depend on hydraulic stabilization for control of radial (side-to-side) movement. Once the engine starts, the shaft rises off its bearing and floats on a wedge of oil. Because the bearings also float in their bosses, static clearances (bearing ID to shaft OD, bearing OD to boss ID) are quite large and it will be possible to rock the impeller from side to side. Thrust bearing (axial) clearance is a more accurate gauge of turbocharger condition. Measure at the hub of either wheel with a dial indicator and suitable extension bar. The wear limit varies from a few thousandths of an inch to more than 0.030 in., for different makes.

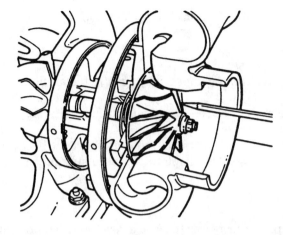

4-23
It is usually easier to make preliminary bearing checks from the compressor end of the turbocharger.

Figure 4-24 illustrates a delicate operation used to remove soft carbon deposits from IHI compressor wheels. With the engine operated at or near full load, the mechanic introduces small quantities of a proprietary product, known as Blower Wash, through the inspection cover. Plain water also can be used. The rate is critical; the cleaning solution can damage the impeller if injected too rapidly. Several washings, repeated at ten-minute intervals, might be required to restore full boost.

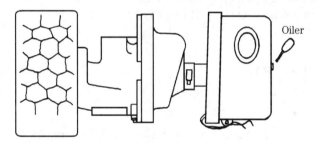

4-24 Carbon accumulations on the blower can be removed by controlled injections of detergent.

The hot section can also be cleaned under power, but this operation is usually confined to large stationary power plants and to aircraft engines. Various detergents are used for this purpose, together with mild abrasives such as ground walnut or pecan shells.

Blue/white smoke Carefully inspect the intake ducting for oil smears that would mean that the source is upstream of the turbocharger, possibly from the air cleaner.

1. Remove the compressor housing from the turbocharger.
2. Clean the inner surfaces of the housing with special attention to the area behind the compressor wheel, adjacent to the seal cartridge.
3. Apply a light coat of Spot-Check type SKD-MF developer to the critical areas. Reassemble.
4. Run the engine at maximum governed speed for five minutes.

5. Open the turbocharger and look for oil streaks on the developer that would indicate seal leakage.

Seal failure might be caused by:

*   Wear (often associated with bearing failure).
*   An abnormally high-pressure differential imposed by a restricted air filter.
*   Restrictions in the return fuel oil line.
*   The practice of using the engine as a brake; for example, when motoring downhill.

Noisy operation   This condition is almost always caused by contact between the rotating members and the turbine case. See Chapter 9 for more information.

Exhaust system

Excessive back pressure costs power, increases the smoke level, and can prevent the engine from idling. Inspect the exhaust plumbing for kinks, flattened or necked down areas, mismatched flanges, and in general, for any evidence of flow disruption.

Detecting internal faults—baffle plates that have come adrift or blisters in double-walled piping—requires a back-pressure test, using a water manometer connected a foot or so downstream of the turbocharger. Specifications are generally available for commercial engines with factory-supplied exhaust systems. For example, DDA Series 60 six- and eight-cylinder engines should develop no more than 3 in. Hg (1.5 psi) on a relatively straight length of pipe, six to twelve inches downstream of the turbocharger.

Starting system

Cranking

Electrical cranking systems are based on a common model—battery, switch, solenoid (which magnetically switches the motor on and engages the pinion gear with the flywheel ring gear), and motor (Fig. 4-25). These elements are always present, although some manufacturers cascade the solenoid with a relay. Turning the switch to "Start" energizes the relay, which in turn energizes the solenoid and starter motor.

The cardinal rule of starting system diagnosis is to begin at the battery. The electrolyte should cover the plates. Check each cell individually with a hydrometer. Full charge electrolyte density varies somewhat with battery make, but 1.260 at room temperature is a good average. More than 0.01 difference between cells indicates some loss of efficiency; 0.05 or more difference means a shorted cell, and if the system is to operate reliably, the affected battery must be replaced. Bring the battery up to full charge. Remove the cables, clean the terminals, reconnect the cable clamps, and grease.

Table 4-4 outlines the basic diagnostic approach, which depends entirely on voltage measurements made between the terminal connections shown in Fig. 4-25 and engine ground. You will need a digital voltmeter and jumper cables to shunt suspect components.

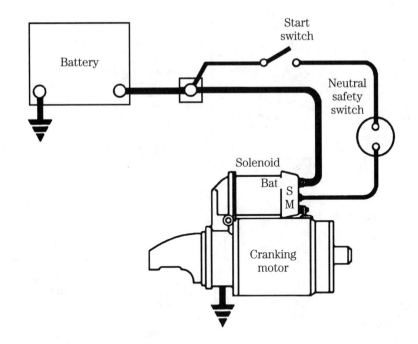

4-25 Typical starting circuit.

I prefer to work backwards from the starter motor; you might find it less confusing to work from the direction of apparent current flow; that is, from battery forward, through the control switch to the solenoid and motor. In any event, make these tests quickly, engaging the starter for no more than 30 seconds at a time and allowing several minutes between engagements for cool down. A short in a cable-fed circuit will pass very high current—in excess of 1000 amps. Be alert for signs of overheating at the cable terminals and motor.

The procedure outlined in Table 4-4 should give you a good start at identifying the problem. More detailed diagnostic procedures, including component-by-component voltage drop tests, are described in chapter 11.

Glow plugs

Early glow plugs were rated for continuous operation at full battery power and remained energized until the engine started. More modern systems generally cycle the plugs on and off or, in the case of Mitsubishi automotive engines, reduce supply voltage until glow plug resistance reaches a preset value, corresponding to near-normal coolant temperature. Thus, complaints of intractability on start-up, persistent detonation, and white smoke can mean a problem in the control module or wiring harness. For example, Fig. 4-26 illustrates a Ford-Navistar module output determination.

The diagnostic procedure described here applies to most "make-and-break" control circuits. More complex, voltage-limiting circuits must be tested as outlined by the manufacturer.

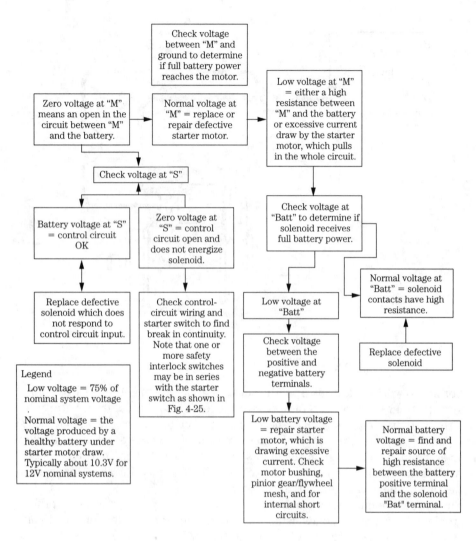

Check voltage between "M" and ground to determine if full battery power reaches the motor.

Zero voltage at "M" means an open in the circuit between "M" and the battery.

Normal voltage at "M" = replace or repair defective starter motor.

Low voltage at "M" = either a high resistance between "M" and the battery or excessive current draw by the starter motor, which pulls in the whole circuit.

Check voltage at "S"

Battery voltage at "S" = control circuit OK

Zero voltage at "S" = control circuit open and does not energize solenoid.

Check voltage at "Batt" to determine if solenoid receives full battery power.

Normal voltage at "Batt" = solenoid contacts have high resistance.

Replace defective solenoid which does not respond to control circuit input.

Check control-circuit wiring and starter switch to find break in continuity. Note that one or more safety interlock switches may be in series with the starter switch as shown in Fig. 4-25.

Low voltage at "Batt"

Check voltage between the positive and negative battery terminals.

Replace defective solenoid

Legend

Low voltage = 75% of nominal system voltage

Normal voltage = the voltage produced by a healthy battery under starter motor draw. Typically about 10.3V for 12V nominal systems.

Low battery voltage = repair starter motor, which is drawing excessive current. Check motor bushing, pinior gear/flywheel mesh, and for internal short circuits.

Normal battery voltage = find and repair source of high resistance between the battery positive terminal and the solenoid "Bat" terminal.

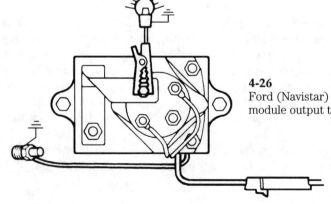

4-26
Ford (Navistar) glow plug control module output test.

1. Locate the glow plug control module.
2. Connect a test lamp to the power input lead. Turn the ignition switch on. The lamp should light. If it does not, find the break by tracing the power path back to the switch. If it does light, go on to the next step.
3. Connect the test lamp to the module output lead and start the engine. On most systems, the light should remain on initially and, as temperature rises, cycle with progressively shorter on times until engine temperature approximates normal. The manufacturer's manual will provide a detailed duty cycle. If the lamp does not function normally, check the power relay and associated wiring. Disconnect the battery and replace the module only after these other possibilities have been eliminated. If the control circuit appears to be functional, repeat the test at the individual glow plug connectors.

Glow plugs can be tested for continuity with a low-range ohmmeter or battery-powered lamp (Fig. 4-27).

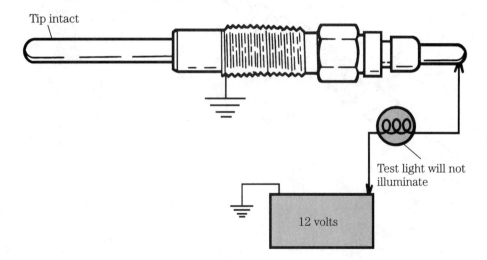

4-27 Glow plug continuity can be determined with an ohmeter or test lamp. But, as pointed out in the text, the voltage available to the glow plug must not exceed its rated voltage.

CAUTION: Test voltage must not exceed the glow plug rating, which for GM, some Navistar, and Navistar 6.9 L Ford engines is 6 V.

Mechanical damage, particularly if it extends to several glow plugs, is evidence of a more fundamental problem. For example, the missing tip on the plug pictured in Fig. 4-28A suggests chronic detonation. (Wire-type glow plug elements will come adrift under the same conditions and batter the piston heads.) Effects of excessive current are shown in Fig. 4-28B.

Cooling system diagnosis cannot be profitably discussed without detailed reference to the components that make up the various systems. Please see chapter 12.

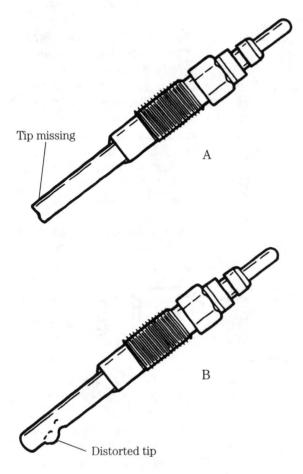

Tip missing

A

B

Distorted tip

4-28
Tip breakage (A) suggest detonation, usually the result of improper fuel or over-advanced injector timing. A melted or blistered tip (B) can mean control circuitry malfunction.

5
CHAPTER

Fuel systems

The fuel system has always been one of the more problematic areas of the CI engine. Fuel must be introduced against cylinder compression at some point before tdc (top dead center). Early engines employed air injection. A small amount of compressed air was mixed with the fuel and injected into the cylinder with it. Besides providing the energy for injection, compressed air helped to atomize the fuel. The disadvantages were several. The compressor—or, as was more often the case, the compressor stages—were parasitic loads, absorbing power without directly producing any. And the weight of the compressors, high-pressure plumbing, and starting tanks meant that the diesel was confined to stationary or shipboard applications.

In 1910 James McKechnie developed the first successful mechanical or solid injector. Development continued in the next decade, with the Robert Bosch pump and injector as one of the most significant breakthroughs. Today all diesel engines employ some means of solid injection, although you might encounter a few ancient examples with the earlier system.

Fuel injectors

The fuel injector—variously called the *spray nozzle, spray valve*, or *fuel delivery valve*—is the final component of the fuel system and, in many ways, the most critical. The injector must deliver a timed and metered spray of fuel to the chamber, and must double as a check valve to prevent compression and combustion pressures from entering the fuel supply lines. In addition the injector is fitted with a return pipe to ensure continuous fuel flow in the circuit.

An injector consists of two main parts—the nozzle holder (called the injector body in Fig. 5-1) and nozzle (or as in Fig. 5-2, the nozzle body). The nozzle mounts the needle valve, also known as the nozzle valve. The nozzle holder secures the assembly to the cylinder head, carries the fuel inlet and spillage fittings, and on some designs, incorporates a high-pressure fuel pump.

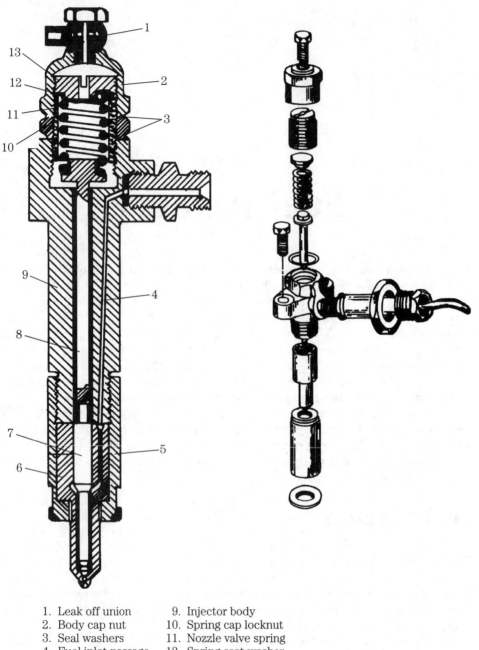

1. Leak off union
2. Body cap nut
3. Seal washers
4. Fuel inlet passage
5. Nozzle cap nut
6. Nozzle
7. Nozzle valve
8. Nozzle valve spindle
9. Injector body
10. Spring cap locknut
11. Nozzle valve spring
12. Spring seat washer
13. Spring cap nut

5-1 CAV pintle-type nozzle in cross section with parts nomenclature (GM Bedford Diesel) and a similar CAV nozzle in exploded view (Lehman Ford Diesel).

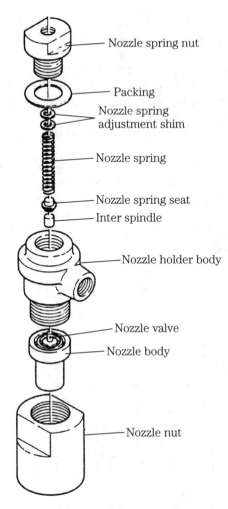

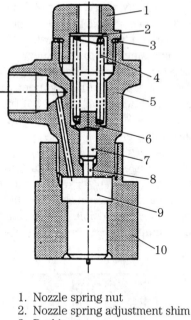

1. Nozzle spring nut
2. Nozzle spring adjustment shim
3. Packing
4. Nozzle spring
5. Nozzle holder body
6. Nozzle spring seat
7. Inter spindle
8. Nozzle valve
9. Nozzle body
10. Nozzle nut

5-2 Yanmar throttling pintle nozzle. Diesel nomenclature has never been completely standardized, as can be seen by comparison of the Yanmar parts lists with the list in Fig. 5-1.

The nozzle should:
- discharge cleanly, without "afterdribble."
- atomize the fuel for easy combustion.
- direct the spray pattern into the far reaches of the cylinder, but not against the piston or cylinder walls.

The spray pattern should have a definite start and stop for consistent atomization and to prevent fuel from accumulating in the cylinder between cycles.

Droplet size is a function of fuel viscosity, delivery pressure, air density, and of the ratio of orifice length to diameter. The more finely divided the fuel, the more quickly it ignites and the smoother the engine runs. But diesel combustion proceeds at uneven rates, due to varying fuel/air ratios in the vicinity of the fuel stream. To obtain oxygen for complete combustion, fuel droplets must penetrate to all regions of the cylinder. The larger (and heavier) the droplets, the better the penetration. In order to satisfy these contradictory requirements—small droplets for rapid ignition, large droplets for complete combustion—the spray pattern consists of a "hard" core surrounded by a "soft" sheath of foglike particles.

The simplest form of nozzle is the *open hole* type, in which the orifice is exposed to cylinder pressure. Tandem check valves, located in the fuel passage above the nozzle, prevent hot cylinder gases from blowing back into fuel line and also help to control dribble at the end of the pump stroke. Open-hole nozzles began to disappear in the 1920s with the advent of solid injection, but can still be found on certain Caterpillar engines.

The *differential pressure nozzle* can be recognized by a spring-loaded nozzle valve, or needle, which seats against the orifice (Fig. 5-3). Fuel collects in a small pressure chamber formed by the recess in the nozzle valve spindle and nozzle valve shoulder. Pressure in the chamber rises until the force on the nozzle shoulder overcomes spring tension. The nozzle valve lifts off the seat, fuel flows out of the pressure chamber, past the nozzle seat, and through the orifice to the cylinder.

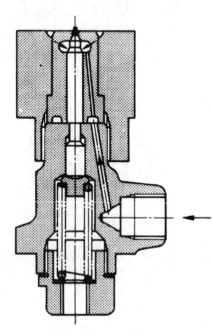

5-3
Fuel enters the nozzle body pressure chamber at the lower end of the nozzle body and generates an axial force against the nozzle shoulder. When this force overcomes spring tension, the nozzle unseats. Fuel is injected through the nozzle orifice and into the engine. When the delivery valve (see Fig. 5-21) closes, fuel pressure abruptly drops and the spring closes the nozzle valve.
Yanmar Diesel

Once the nozzle valve opens, its full cross-sectional area—the annular shoulder and tip—comes under fuel pressure. Because force is a function of pressure acting on area, less pressure is required to hold the needle open than to unseat it; hence the term *differential pressure nozzle*.

The size of the orifice determines the fineness of the spray. The smaller the diameter, the greater the atomization. But if the hole is too small it will not be able to deliver sufficient fuel. Multihole nozzles overcome this difficulty because the holes can be drilled small enough to atomize the fuel, and the number can be large enough to provide sufficient delivery. Depending on the application, the nozzle might have as many as 16 orifices, with diameters as small as 0.006 in. The spray pattern is usually symmetrical, although some chamber shapes and nozzle locations demand a canted pattern.

Some nozzles feature a pintle or pin that retracts into the orifice during discharge. The pintle is an extension of the nozzle valve. When retracted it releases a hollow, cone-shaped spray at an angle of up to 60° from vertical (Fig. 5-4). Pintle-type nozzles require, on the average, less maintenance than the multihole types, because the shuttle action of the pintle discourages carbon buildup and because the orifice diameter is larger.

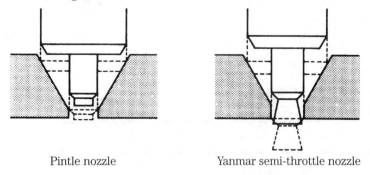

Pintle nozzle   Yanmar semi-throttle nozzle

5-4 Pintle (left) and throttling pintle nozzles. _{Yanmar Diesel}

A variant of this design is the throttling pintle, employed by Mercedes-Benz and others. The pintle has a tapered section. In the closed position the orifice is completely blocked, and no fuel flows. Under part pressure, during the beginning of the pump stroke or when the distributor valve has just initiated delivery, the pintle retracts until its small section is in the nozzle orifice. It is unseated at this point, but fuel delivery is restricted by the diameter of the pintle. At full delivery pressure the pintle retraces clear of the orifice, allowing maximum fuel flow. Other throttling pintles work in the reverse manner—that is, the pintle is fully retracted under zero or less than opening pressure, and is pushed outward as fuel pressure increases.

Fuel delivery systems

Solid injection systems (the term *solid* distinguishes these systems from the air injection system originally used) can be classified into three broad categories.

Common-rail Common-, or third-rail, systems employed a continuously operating pump to pressurize the fuel header and injectors (Fig. 5-5). Injectors were opened mechanically by a camshaft, which on more sophisticated designs operated through wedges to vary the duration of injection with engine load. Most common-rail systems included an accumulator in parallel with the header. The accumulator averaged pump impulses so that system pressure remained a fairly constant 5000 psi or so.

These systems were an outgrowth of air injection technology, substituting pressurized fuel pressure for compressed air. However easy this might have made the transition 70 years ago, common-rail injection had serious shortcomings. The sys-

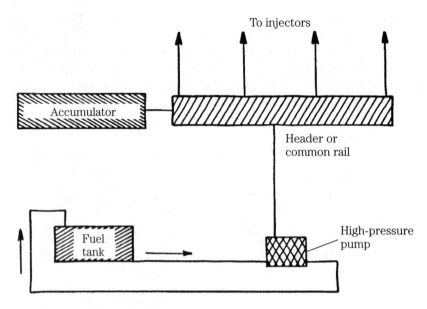

5-5 Common- or third-rail systems.

tem, wedges and all, was not adaptable to high-speed engines, and to judge from contemporary accounts, leaks were endemic. Headers leaked and injectors dribbled fuel into the combustion chamber throughout the whole cycle.

Common-rail systems were conceptually obsolete by 1920, but a few of these engines continued to be manufactured and are occasionally encountered today.

Pump-controlled systems Contemporary practice is to deliver fuel to the nozzles in small, high-pressure lots, timed to engine requirements and metered according to load. Injection, or jerk, pumps are classed according to location and distribution arrangements. Each injector can be fitted with its own cam-operated pump, in which case the assembly is known as a *unit injector*. The alternate approach is to supply all injectors from a central pump, mounted at some remote location on the engine and connected to the injectors by piping. *In-line* pumps dedicate a plunger to each injector; *distributor-type* pumps employ a single plunger or plunger set, and apportion fuel to the various injectors by means of a rotary valve.

Unit injectors Figure 5-6 illustrates a typical DDA fuel circuit. The fuel pump (more commonly called a lift or transfer pump) provides low-pressure fuel to the high-pressure injector pump assemblies.

The most immediate advantage of this system is the containment of high-pressure fuel within the injectors. Injection pressures can be very high, on the order of 20,000 psi. Eliminating the piping between the injectors and pump also eliminates pressure waves.

When the injector opens, a low-pressure wave moves through the fuel on the discharge side of the pump. A high-pressure wave is generated when the pump column encounters the injector nozzle. Even open, there is enough of a restriction to send a high-pressure wave back to the pump at speeds approaching 5000 ft/second. These waves can—depending on the length and shape of the plumbing—reinforce each other to unseat the nozzle valve, thereby injecting fuel at the wrong time in the

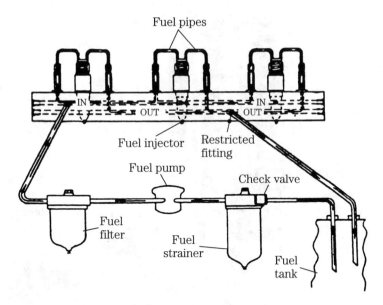

Fuel pipes

IN OUT IN OUT

Fuel injector

Restricted fitting

Fuel pump

Check valve

Fuel filter

Fuel strainer

Fuel tank

5-6 Unit injectors combine individual high-pressure pumps and nozzle holders. Detroit Diesel

cylinder, or they can delay injection. The whole problem is bypassed by mounting the pump as a unit with the injector.

In addition to parts found on more conventional injectors, the unit injector has a mushroom-shaped cam follower, a return spring, a pump plunger, oil supply reservoirs, and one or more check valves. The pump plunger is moved up and down by the engine camshaft, acting through pushrods and rocker arms. The control rack is a bar with teeth milled on one edge that engage a gear on each injector. The rack is connected to the throttle and to the governor, and its position determines the amount of fuel delivered to the engine.

Metering is achieved by varying the effective stroke of the pump. The plunger reciprocates the same distance each time it is tripped by the cam. But the effective stroke is a matter of the position of the fuel inlet and discharge ports relative to helical reliefs on the flank of the plunger.

Fuel enters the injector at the filter and passes to the supply chamber (Fig. 5-7). As the pump plunger moves down, part of the fuel is routed back to the supply chamber through the lower port, until the port is closed off by the edge of the plunger.

The plunger has an internal oil gallery that effectively shunts the pump as long as fuel trapped in the helix can escape. Its escape route is through the upper port, whose timing is a function of the contour of the helix. The lower port, which is blanked off by the plunger diameter once the downward stroke begins, plays no further part in the story.

With both ports closed, the fuel trapped below the plunger has no place to go except past the check valve and out the spray nozzle. The disc-shaped check valve, shown in profile in Fig. 5-7, is a fail-safe for the needle valve. Should the needle not seat because of dirt or carbon, the check valve will prevent air from entering the fuel supply.

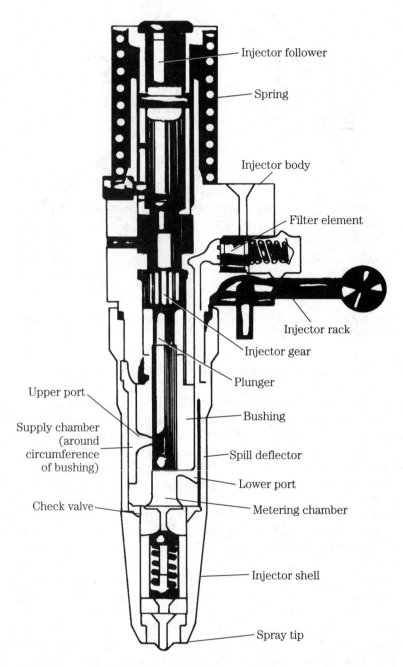

Injector follower

Spring

Injector body

Filter element

Injector rack

Injector gear

Plunger

Upper port

Bushing

Supply chamber
(around
circumference
of bushing)

Spill deflector

Lower port

Check valve

Metering chamber

Injector shell

Spray tip

5-7 Cutaway of a Detroit Diesel unit injector. EMD locomotive and marine engine injectors are similar in construction.

The position of the control rack determines the timing of the upper port because rotating the plunger changes the contour of the helix. At no injection (rack full out), the upper port remains open until after the lower port is uncovered, as shown in Fig.

5-8. The pump idles uselessly, alternately delivering fuel through the lower port and then through the upper port. No pressure is developed below the plunger, and no fuel is delivered to the spray head. At full load the contour of the helix is advanced (rack full in), and the upper port is closed almost as soon as the lower port, giving the fuel no alternative but injection.

Part of the reason for the relative complexity of this particular unit injector is the need for lubrication and cooling. Oil is in constant circulation around and through critical parts.

Distributor pumps

Distributor pumps develop pressure through a single or paired set of plungers, which are then apportioned to the injectors by means of a distributor. Various fuel distribution schemes have been tried in the past, but most modern pumps discharge from one side of the plunger, which rotates to align the discharge point with a fixed distributor head.

The major technical advantage of a distributor pump is that each cylinder receives precisely the same amount of fuel. The economic advantage is that a single plunger or plunger set costs less than the alternative scheme of individual plungers for each cylinder. Consequently, these pumps are widely used on automobiles, light trucks, and small engines.

Yanmar system

The newly developed Type VE pump, illustrated in relation to the total fuel delivery system in Fig. 5-9 and in more detail in Fig. 5-10, combines transfer and high-pressure pump functions in a single unit. A cam disk, driven from the engine by gear or Gilmer belt (this must be the first application of toothed-belt drive to injector pumps), is pinned to the plunger, which has a double motion. Lobes on the disk cause the plunger to reciprocate, alternately drawing fuel and pressurizing it. The plunger also rotates with the disk to align its discharge groove successively with each of the four injector ports on the pump body. Figure 5-11 illustrates the plunger at the moment of discharge through one of the injector ports.

The same drawing shows the control sleeve, which limits the plunger's effective stroke. The distance the plunger moves is fixed by the cam lobes; the effective stroke is determined by the position of the control sleeve, also shown in Fig. 5-12. Moving the sleeve to the left uncovers the spill port, before the beginning of injection, reducing the amount of fuel delivered. Conversely, moving it to the right blocks the spill port, so that more of the charge is delivered.

The pump also includes a timer, which initiates injection early by advancing the plunger carrier assembly relative to the cam plate during high-speed operation. Like most such timers, the Yanmar unit takes its cue from the increase in pump pressure at speed.

The magnet valve, or solenoid (shown in Figs. 5-11 and 5-12) is normally closed. When voltage is applied, the plunger moves down to block fuel delivery and stop the engine. Automotive and small diesel engines generally have this feature, which can be supplemented by a manual cutoff.

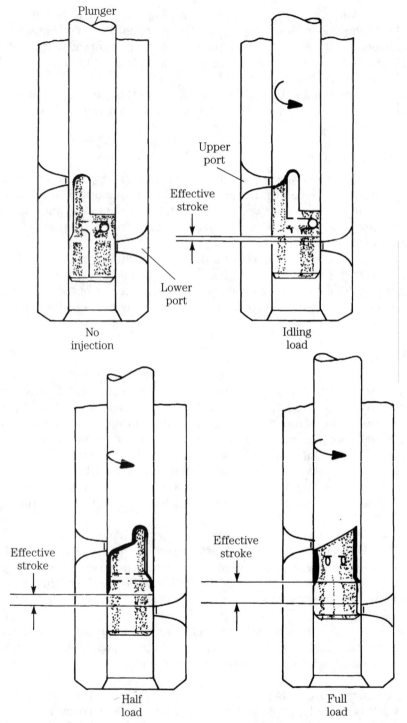

Plunger

Upper
port

Effective
stroke

Lower
port

No
injection

Idling
load

Effective
stroke

Effective
stroke

Half
load

Full
load

5-8 Pump plunger positions from zero delivery to full load. Unlike most contemporary designs, this plunger stroke has a variable beginning and constant ending.

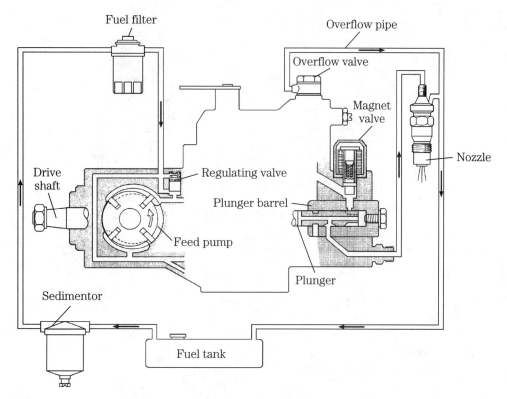

5-9 Typical small-engine fuel system. Yanmar Diesel

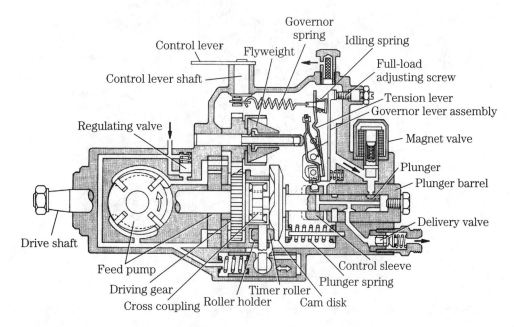

5-10 Yanmar type VE injection pump.

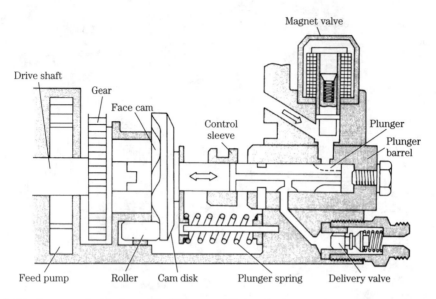

5-11 Cross-sectional view of the VE pump. Notice the integration and compactness of the design. Yanmar Diesel

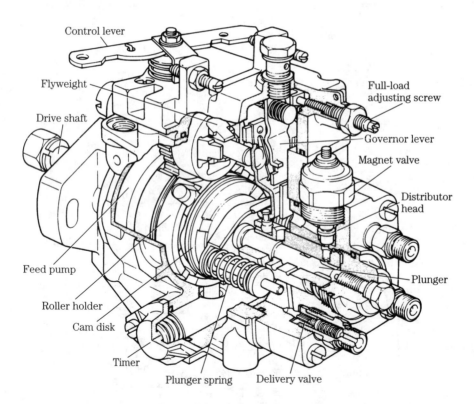

5-12 VE plunger reciprocates in the pumping function and rotates as a distributor.
Yanmar Diesel

CAV system

Figure 5-13 is a schematic of the CAV system, which combines the high-pressure pump under the same cover as the fuel distributor, transfer pump, governor, and regulator.

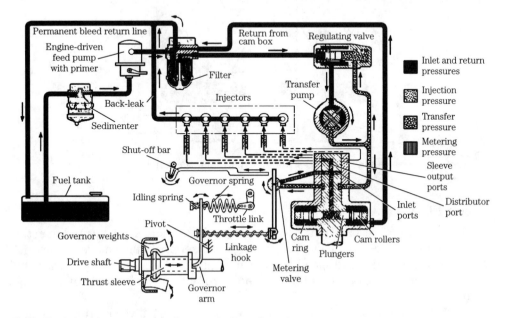

5-13 Fuel distribution system for a typical truck engine. GM Bedford Diesel

The pump consists of a pair of opposed plungers that are actuated by cam rollers. On the inlet stroke the plungers move outward under pressure from fuel delivered through the metering port (see Fig. 5-14). As the rotor turns, the metering (or inlet) port is blanked off by the rotor body. At the same time, the plungers make contact with the cam lobes and, as the cam rotates, are forced inward. Fuel passes through the distributor port, and to one of the injectors through the outlet port. There are as many inlet and outlet ports as there are engine cylinders.

The amount of fuel injected per stroke is determined by the regulating valve (upper right in Fig. 5-13), which is controlled by the throttle and governor. At low speeds, the regulator reduces fuel pressure and, therefore, the volume of fuel delivered by the vane-type lift pump. Because the opposed plungers in the high-pressure pump are driven apart by incoming fuel, their outward displacement is determined by the amount of fuel passed by the regulating valve.

The contour of the cam provides relief of injection pressure near the end of the delivery cycle, and helps control afterdribble. Timing of individual engine cylinders (relative to the datum provided by cylinder No. 1) is determined by accurate machine work on cam lobes and outlet ports.

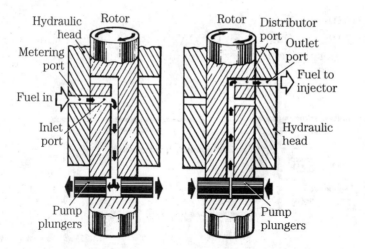

5-14 Double-piston distributor-type pump operation. GM Bedford Diesel

Bosch EP/VA pump

The Bosch EP/VA series pump is, in some respects, more sophisticated than the CAV unit. It is intended to be used in passenger car service, which makes severe demands on fuel metering and speed regulation. Figure 5-15 is a schematic of the main circuits. Other features such as the governor, hydraulic advance, and deferred idle are not shown.

Each upward stroke of the piston (shown at No. 1) involves three distinct events. The fuel transfer pump feeds the piston by means of two pipes, the back-pressure stage (2) and the regulator stage (3). The piston compresses the fuel and sends it to the injectors and high-pressure chambers (6) of the regulator shuttle (8). The compression chamber (2) is opened to the return circuit by the regulator shuttle, which moves to the right by virtue of regulator pressure and metering.

Metering is achieved by means of the shuttle. The main piston (1) moves upward, impelled by the pressure of the fuel passing through two parallel canals and emptying behind the shuttle. The shuttle moves to the right and controls the final injection by effectively varying the capacity of the compression chamber (2) and the diameter of the fuel-return port (11). On the downward movement of the piston the shuttle moves to the left under spring pressure (10) and closes the fuel-return port (11). Fuel is forced back through the two parallel lines connecting the piston and shuttle. The one-way valve (5) closes. The return of the shuttle spring pressure is a function of the opening of the accelerator valve (4). The amount of fuel injected is a function of the stroke of the shuttle, which in turn depends on the position of the accelerator valve.

Compensation for load and fixed-throttle speed variations is automatic. As the speed increases the piston moves upward more quickly. The shuttle moves back earlier, shortening the stroke and reducing the fuel volume per delivery. If the load increases the opposite happens, and more fuel is injected until new equilibrium is established.

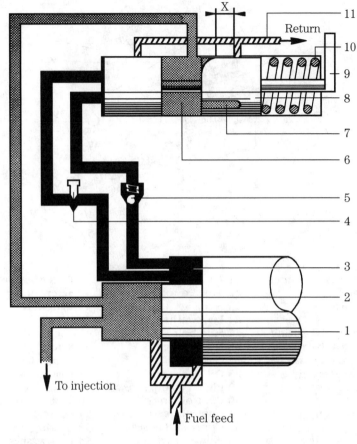

X

Return

11
10
9
8
7
6
5
4
3
2
1

To injection

Fuel feed

1–Main piston
2–Back-pressure stage (compression chamber)
3–Regulator stage
4–Speed valve (accelerator)
5–Non return valve
6–High-pressure chamber
7–Stop slot
8–Regulator shuttle
9–Stop lever
10–Spring
11–Canal returning fuel to reservoir

Three main hydraulic circuits

▨ Low-pressure feed

▦ High-pressure distribution

▮ Metering at medium pressure

5-15 Bosch pump schematic. Peugeot

In-line pumps

Another approach is to provide a separate plunger for each injector (Fig. 5-16). Plungers share a common housing and are driven by an integral camshaft that, together with the governor, roller cam followers, and shaft bearings, receives lubrication from the engine. (Pumps built as late as the 1950s had integral oiling systems).

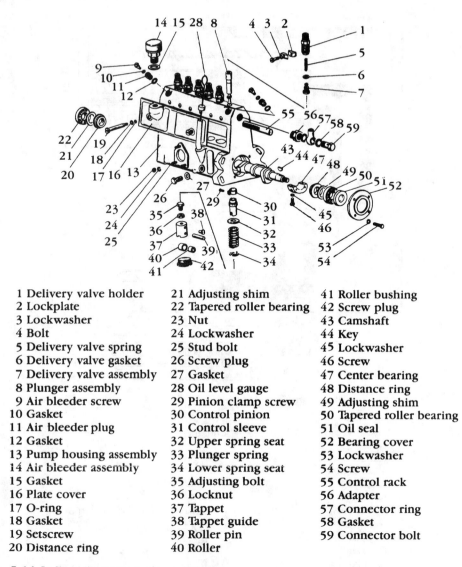

1 Delivery valve holder	21 Adjusting shim	41 Roller bushing
2 Lockplate	22 Tapered roller bearing	42 Screw plug
3 Lockwasher	23 Nut	43 Camshaft
4 Bolt	24 Lockwasher	44 Key
5 Delivery valve spring	25 Stud bolt	45 Lockwasher
6 Delivery valve gasket	26 Screw plug	46 Screw
7 Delivery valve assembly	27 Gasket	47 Center bearing
8 Plunger assembly	28 Oil level gauge	48 Distance ring
9 Air bleeder screw	29 Pinion clamp screw	49 Adjusting shim
10 Gasket	30 Control pinion	50 Tapered roller bearing
11 Air bleeder plug	31 Control sleeve	51 Oil seal
12 Gasket	32 Upper spring seat	52 Bearing cover
13 Pump housing assembly	33 Plunger spring	53 Lockwasher
14 Air bleeder assembly	34 Lower spring seat	54 Screw
15 Gasket	35 Adjusting bolt	55 Control rack
16 Plate cover	36 Locknut	56 Adapter
17 O-ring	37 Tappet	57 Connector ring
18 Gasket	38 Tappet guide	58 Gasket
19 Setscrew	39 Roller pin	59 Connector bolt
20 Distance ring	40 Roller	

5-16 In-line injector pump in exploded view. Marine Engine Div., Chrysler Corp.

Splitting the pumping function among several plungers increases costs—to hold pressure, the plunger/barrel assemblies must be finished to tolerances measured in wavelengths of light—but enhances durability. The design also allows higher pressures than can be achieved with a single plunger.

Metering

In-line pumps of the class we are concerned with are constant stroke devices. The plungers move the same distance with each revolution of the camshaft. Fuel delivery is adjusted by moving the control rack (Fig. 5-17) in and out.

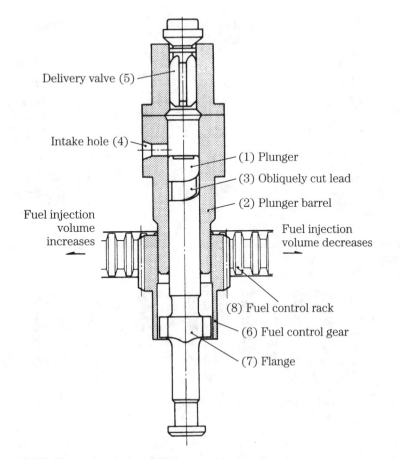

Delivery valve (5)

Intake hole (4)

(1) Plunger

(3) Obliquely cut lead

(2) Plunger barrel

Fuel injection
volume
increases

Fuel injection
volume decreases

(8) Fuel control rack

(6) Fuel control gear

(7) Flange

5-17 Plunger, barrel, and rack assembly.

How this is typically done is shown in Fig. 5-18. At bottom center, the plunger uncovers the inlet port. Fuel enters the pressure chamber above the plunger. The plunger rises, initially pushing fuel back out the inlet port, and, moving further, covers the port. The plunger continues to rise, building pressure in the fuel trapped above it. The delivery valve opens, and a few microseconds later, the injector discharges. Fuel continues to flow until the annular groove milled along the side of the plunger uncovers the inlet port. At this point, pressure bleeds back through the inlet port and injection ceases. Because of the shape of the groove, rotating the plunger opens the inlet port to pressure earlier or later in the plunger stroke.

American Bosch, Robert Bosch, and CAV barrels are drilled with a second port above the inlet port to accept spillage during part-throttle operation. Figure 5-19 illustrates the metering action of a CAV pump. Fuel enters the barrel at A and continues to flow until the plunger movement masks the two ports (inlet shown on the left, spill port on the right). At full load, pressure-bleed down through the spill port is delayed until the plunger approaches the end of the stroke, as shown in drawing C. You might wonder why the spill port opens before top dead center, when a few more degrees of cam movement would reverse the plunger and lower injection pressure. The

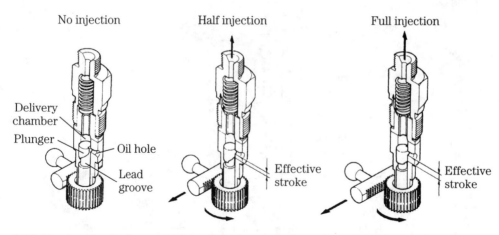

No injection   Half injection   Full injection

Delivery chamber

Plunger

Oil hole

Lead groove

Effective stroke

Effective stroke

5-18 Constant-beginning, variable-ending plunger action, with surplus fuel exiting the inlet port.

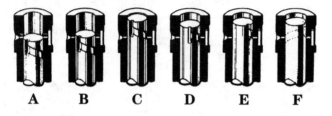

A   B   C   D   E   F

5-19 A variation of the principle illustrated in Fig. 5-18, using a spill port to accommodate the unneeded fuel.
GM Bedford Diesel

reason is that cam-driven plungers behave like pistons, accelerating at mid-stroke and slowing as the dead centers are approached. Opening the spill port early, while plunger velocity and pressure rise are rapid, terminates injection far more abruptly than if the port remained closed until the plunger reached tdc.

In drawing D, the annular groove has rotated to the half-load position. The effects of further rotation are shown at E, which represents idle, and at F, shutdown.

However the porting is arranged, the stroke has a constant beginning and variable ending for pumps of the type this book is concerned with. Some marine and large stationary engine pumps meter fuel delivery at the beginning of the stroke.

The mechanism by which rack movement is transmitted to the plungers varies with the manufacturer, but always includes an adjustment to equalize delivery between plunger assemblies. At a given rack position, each cylinder must receive the same amount of fuel.

Delivery valves

Delivery valves, installed between the pump and the injector feed pipes, create the sudden pressure drops that signal the end of injection (Fig. 5-20). Design details vary, but all include a spring-loaded check valve and a piston. The piston for the

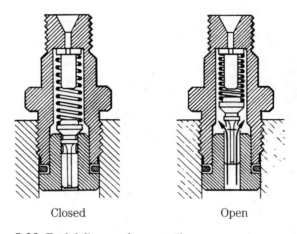

Closed   Open

5-20 Fuel delivery valve operation. <small>GM Bedford Diesel</small>

valve shown is located in the valve bore as an extension of the tulip-shaped check valve. The check valve opens under pump pressure and is closed by its spring when that pressure relaxes. Closure is critical. The check valve moves toward its seat, and the piston, as it were, runs interference. It displaces fuel out of the valve body and back into the pump. Pressure in the fuel line above the piston drops abruptly. This sudden loss of pressure causes the injector needle valve to snap shut.

Fuel supply pumps

High-pressure pumps have poor suction capability and must be primed by the device variously known as the fuel supply, transfer, or lift pump. These positive-displacement devices operate more or less independently of fuel viscosity, pressure, or temperature. The geared type consists of a pair of meshed gears rotating inside of a housing. Pressure is derived from the action of the meshed gear teeth, which prevents oil from passing between the gears and, instead, forces it around the outside of each gear. These pumps generally include a pressure regulator in the casing.

Vane-type pumps consist of one or more pairs of sliding vanes in an eccentric housing. Pressure is developed as the vanes turn and wedge the fuel against the progressively narrower sides of the housing. Fuel is expelled through a discharge port on the periphery of the housing.

The diaphragm pump is by far the most popular for small engines. An AC design is shown in exploded view in Fig. 5-21. It is driven from the engine cam or from a lobe on the injector pump cam. Note the provision for manual priming.

Figure 5-22 illustrates the piston pump used on Chrysler-Nissan SD22 and SD33 engines. It is integral with the high-pressure pump and driven from the same cam. The operation of the pump is slightly unorthodox in that fuel is present on both sides of the piston, thus eliminating the possibility of air lock.

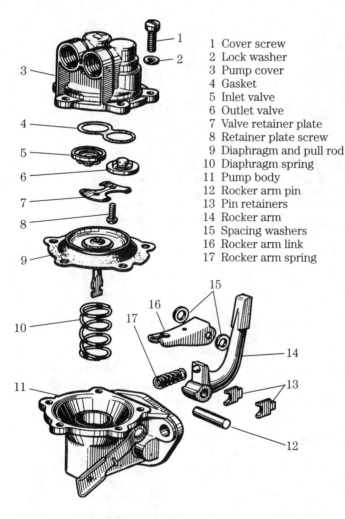

1	Cover screw
2	Lock washer
3	Pump cover
4	Gasket
5	Inlet valve
6	Outlet valve
7	Valve retainer plate
8	Retainer plate screw
9	Diaphragm and pull rod
10	Diaphragm spring
11	Pump body
12	Rocker arm pin
13	Pin retainers
14	Rocker arm
15	Spacing washers
16	Rocker arm link
17	Rocker arm spring

5-21 AC diaphragm-type fuel supply pump. GM Bedford Diesel

Figure 5-23 diagrams its operation. On the downward movement of the piston the discharge-side check valve (No. 3) closes, and fuel goes up the outer chamber (B) to the high-pressure pump. The intake-side valve (4) is open. At position II the piston is forced up by the rotation of the cam. Fuel enters the outer chamber (B) through the check valve (3) and the passage on the intake side. Discharge commences at position III.

Fuel filters

Fuel oil doubles as a lubricant for the pumps and injectors and must be compatible with parts whose tolerances are gauged in hundred-thousandths of an inch. Refiners take pains to keep diesel oil free of impurities, but contaminants can enter the

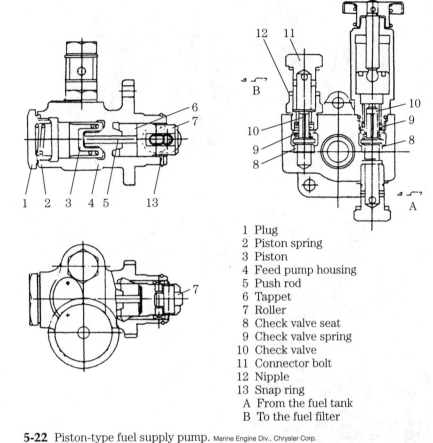

1 Plug
2 Piston spring
3 Piston
4 Feed pump housing
5 Push rod
6 Tappet
7 Roller
8 Check valve seat
9 Check valve spring
10 Check valve
11 Connector bolt
12 Nipple
13 Snap ring
A From the fuel tank
B To the fuel filter

5-22 Piston-type fuel supply pump. Marine Engine Div., Chrysler Corp.

fuel at any point from distillation to final tankage. Filters are vital to efficient and economical operation (Fig. 5-24).

Most engines incorporate at least three stages of filtration. The filters are the full-flow type, in series with the fuel line. A bypass valve is usually present to keep the fuel flowing if the filter clogs. In many applications it is better to risk scoring the pump than to suffer unexpected shutdown. The filter stages are progressive; the primary stage is usually rather broad-gauged and traps particles larger than 0.005 in. or so. The secondary stage, which is located between the supply and the injector pumps, has a pleated paper element that filters down to 25 microns (0.0009 in.) or finer. Where water is a problem, as in marine or occasional-use applications, a separator is fitted. It might be a glass or aluminum blown upstream of the supply pump, or it might be integral with the primary filter. In addition, some engines feature filters at the injectors to give added protection to these sensitive components. Figure 5-25 illustrates a typical fuel supply system.

Filter casings have bleeder valves mounted high to purge the air from the lines, and many also incorporate return valves to divert excessive fuel back to the tank (Fig. 5-26).

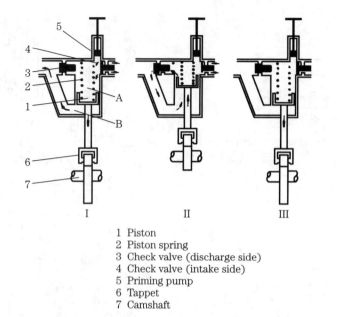

1 Piston
2 Piston spring
3 Check valve (discharge side)
4 Check valve (intake side)
5 Priming pump
6 Tappet
7 Camshaft

5-23 Fuel pump operation. Marine Engine Div., Chrysler Corp.

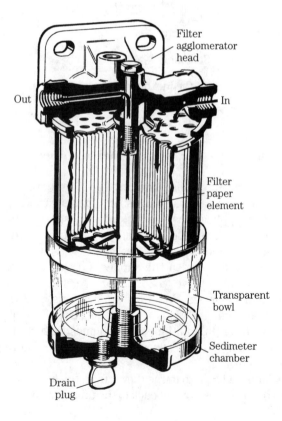

5-24
Cartridge-type fuel filter.
GM Bedford Diesel

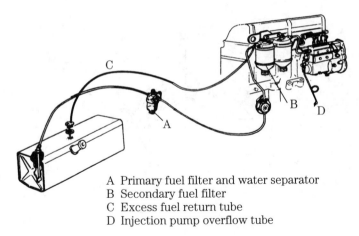

A Primary fuel filter and water separator
B Secondary fuel filter
C Excess fuel return tube
D Injection pump overflow tube

5-25 Typical boat-engine fuel system with primary and secondary filtration. Lehman Ford Diesel

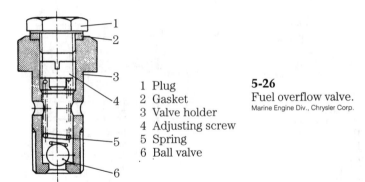

1 Plug
2 Gasket
3 Valve holder
4 Adjusting screw
5 Spring
6 Ball valve

5-26
Fuel overflow valve.
Marine Engine Div., Chrysler Corp.

Refinements

The basic fuel system—injectors, high-pressure pump, transfer pump, and staged filters—has been described. As fuel costs have climbed and diesels have moved into territory hitherto occupied by SI engines, certain refinements to the fuel system have been introduced. The most significant of these are intended to improve the engine's flexibility at either extreme of the rpm scale.

Fuel timer

As mentioned earlier, current designs employ a fuel timer. The timer is integral with the injector pump and operates to advance the pump cam at high speeds.

The timer can be actuated hydraulically (Fig. 5-27A) or mechanically (Fig. 5-27B). The mechanical timer is reminiscent of the centrifugal advance mechanism employed on ignition distributors. As engine speed increases, the flyweights pivot outward against their restraining springs. This movement is transmitted to the pump camshaft and turns the split shaft against the direction of its rotation. The pump discharges earlier.

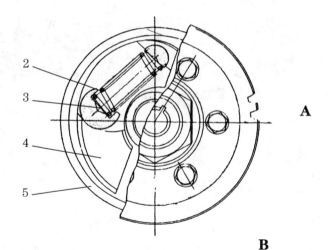

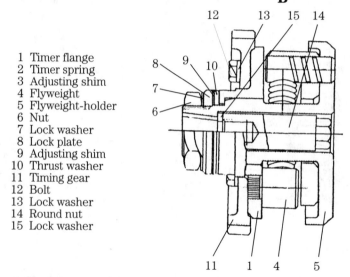

1 Timer flange
2 Timer spring
3 Adjusting shim
4 Flyweight
5 Flyweight-holder
6 Nut
7 Lock washer
8 Lock plate
9 Adjusting shim
10 Thrust washer
11 Timing gear
12 Bolt
13 Lock washer
14 Round nut
15 Lock washer

5-27 Yanmar hydraulically controlled fuel timer (A) advances in response to increased transfer pump pressure. Chrysler-supplied unit (B) sense engine speed with flyweights.

The timer should be checked for freedom of movement and maximum advance. The model shown here is allowed 5–7.5 degrees advance, depending on application. Travel limit is determined by shims.

Maintenance & repair

The fuel system is most critical from the viewpoint of preventive maintenance. Failure to follow a maintenance and inspection schedule will result in time wasted in shutdowns, and can cause expensive repairs. Filter elements are cheaper than injector pumps.

Bleeding & priming

Whenever the system has been opened or whenever the tank has run dry, the system must be bled. Some fuel will escape in the process. Place a drip pan under the engine and be sure that the wiring harness is clear of the spillage. Fuel oil softens insulation.

Bleeder screws are located in the upper filter bodies (Fig. 5-28) and in one or more locations on the accessible side of the injector pump (Fig. 5-29). Carefully clean the screw tops and adjacent areas to prevent foreign matter from entering the fuel supply. A hand-operated priming pump is incorporated with the injector pump or integral with the transfer pump. Working from the tank end forward, crack the bleeder screws, and tighten as fuel flows in an unbroken stream. It might be necessary to break the high-pressure lines at the injector pumps. Continuing to keep pressure up (10 psi is adequate), crack one union at a time, and hold until the bubbles disappear. Run the engine for at least 10 minutes to ensure that the fuel system is completely purged.

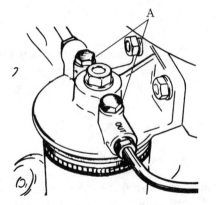

5-28
Fuel filter bleed screws.
Lehman Ford Diesel

1   Injector pipe clamp
2   Union nut
3   Injector pipe clamp
4   Delivery valve holder
5   Valve holder clamp
6   Bleed screws

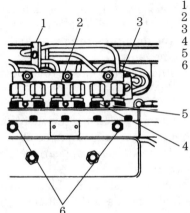

5-29
Injection pump bleed screws.
Ford Industrial Engine and Turbine Operations

System purge

Flushing the fuel system is necessary in the event of massive contamination. Water traps and filters are intended to remove minor contamination incidental to fuel storage and handling, and cannot cope with large amounts of water (especially when present as an emulsion), solids, or bacterial growth.

With the battery disconnected and fire extinguisher ready, follow this procedure:
1. Drain as much fuel as possible from the tank.
2. Wash the tank out with clean diesel fuel, removing the tank if necessary.
3. Working from the suction side of the injector pump to the tank, disconnect each system component (filters, water/fuel separator, and so forth) and blow the connecting lines out with a mist of air and diesel fuel (Fig. 5-30).

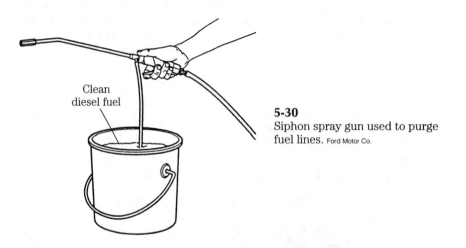

Clean
diesel fuel

5-30
Siphon spray gun used to purge fuel lines. Ford Motor Co.

4. Disassemble, clean, and reassemble each component in turn.
5. Remove the fuel return line, and blow it out as in Step 3 (Fig. 5-31).
6. Make up the return line circuit at the injectors, pump, or pump filter, but leave the tank connection open. Slip a length of flexible hose over the discharge end of the line to direct fuel into an approved (fire safe) container, which should hold several gallons. Pour enough clean diesel fuel into the container to cover the end of the hose. The purpose is to prevent contaminated fuel still present in the high-pressure pump and injector bodies from being recirculated to the tank.
7. Make up the fuel delivery lines, with careful attention to achieving leakproof connections.
8. Fill the tank with clean fuel.
9. Prime the injector pump, and bleed the system as previously described.
10. Crack the high-pressure pipe connection a half-turn or so at each injector, and crank the engine (Fig. 5-32). If fuel does not appear, bleed the injector pump as per manufacturer's instructions.
11. Operate the engine for 15 minutes or so at slow idle. Shut down the engine, reconnect the return line to the tank, and discard the fuel caught in the container.

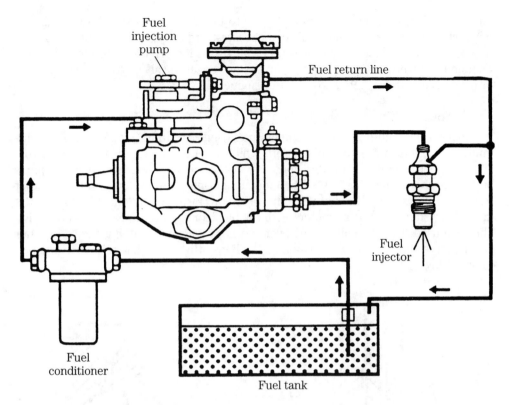

Fuel
injection
pump

Fuel return line

Fuel
injector

Fuel
conditioner

Fuel tank

5-31 Fuel return line recycles unused fuel to tank. Ford Motor Co.

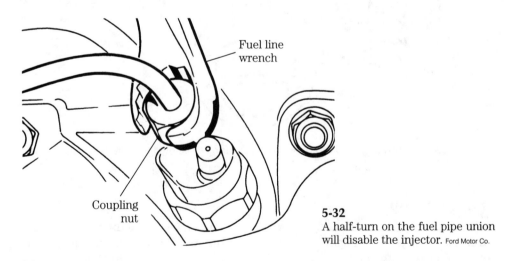

Fuel line
wrench

Coupling
nut

5-32
A half-turn on the fuel pipe union
will disable the injector. Ford Motor Co.

Filter service

Sediment chambers should be drained daily or prior to start-up. To replace the
filter, drain the fuel out the petcock. Remove the mounting bolt. Clean the shell in
fuel oil and dry with a lintless rag or, preferably, compressed air. Density-type ele-

ments should be immersed in fuel oil prior to installation to expel trapped air and make starting easier. Fill the shell to about two-thirds capacity with fuel oil. Install new gaskets at the recess and, if present, at the union. Tighten the cover nut just enough to prevent air leaks. Do not jam the nut because this might distort the shell or cover. Bleed the system and start the engine.

Fuel lines

Inspect the high-pressure fuel pipes and fittings carefully (Fig. 5-33). Overtightening the unions will cause distortion, which might be seen on the pipe ferrule (flange). Deep scratches or abrasions can develop into cracks. Replace any pipes and fittings with these defects. When installing, tighten fingertight and check that the pipe is centered as it emerges from the union. Firm up and check the lay of the lines. They should not be in contact with the block or any other part. When retaining clips are used, the clips should fit over the lines without straining them. In other words, there should be no pretension on the lines other than that offered by the terminal fittings.

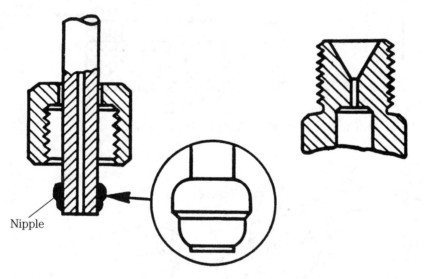

Nipple

5-33 Pipe fitting should be concentric with nipple as shown by wear patterns.
GM Bedford Diesel

Transfer pump

Periodically (every 200 hours or as per maintenance schedule) remove the cover and clean the fuel chamber of diaphragm pumps (Fig. 5-34). Examine the diaphragm for tears, cracking, and loss of elasticity. Piston- and gear-type pumps do not require routine inspection and should not be disturbed short of failure or overhaul. In most instances failure will come about because of a clogged valve. The valve can be lapped to fit, or replaced if that is impractical. GM rotary pumps might fail at the drive shaft or coupling, though this is rare. You can determine if the pump is

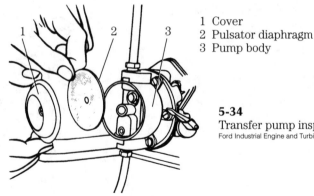

1 Cover
2 Pulsator diaphragm
3 Pump body

5-34
Transfer pump inspection.
Ford Industrial Engine and Turbine Operations

turning by inserting a fine wire through one of the drain ports. The wire will vibrate when struck by the gears.

Pump delivery rate and pressure specifications are not always available. You can get a rough idea of pump efficiency by disconnecting the line between it and the high-pressure pump. Collect the spill over in a glass jar, and look for bubbles, which indicate that the pump is losing its prime by sucking air. If this is the case, check all fittings and lines on the suction side of the pump. Pay particular attention to strainers or filters between the pump and tank.

Injectors

The best maintenance policy is to replace fuel injectors at periodic intervals, before problems develop. Solids accelerate tip, seat, and valve stem wear and encourage carbon formation; water is both a poor lubricant and a corrosive that attacks all polished parts; heat solidifies the fuel into varnish and carbon, which slows valve action and distorts the spray pattern.

Solids can be controlled by the simple expedient of checking the filters. Water intrusion can be minimized by testing the fuel before use, installation of a water separator, and by topping the tank off on a daily basis. Heat can be kept within reasonable limits by keeping a gentle hand on the throttle, monitoring exhaust stack temperatures, giving routine cooling system service, and paying careful attention to injector seating.

Removal Disconnect the return line. Using a flare nut wrench, remove the fuel pipe union at the injector and loosen the pump union. Disengage the stay clips that are sometimes fitted to the piping, and without stressing the pipe, swing it out of the way. Cap the open end.

Before a unit injector can be withdrawn, it is necessary to disassemble the associated rocker arm shaft and disengage the rack or, as is becoming more frequently the case, the wiring harness. Figure 5-35 illustrates the DD Series 60 (electronic) injector tube configuration. Other injectors thread into the block (Fig. 5-36) or else mount with a two-bolt flange (as shown back in Fig. 5-1).

Standard design practice is to seat the injector against a sleeve, or tube, which is exposed to coolant on its OD (Fig. 5-37). The sleeve acts as a sink, dumping heat from the spray tip into the coolant. Sleeve temperatures are relatively high, and fuel

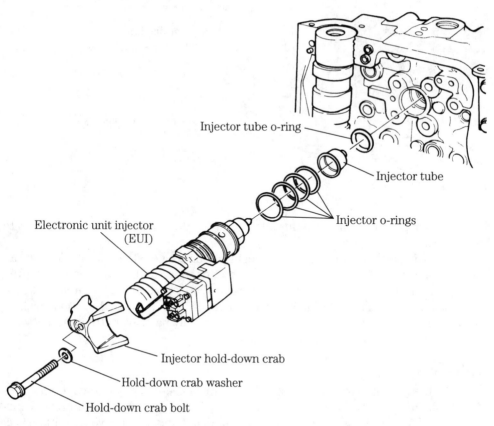

5-35 Series 60 unit injector mounting configuration. <small>Detroit Diesel</small>

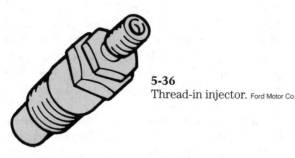

5-36
Thread-in injector. <small>Ford Motor Co.</small>

that leaks past the O-ring or copper ring seals will carburize and bind the nozzle holder to the sleeve. In that unhappy event, a special puller is required to remove the injector without damaging the injector and, quite possibly, the sleeve.

Sleeves Once the injector is withdrawn, clean the sleeve ID with the appropriate factory reamer or, lacking that, with a rotary wire brush. Smear the sleeve ID with grease to keep as much carbon as possible from falling into the combustion chamber. Wipe the grease off and carefully inspect the tube for coolant leaks, cracks, and pronounced chatter marks on the injector seat (Fig. 5-38). Sleeve re-

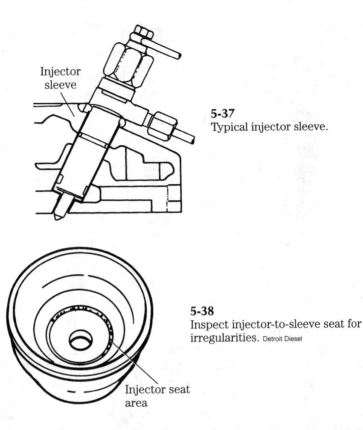

5-37
Typical injector sleeve.

5-38
Inspect injector-to-sleeve seat for
irregularities. Detroit Diesel

placement is best left to a diesel machinist with a reputation for careful work-manship. While sleeves for some engines can be installed without dismantling the cylinder head, one always feels better about the job when the head is removed for pressure testing.

Injector tests See chapter 4 for a description of the four standard injector tests—spray pattern, opening pressure, dribble, and chatter.

Service procedures Most injectors can be disassembled for cleaning and inspection; some ⅜-inch "pencil" injectors and a number of foreign automotive injectors are throwaway items, serviced as complete assemblies. Although it is possible to rebuild injectors in the field, most shops limit their activity to testing and tip cleaning. Figure 5-39 illustrates a Ford-supplied nozzle cleaning kit, which includes a special solvent.

Replacement injectors should be cycled in the test stand both to check performance and to flush out the rust inhibitor.

Installation Mount the injector with new copper gaskets and, where applicable, O-rings. On injectors with two-bolt mounts, it is good practice to loosely make up high-pressure fuel pipes before the mounting bolts are drawn down. This precaution will help distribute stresses in the lines. Tighten the injector and fuel lines to specifications, and (with the control rack pulled out) crank the engine. Check for compression leaks around the injector body. If there is any doubt about the integrity of the seal, verify with one of the aerosol powders sold for this purpose. Most com-

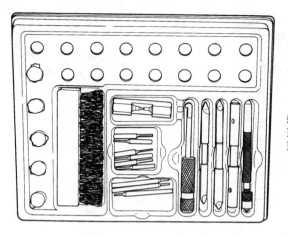

5-39
Ford-supplied injector cleaning kit.

pression leaks can be cured by loosening and retorquing the mounting bolts. Start the engine and look for leaks at the fuel pipe connections.

Fuel injection pumps

Field service is normally limited to timing the pump and replacing it as a unit.

Timing Drills vary all over the map, but the purpose of the exercise is to synchronize No. 1 plunger fuel delivery with No. 1 cylinder. Once this is done, pump geometry can be trusted to provide fuel to the remaining cylinders at the proper time.

Static timing refers to the alignment of the pump reference mark—indicating the onset of fuel delivery—with a timing mark stamped on the drive member. Figure 5-40 illustrates timing marks on a typical Bosch pump; Stanadyne and other manufacturers employ internal marks, visible through a window in the pump mounting flange. Some Navistar pumps are timed before installation by aligning match marks on the drive gears.

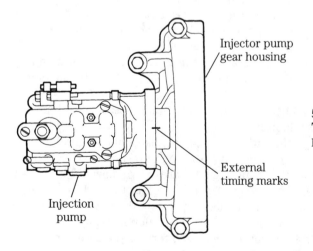

Injector pump gear housing

External timing marks

Injection pump

5-40
Timing marks—Bosch distributor pump. Ford Motor Co.

Perhaps a more accurate method, and one used by Mercedes-Benz, involves substitution of a length of clear plastic tubing for No. 1 fuel pipe. The mechanic slowly bars the engine over, adjusting the pump position until the fuel level in the tube just begins to climb at the instant engine timing marks align (Fig. 5-41).

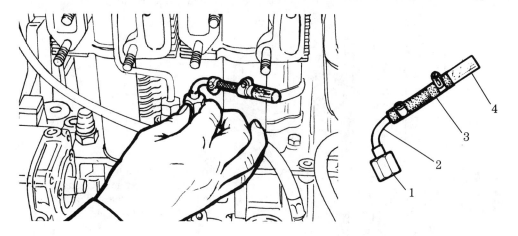

5-41 Transparent section (4) of fuel pipe adapter indicates plunger movement for accurate timing. Lombardini

An approximately equivalent measurement is the free height adjustment, necessary to uniformly align some unit injectors with the actuating mechanism. Figure 5-42 illustrates the DDA gauge, which indexes to a pilot hole in the injector body and, when the adjustment is correct, makes a light interference fit with the top of the injector follower.

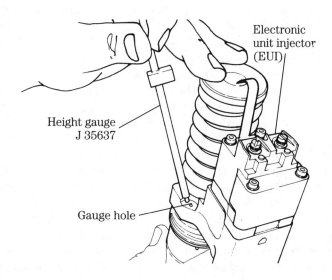

5-42 Height gauge used to "time" unit injectors. Detroit Diesel

Remote pumps can also be timed by using a dial indicator to measure plunger lift, as shown in Fig. 5-43. When this method is recommended, the manufacturer will indicate the amount of plunger travel necessary to initiate injection. The location of the pump timing mark can be adjusted accordingly. In like manner, measurement of piston height at the onset of injection can detect small inaccuracies in engine timing mark placement.

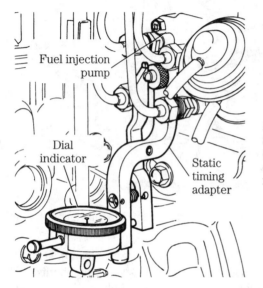

Fuel injection pump

Dial indicator

Static timing adapter

5-43
Static timing drill for two distributor-type pumps, using a dial indicator to measure plunger movement. Ford Motor Co.

Static timing establishes a baseline, adequate for most engines, but is by no means the last word in precision, at least as far as remote pumps are concerned. Careful mechanics supplement static timing with another method of *dynamic* timing.

Ford's Rotunda Timing Tester (PN 078-00100) and equivalent tools combine a luminosity probe with an accurate tachometer. The probe screws into the cylinder head in place of No. 1 glow plug to detect the moment of combustion; the tachometer triggers off a magnetic strip affixed to the harmonic balancer.

Such instruments compensate for the variables that can affect timing. These include fuel and air quality, pump wear, engine speed, and the inevitable delay between the onset of pump pressure and injector opening. By making a series of readings at different speed settings, the mechanic can verify timer function and, in some cases, recalibrate the timer. Unfortunately these tools are neither inexpensive nor adaptable to all engines. Some enterprising inventor should look into the possibility of using knock sensor technology (developed for computer-controlled SI engines) as a timing indicator.

The Sun timing light, shown in Fig. 5-44, is a relatively simple, reliable tool, used widely in the industry. A transducer, clamped to No. 1 injector pipe, triggers the strobe in response to pipe expansion under pressure. Pressure rise precedes ignition, but the difference is small enough to be ignored for all but the most highly stressed engines.

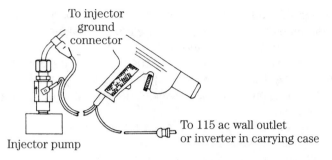

To injector
ground
connector

To 115 ac wall outlet
or inverter in carrying case

Injector pump

5-44 Sun timing light triggers from fuel pulses in the fuel pipe. Sun Electric Corp.

Power is taken from the engine's 12- or 24-V system or from a 120-Vac line. The engine is run at idle with the stroke light illuminating the crankshaft timing marks. A dial on the instrument shows timing advance in crankshaft degrees. A built-in tachometer allows the mechanic to set the governor and plot the timing advance curve.

Removal/installation Prior to removal, scrub the pump body and mounting flange area with solvent and blow dry. Bar the engine over to align the timing marks. Disconnect the cap fuel and oil lines, and clean the pump casing a second time, now that most surfaces are accessible. Remove the subassemblies, including the drive coupling, transfer pump, fuel timer, and attached filters.

Mount the pump, aligning the timing marks. Make certain all lash is removed from the drive gear train; Peugeot catalogs a special tool for this purpose. Once the pump is installed and connected, retime it and bleed the system.

The following is a brief introduction to the repair process, as applied to Bosch-pattern in-line pumps. This material is included for purposes of general information and should not be taken as an instructional text.

Inspection & repair

The disassembly drill for the pump proper varies with makes and models. Normally, the first step is to remove the delivery valves, marking them or arranging them on the bench to ensure proper assembly. Rotate the cam so that each tappet is in its *up* position. Shim the tappets and withdraw the cam. The tappets are removed from below. The tappet bores might be covered by the lower plate or by threaded plugs. Mark or arrange the tappets sequentially. Withdraw the plungers from the top with the aid of a length of hooked wire or expansion forceps. *Do not touch the lapped surfaces with your fingers.*

Inspect the plungers and bore for wear with a powerful magnifying glass. Scratches deep enough to be felt mean that the plunger must be replaced. This kind of wear comes about because of poor filter maintenance. Dulled rubbing surfaces are caused by water or acid contamination. Polish lightly to restore the sheen. Uneven wear patterns indicate that the bore has been warped from overtightening the fuel delivery valve. Wear in the helix profile destroys pump calibration.

A plunger should fall of its own weight when the barrel is held 45 degrees from vertical. Gummed or varnish-covered plungers can make governor action erratic and accelerate wear on the helixes and rack.

New plungers must be handled with the greatest care. Soak in kerosene or other solvent to dissolve the preservative, and dip in fuel or test oil. Assemble without leaving fingerprints on the lapped surfaces. Be sure that the plunger mates with the control mechanism.

Clean the delivery valves and make a functional test by depressing the plunger and releasing it as shown in Fig. 5-45. It should hold air pressure. Most valves cannot be resurfaced and must be replaced as an assembly.

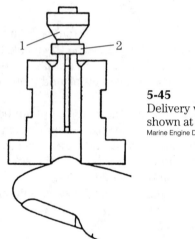

5-45
Delivery valve test. Check valve shown at 1, piston at 2.
Marine Engine Div., Chrysler Corp.

With a micrometer, check the cam against the original specifications. Inspect the control rack for wear. Figure 5-46 illustrates, in exaggerated fashion, the knife-edged wear pattern. Check the rack bushings for wear and ovalness by fitting a new control rack. Comparison of the sliding fits will show the state of the bushings. New bushings must be reamed to fit.

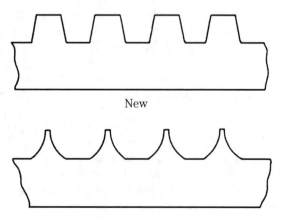

New

5-46
Control rack wear. GM Bedford Diesel

Worn-knife edge

Tappet height fixes the stroke of the pistons and influences the pump output. Initially all tappets should be set to the same height, either by means of shims or integral adjustment screws. Before the pump is installed on the engine it should be calibrated and the individual elements phased. This job requires a machine sensitive to cam angle and to output.

Assemble the pump with new gaskets, oil seals, O-rings, torquing the fuel delivery valve holders, and other critical parts to specs. Some manufacturers use adhesive in conjunction with the gaskets to make a secure joint. Scrape the old joints clean and use the brand of adhesive originally specified.

6
CHAPTER

Mechanical governors

The majority of diesels employ centrifugal flyball governors, which are integral with the injector pump. Operation differs little from the flyball governor invented by James Watt for his steam engines. But instead of opening and closing a steam valve, the diesel governor rotates the injection pump plungers to deliver more or less fuel per stroke. Pneumatic governors respond to air velocity entering the manifold, which in turn is a function of piston speed.

No governor can act instantaneously—the engine senses and responds to a load before the governor can react. Coarse regulation is adequate for most installations and might result in speed change peaks of 10 percent or so. Fine regulation cuts that figure in half, for maximum of 2.5% over or under the desired rpm.

Service work on governors consists of adjustment (particularly of the high- and low-speed limits), cleaning, and parts replacement. Bearings and pivots wear, coarsening the regulation. And occasionally a diaphragm will fail because of leaks or age hardening.

Centrifugal governors

Before any work more demanding than adjustment is undertaken, you should try to understand how the mechanism operates. We will consider two rather complex governors in this section. They are similar to others in all but small detail. The trick is to concentrate on what typewriter mechanics call the *chain of motion*, or how movement of one part is transferred to others.

CAV governor

The CAV governor is quite conventional, but it does have a manual override, which complicates matters. The lever (1 in Fig. 6-1) pivots on the axle (2). The whole assembly pivots at point 12 on command from the throttle (8).

In Fig. 6-1 the engine is running at about half-speed. A change in speed will be reflected by the flyweights (6). Should the engine accelerate, the weights will press outward against their springs (13) and move the lever and rack to the right, reduc-

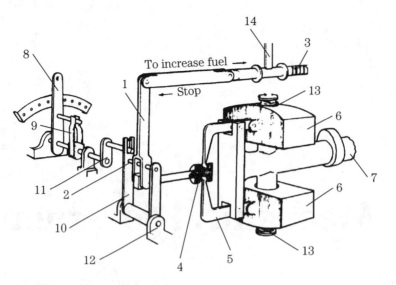

6-1 CAV midspeed setting.

ing fuel delivery. Under load the engine slows, and the weights, impelled by their springs, move inward, causing the throttle to open. Now let us increase speed by moving the control level (8) a few clicks to the right, as shown in Fig. 6-2.

The yoke (10) pivots on its axle (12), thus moving the lever (1) closer to the flyweights. To keep the geometry constant, the lever is kicked to the right by the spool shaft (4). Consequently, more fuel leaves the pump, and the engine accelerates. When centrifugal force is balanced by the counterweight springs, speed will stabilize, but at a higher rpm than would be the case in the previous drawing.

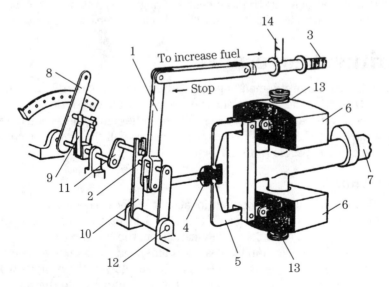

6-2 CAV three-quarter-speed setting.

It is possible that the rack will be opened until it contacts its stop. Additional movement of the control lever could be created only by forcing the weights apart. Rabbit-ear springs (9) are provided to protect the mechanism.

To reduce the speed the control lever is moved to the left, and it brings the lever (1) and yoke (10) with it. Regardless of engine speed the weights describe the same track.

CAV repairs Governor malfunctions—hunting, sticking, refusal to hold adjustments—can usually be traced to binding pivots. Most of the time the pivots bind because of dried oil and dirt accumulations. After long service the pivots and bearings might wear enough to require replacement. The drawing in Fig. 6-3 shows most of the 200-odd parts that make up a CAV governor, and can serve as our reference in this discussion. The bearings to check are numbered as 223, 225, and 226. Check the bushings on the bellcrank levers (223 and 224) and the element support sleeve (231). Bushings should be pressed into their bosses and reamed. Use a factory-supplied reamer for best accuracy.

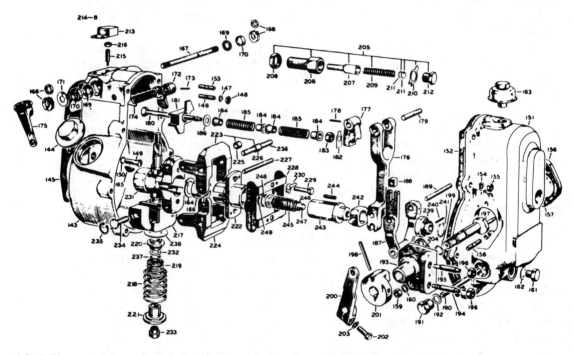

6-3 CAV governor in exploded view. Refer to the text for parts identification.

The pivot pins, shown at bottom left as 234, fit into bores on the sides of the weights. Check these bores for wear. Weights are sold in matched pairs.

Further disassembly requires that the oil be drained and all external connections opened. Remove the filler and oil level plugs (144 and 161). You will observe two hex nuts (191) secured with safety wire. Cut the wire and remove the nuts. The timing lever and fulcrum pin (190) can now be withdrawn. Next, remove the pair of nuts (159) that hold the control lever (200) and flange assembly (193) in place. Remove the assembly from the end cover.

The end cover is held by eight screws. Once they're off you will be able to get to the vitals of the governor. At this point disassembly becomes complex and promises that assembly will be equally so. Lay the parts out on a paper-covered bench, in rows from left to right. When you put the mechanism back together, follow the pattern exactly as if you were reading a newspaper.

The control rod linkage (177) is unscrewed from the control rod, and the assembly withdrawn by rotating it 90 degrees and disconnecting the yoke-shaped lever. The governor weights carry two screws (229). Remove them and then observe that the plate (228) is secured by a pin. Rotate the shaft so that the pin can be punched free. The governor linkage now can be separated from the weights.

Next, remove the guide bush from the end of the camshaft. The bush floats freely and should lift right out. Behind it you will find the governor weight assembly nut (166). To remove this nut, you need a CAV tool, which can be ordered from your distributor as part number 7044-112B. Another tool, No. 7044-8, is needed to pull the weights. Bottom the puller in the threads before applying force; otherwise, the threads might strip and you will be in real trouble.

Remove the key (150) from the camshaft keyway and the linkage at the control rod. The rear cover (151) is next, along with the auxiliary idle stop (205). A spring (209) and a plunger (207) lurk under the plug (212); it is rarely necessary to disturb these parts.

The weights are usually treated as an assembly. If you need to replace them (as in the case of worn pivot holes), place the weights in a brass-jawed vise and compress the spring by exerting force on the upper spring plate (221). Remove the adjusting nut with a CAV No. 7044-65 on each weight. Remove the spring clips (235) and their pins (234). You can now lift off the two bellcranks (223 and 224). The next step is to remove the excess-fuel hood (213), the excess-fuel adjusting screw (215), spindle stop lever (175), and the spring clip (168). Replace this clip with a new one on reassembly. The pawl (172) is located by a pin (173). Drive the pin clear and withdraw the spindle from the housing. Discard all seals.

CAV assembly Begin assembly with the bellcranks (223 and 224), using the pivot pin to locate them on the element support sleeve (231). Fit the weights (217) to the bellcranks with the pins (234) and fit the pin clips (235). Replace the excess-fuel-device seals (170) and assemble the pawl (172) and pawl spring (174). Hold them in position while you fit the spindle (167) into the housing. Pawl and spindle go together with a pin (173). With new clips on the spindle, mount the control lever (175). All parts should move freely at this and subsequent stages of assembly. If something binds, find out why before you proceed.

Fit the control linkage to the control rod and tighten the tab. Next, insert the key (165) on the camshaft and fit the weight assembly over the shaft. The spring washer and nut should be tightened to 500 in.-lb. with CAV No. 7044-112B. The brass element guide (222) has its slotted end nearest the pump camshaft. Screw the control link assembly onto the control rod. Rotate the governor weight assembly and camshaft to give access to the bellcrank pin boss. Insert the pin (227) and secure with the retaining case (228). Tighten the screws and secure with the locking tabs provided. With a new gasket on the rear housing cover, assemble and tighten the eight nuts in a criss-cross progression. The timing lever and fork pivot pin (190) go through the lower pair of holes in the rear cover. Use new sealing washers (192). The capnuts are threaded to the timing lever and fork pivot, and not to the housing cover.

Fit the lever and flange assembly (193) with a new gasket. Note the way in which the eccentric on the camshaft adjusting lever (197) fits into a hole on the block (188). Tighten the fasteners, being sure to include the lockwashers. Finally, re-place the oil filler plug and drain plug. Use a new gasket on the drain. Mount on the engine and adjust as required.

Simms governor

Although the Simms governor is a bit simpler than the CAV, it operates on the same principle; a force acting to pivot the weights outward is balanced by springs holding the weights to the shaft. Centrifugal force acts to pivot the weights outward in proportion to shaft speed. This force is counteracted by spring tension. An over-ride is provided for throttle control at speeds above idle and below maximum.

Figure 6-4 is a schematic of its operation. At start, the fuel rack is all the way in. (The control lever position is shown as the solid line.) The governor weights will be pressed inward. When the engine starts and picks up speed, the weights are flung outwards and cam against the hub sleeve, moving it horizontally. The hub sleeve moves the control lever to the position indicated by the dotted lines in the drawing. The rack will be moved toward the "no fuel" position. To avoid abrupt movement, Simms engineers have fitted two springs between the crank lever and throttle lever. The smaller of the two controls idle speed, and the heavier spring regulates high speed. The heavy spring (Fig. 6-5) is mounted in an elongated slot at its upper end and does not come into play at low speeds.

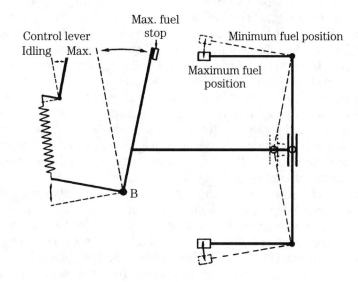

6-4 Simms governor action.

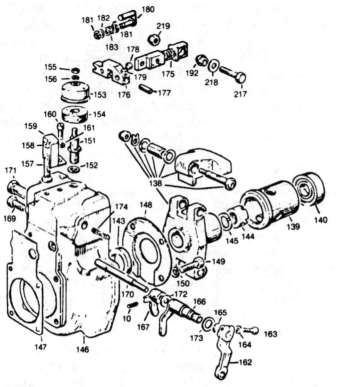

6-5 Governor elements.

As the engine starts, the fuel rack tends to cut off the fuel supply, and the engine slows. The governor weights generate less centrifugal force, and the springs then are able to pull the rack out toward the direction of more fuel. Operation is automatic, and the engine rpm is stable at whatever speed the operator has chosen.

The excess-fuel button opens the excess-fuel catch and allows the control lever to move forward beyond its normal limits during cold starts. Once the lever moves away from the catch, it locks; maximum speed is limited by stops on the lever.

Disassembly This governor is, as stated earlier, simpler than most. But care should be taken to understand the function and interrelationship of the parts as they are disassembled. Scrupulous standards of cleanliness should be maintained. Replace all gaskets and seals, and treat the springs with respect. Do not stretch or bend them. Parts numbers are keyed to Fig. 6-6.

Begin by undoing the bolt securing the control lever (206) to the shaft (197). Remove the lever. The governor rear casting is fastened to the front casting by six bolts. Separate the two castings enough to remove the Nyloc nut (219) and bolt (217) that secure the lever (209). The rear casting is free and can be set aside.

Remove the sleeve (Fig. 6-5, 139) with its bearing and yoke (Fig. 6-6, 141) from the weight assembly. Remove the control rod link from the control rod. It is held by a screw. Then remove the pin from the maximum-fuel stop lever. Now take out the

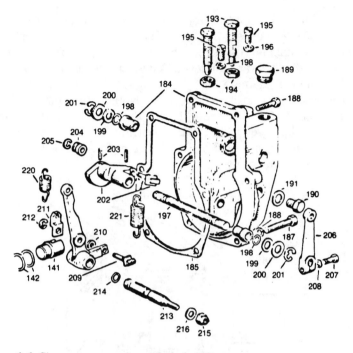

6-6 Simms governor in exploded view.

bearing (171) and shaft (170). Remove the spring clip and stop assembly (166). The stop lever is now free.

The weight assembly must be removed with special tools. Hartridge No. 87744 is needed to loosen the retaining nut (144), and Hartridge puller No. 7044-48 to remove the weights.

The front half of the mechanism is held by four bolts. To disassemble further, remove the lower spring plate (211). It is held by an E-clip. Observe the lay of the springs for future reference; it is possible to assemble them wrong. Carefully remove the governor springs (221 and 220). Next, remove the Nyloc nut (215) and crank lever fulcrum shaft (213), along with the lever (209). Undo the full-speed and idle stops. The control lever cross-shaft is held by taper pins. Support the shaft and drive the pins out. Remove the arm (206) and the E-clips. Observe the number and location of any shims on the shaft.

Assembly Insert the cross-shaft (197) of the control lever in the governor rear half and assemble the stop control (206) and spring arm (202). Replace the taper pins (203), E-clips, and shims. If you have rebushed or replaced the cross-shaft, the end float should be adjusted by adding or subtracting shims to between 0.05 and 0.25 mm. Thread the high-speed and idle stop screws into their bosses. The control stop can be fitted backwards if you are not observant.

Thread the fulcrum shaft (213) of the crank lever into the casting and assemble the crank lever (209) together with it. Use a new Nyloc nut (215) on the shaft. Place the governor springs in position and locate the plates on the small spring with E-clips.

Turn to the front casting. Fit the stop lever (166) and locate with a washer (173) and spring clip (165). The excess-fuel shaft (170) goes on the opposite side of the housing, along with maximum stop lever (167) and washer (172). The shoulder of the shaft should extend through the stop lever. Secure the shaft with its pin and replace the spring (174) and the bearing (171). The bearing should be torqued to 16 ft.-lb. The front casting is assembled with a new gasket. Note that the oil baffle (144) is part of the package; torque to 6 ft.-lb. Secure the link to the control rod and tighten.

To replace the weights you will need the Hartridge No. 87744 again. Replace the hub. If parts (hub, bearing, sleeve, fork) have been replaced, check the dimension at x in Fig. 6-7 and shim accordingly.

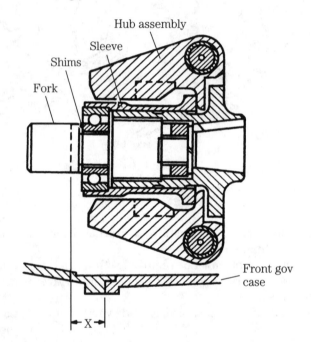

Governor type	Dimension X
B N Z—weights closed	13.7 – 13.9 mm
A—weights closed, with excess fuel device in governor	13.7 – 13.9 mm
A—weights closed and no excess device in governor	10.7 – 10.9 mm
M—weights open	19.5 – 19.7 mm

6-7 Critical dimension on Simms governor.

The telescopic link connects to the crank lever in the rear casting. Torque the castings together with the six bolts to 6 ft.-lb. Replace the control lever (206) on the lever shaft (197).

Engine speed is a function of pump delivery as much as governor adjustment. After you are satisfied that the pump is delivering as it should, you can adjust the idle and high-speed stops to the engine. But if these stops are factory sealed they should not be disturbed.

RSV governor

The RSV governor requires some explanation, although it employs the familiar principle of flyweights acting in opposition to spring tension. Figure 6-8 shows the basic arrangement of the parts. The fuel rack (1) extends to the plungers and determines delivery per stroke. The floating lever (3) is suspended, pendulum fashion, at the pivot (B) and moves the fuel rack through the connecting link (2). The guide lever (4) is adjusted by the operator to determine speed. The shifter (5), which other manufacturers would describe as the *sleeve*, converts the angular motion of the weights (6) to fore-and-aft movement. The weights are driven by the camshaft (7).

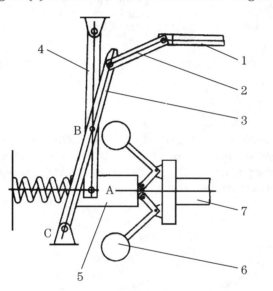

1 Control rack
2 Connecting link
3 Floating lever
4 Guide lever
5 Shifter
6 Counterweight (flyweight)
7 Camshaft

6-8 Simplified drawing of the RSV governor.

As speed increases, point A moves to the left in response to the displacement of the weights away from the camshaft. The shifter moves the lower end of the floating lever (A) to the left, and in so doing, moves the shared fulcrum (B) to the left. This movement causes the floating lever to retract the rack, reducing fuel delivery. Under load the mechanism moves to the right, reversing the procedure.

The drawing at Fig. 6-8 should be consulted as we get further into the discussion, because it is easy to lose sight of the fundamental principles involved under different speed and load conditions.

The next series of drawings are more realistic because they include all the important functional parts. Be patient and bear with me in this long and somewhat tedious explanation. Unless you understand how the mechanism works—either from the drawings or in conjunction with an actual governor—repair attempts are almost hopeless.

Figure 6-9 shows the governor at "start." The control lever (11) is set to the excess-fuel position; it moves the swivel lever to the right and stretches the governor spring (2). One end of the spring is connected to the tension lever (7) and brings the high-speed stop. When the engine starts, the weights are forced outward, overcoming the weak tension of the starting spring (10), and the shifter retracts until it comes into contact with the tension lever.

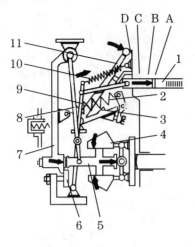

1 Control rack
2 Governor spring
3 Swivel lever
4 Counterweight
5 Shift and sleeve
6 Full-load stopper
7 Tension lever
8 Guide lever
9 Floating lever
10 Starting spring
11 Control lever
A Start
B Full load
C Idling
D Stop

6-9 Start position.

During idle the governor spring is slack (Fig. 6-10). It exerts little force on the tension lever. The weights are free to move outward, even at low rpm, and cause the tension lever to contact the auxiliary idler spring (8). The floating lever is moved to the idle position, severely restricting fuel delivery.

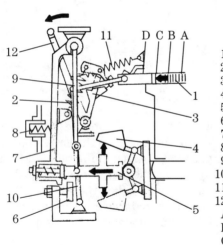

1 Control rack
2 Governor spring
3 Swivel lever
4 Counterweight (flyweight)
5 Shifter and sleeve
6 Full-load stopper
7 Tension lever
8 Auxiliary idler spring
9 Guide lever
10 Floating lever
11 Starting spring
12 Control lever
A Start
B Full load
C Idling
D Stop

6-10 Idle position.

At maximum rpm the governor spring pulls the tension lever (Fig. 6-11) to the high-speed stop and moves the rack to full power. Should the engine speed fall off from this rpm, force on the weights will decrease, and the governor spring will cause the rack to deliver more fuel.

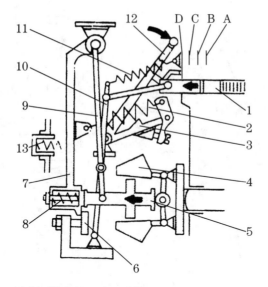

1 Control rack
2 Governor spring
3 Swivel lever
4 Counterweight (flyweight)
5 Shifter and sleeve
6 Full-load stopper
7 Tension lever
8 Balance (Angleich) idler spring
9 Guide lever
10 Floating lever
11 Start spring
12 Control lever
13 Auxiliary idler spring
A Start
B Full load
C Idling
D Stop

6-11 High-speed position.

Other features of this governor include an *angleich* (balance) spring as an override to provide more power at low rpm. The stop lever can be actuated at any speed, regardless of the position of the weights or other levers.

Disassembly Using Fig. 6-12 as a guide, begin disassembly by separating the governor from the injector pump. Remove the *angleich* spring (32) and the auxiliary idler spring (58). Remove the lockscrew (51) from the housing. Raise the swivel lever (13) and lift out the tension lever (27) and spring (14). Now remove the sleeve (21) and guide lever (25). With the swivel lever raised, remove the floating lever (28) from the guide lever.

If required, the bearing (22) can be pressed from the sleeve (21); support as shown in Fig. 6-13. Be extremely careful not to deface the surface of the shifter in the press operation. Next, loosen the lock bolt and dismantle the control lever (38), along with assorted washers and shims (42 and 43). Do not remove the swivel lever (13) unless it is absolutely necessary. It is secured by snap rings. For governors so equipped, remove the lever (secured by a bolt on the lever shaft). The next step is to remove the end cover (65) and *angleich* spring assembly (31, 32, 32, and 34). It is held by a snap ring.

Turn the bearing (22) by hand. It should turn freely and without undue noise. Check the swivel lever (13) bushings for excessive clearance; the lever should turn freely. Do the same for the pin bushing clearance at the weights (18); replace as needed. The spring should be tested by machine to determine if it has lost tension in service. Clean all parts thoroughly and lubricate prior to assembly.

Assembly If a new bearing (22) has been installed, the sleeve (21) will have to be shimmed. Your dealer has the parts and instructions. The lockscrew (51) is torqued to 8 ft.-lb. The distance from the pushrod (31) to the tension lever (27) must be 1.5 mm, or 0.0587 in., as shown in Fig. 6-14. Use new gaskets and O-rings.

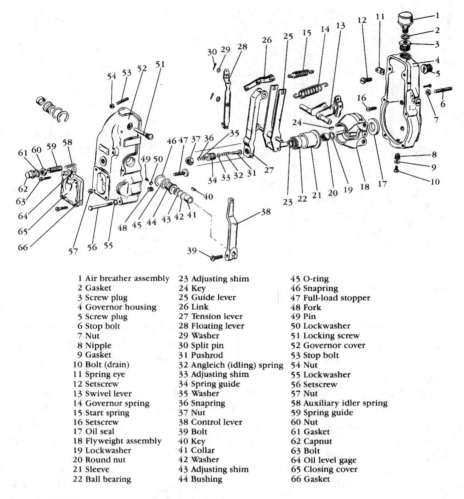

1 Air breather assembly	23 Adjusting shim	45 O-ring
2 Gasket	24 Key	46 Snapring
3 Screw plug	25 Guide lever	47 Full-load stopper
4 Governor housing	26 Link	48 Fork
5 Screw plug	27 Tension lever	49 Pin
6 Stop bolt	28 Floating lever	50 Lockwasher
7 Nut	29 Washer	51 Locking screw
8 Nipple	30 Split pin	52 Governor cover
9 Gasket	31 Pushrod	53 Stop bolt
10 Bolt (drain)	32 Angleich (idling) spring	54 Nut
11 Spring eye	33 Adjusting shim	55 Lockwasher
12 Setscrew	34 Spring guide	56 Setscrew
13 Swivel lever	35 Washer	57 Nut
14 Governor spring	36 Snapring	58 Auxiliary idler spring
15 Start spring	37 Nut	59 Spring guide
16 Setscrew	38 Control lever	60 Nut
17 Oil seal	39 Bolt	61 Gasket
18 Flyweight assembly	40 Key	62 Capnut
19 Lockwasher	41 Collar	63 Bolt
20 Round nut	42 Washer	64 Oil level gage
21 Sleeve	43 Adjusting shim	65 Closing cover
22 Ball bearing	44 Bushing	66 Gasket

6-12 RSV governor in exploded view.

Adjustment Adjustments for this governor are not simple—at least, not simple if you want the performance and responsiveness that this device is capable of. You will need a test bench with a variable-speed drive and dial indicator to determine rack movement as a function of rpm. With the injector pump on the machine, match the indicator with zero extension of rack. Operate the control lever and determine that full stroke is 21.0 mm, or 0.827 in. The rack should move smoothly and should be impelled to maximum fuel injection by the force of the starting spring. The stop bolt bearing against the control lever should be set to correspond with the 0.5–1.0 mm (0.0197–0.039 in.) control rack position.

Begin with the high-speed adjustment. Remove the end cover. You will see the full-load stop screw (Fig. 6-12, No. 47) at the lower part of the unit. Loosen its locknut and turn the screw so that the rack position corresponds to A in Fig. 6-15 at pump speeds between B and C. To increase the fuel delivery—i.e., lengthen the in-

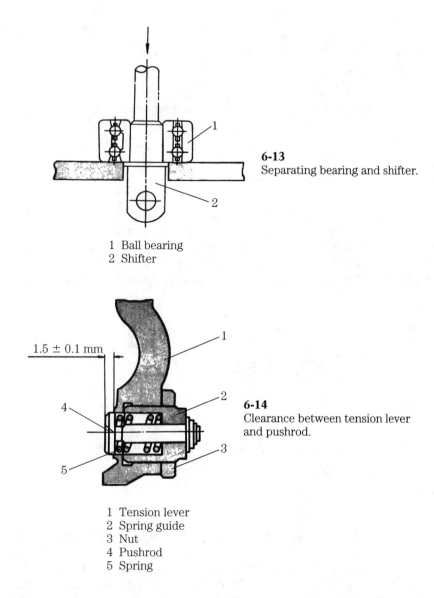

6-13
Separating bearing and shifter.

1 Ball bearing
2 Shifter

1.5 ± 0.1 mm

6-14
Clearance between tension lever
and pushrod.

1 Tension lever
2 Spring guide
3 Nut
4 Pushrod
5 Spring

ward movement of the rack—turn the screw clockwise. The maximum speed is represented by point G. Adjust the stop (Fig. 6-12, No. 6) to obtain the requisite rack position at this rpm.

Next we have the matter of governor sensitivity. Referring again to Fig. 6-15, G represents maximum no-load speed. Point E represents the rated speed. The difference between E and G is expressed as a percentage. Subtract the rated speed from the maximum no-load speed and divide the remainder by the rated speed. For example, if the rated speed were 3200 rpm and no-load speed reached 3400, we would divide the difference, 200 rpm, by 3200 and multiply by 100 to convert the answer to a percentage. Variation would be 6.25%, which is quite acceptable for most applications.

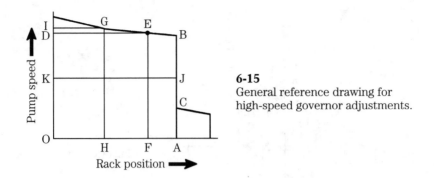

6-15
General reference drawing for high-speed governor adjustments.

If the variation is beyond specs, remove screw plug (Fig. 6-12, No. 3) from the top of the governor casting and turn the adjusting knuckle screw. Tightening the screw reduces the percentage of variation; loosening it increases the variation. At the same time the position of the screw influences rated speed by changing governor spring tension (Fig. 6-16). Rated speed must be reset after this adjustment.

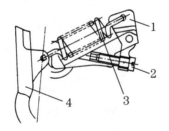

1 Adjusting knuckle
2 Screw
3 Governor spring
4 Tension lever

6-16
Speed variation adjustment.

CAUTION: Do not unscrew the adjustment more than 20 clicks (4 clicks to the turn).

Most RSV governors include an *angleich* spring assembly. The pump-rpm, rack-position curve looks like the one in Fig. 6-17. The *angleich* spring is responsible for the dogleg from B to C. Set the control lever to maximum speed and operate at F rpm. Tighten the *angleich* spring assembly with the appropriate wrench. You can fabricate one in the shop or purchase Chrysler's tool number 47916-212. Tighten until the control rack moves to H. Tighten the locknut to hold the adjustment.

To check the operation of the spring, mount the pump assembly on the engine. Slowly accelerate from speed D. At rpm E the rack should move from H to G.

The idle adjustment is made at the after end of the governor. The idler spring guide, shown as 59 in Fig. 6-12, can be threaded in and out to adjust the tension on the idle spring. To adjust, set the control lever to the stop position, so that the control rack is at B. Set the pump speed to D and adjust spring tension to move the rack to C. Do not overtighten the spring. The idle adjustment curve is shown in Fig. 6-18.

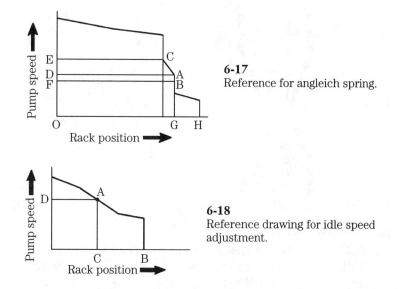

6-17
Reference for angleich spring.

6-18
Reference drawing for idle speed adjustment.

The rack-position, pump-rpm relationship varies between engine makes, models, and to some extent, governor model. No attempt has been made here to include this information, because it is available from dealers and diesel pump specialists.

Pneumatic governors

Pneumatic, or *flap valve*, governors are generally less expensive than the more sophisticated centrifugal types, but are not as progressive in their action (although some engineers would argue that the difference is academic).

These governors operate on the *venturi* principle. When a moving fluid encounters a restriction, its velocity increases. And, in accordance with the principle that we cannot get something for nothing, its pressures falls. Figure 6-19 illustrates the classical venturi—a smoothly narrowing restriction in a bore. The numbers are arbitrary and could express velocity in feet per second, miles per hour, or any other measure. Pressure could be expressed as psi, atmospheres, or inches of mercury. What is interesting about the venturi is that each one has a *constant*. In this case, the constant is 20. If we multiply pressure times velocity at any point in the pipe the answer is always the same. Of course, in the real world, things are not so neat, and our constant tends to be fuzzed by turbulence, air resistance, and other factors.

The vacuum the venturi draws is a function of air velocity through it. The faster the column of air moves, the more vacuum, or negative pressure. Because air velocity in a diesel engine is dependent on piston speed, we can use venturi-induced vacuum to monitor rpm. The venturi need not be a streamlined restriction such as the one shown in Fig. 6-19; any restriction, such as that provided by the edge of the flap valve, will do.

Vacuum is conveyed, by means of a tube, from the venturi to a diaphragmed chamber at the governor. The diaphragm is spring-loaded and free to move fore and aft in the housing. The movement is transferred to the fuel by means of a link.

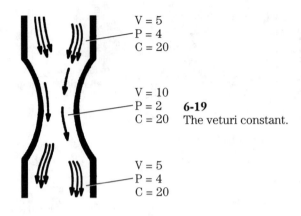

V = 5
P = 4
C = 20

V = 10
P = 2
C = 20

6-19
The veturi constant.

V = 5
P = 4
C = 20

MZ pneumatic governor

Refer to Fig. 6-20 for a structural view of this device. The air intake forms the primary venturi (2), with a secondary near the flap or *butterfly* valve edge (3). A tube bleeds vacuum from these venturis to the left, or low-pressure, side of the diaphragm housing. A second tube brings filtered air to the atmospheric side of the chamber. The diaphragm (8) separates these two halves of the housing. It is loaded by the mainspring (7) and connected to the fuel rack.

As long as the flap value remains stationary in the air intake bore, vacuum in the left side of the chamber is constant. The diaphragm takes a position determined by the spring and the pressure differential on either side of it. Should the engine speed up, vacuum will increase in the left chamber, and the diaphragm, impelled by atmospheric pressure on the right, moves to the left. The mainspring is compressed and the rack is pulled out, reducing fuel delivery. If the engine decelerates, as under load, vacuum decreases and the diaphragm shifts to the right under spring tension. The rack moves in and more fuel is provided. Again the system stabilizes at the predetermined speed fixed by the angle of the flap valve in the air intake.

At full power the valve is open, causing less vacuum to be generated in the venturi, with a consequent increase of fuel delivery as the mainspring pushes the diaphragm to the right. At intermediate valve positions there is correspondingly less fuel delivery. There is no direct connection between the throttle and pump; all signals are delivered by means of pressure differentials in the diaphragm chamber.

The stop lever (23) has two adjustments: The stop bolt (25) limits the maximum fuel delivery during cold cranking, and the stroke setscrew (24) limits fuel once the engine starts. The fuel enrichment provision is by means of a spring that temporarily overrides the stroke setscrew adjustment.

Figure 6-21 shows this relationship graphically. If maximum injection is set at A for best low-speed running, it will increase to B at power; and if we adjust output for best high speed (B), the engine will be starved at low speed, as indicated by A. Ideally the mechanism should be set to A and progress to B at high speed.

The *angleich* mechanism is shown in detail in Fig. 6-22. It acts to reduce fuel delivery at low rpm by moving the fuel rack further out than would otherwise be the case. At low speeds vacuum in the right chamber is low; the diaphragm moves to the

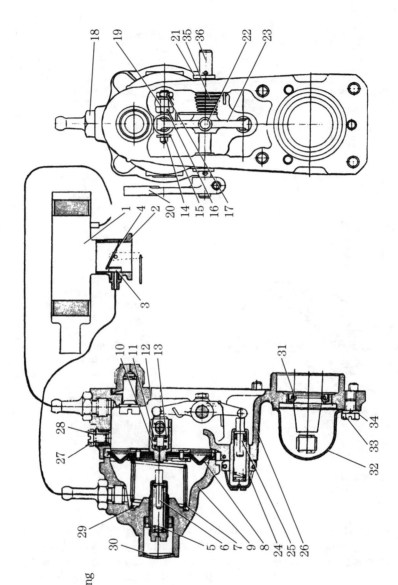

1 Air cleaner
2 Auxiliary venturi
4 Butterfly valve
5 Idler spring
6 Auxiliary idler spring
7 Mainspring
8 Diaphragm
9 Diaphragm housing
10 Adjusting shim
11 Balance (Angleich) spring
12 Pushrod
13 Adjusting shim
14 Cotter pin
15 Washer
16 Connecting bolt
17 Nut
18 Tubing connection
19 Control rack
20 Lever
21 Washer
22 Screw
23 Stop lever
24 Stroke setscrew
25 Stop bolt
26 Governor housing
27 Bolt
28 Gasket
29 Adjusting shim
30 Endplate
31 Oil seal
32 Cap
33 Bolt
34 Lockwasher
35 Split pin
36 Shaft

6-20 The MZ governor.

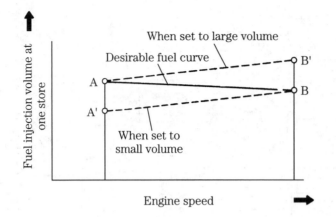

6-21 Fuel injection curves.

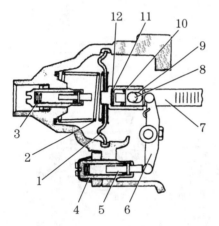

1 Diaphragm
2 Governor spring
3 Auxiliary idler spring
4 Stroke setscrew
5 Stop bolt
6 Stop lever
7 Control rack
8 Connecting bolt
9 Pushrod
10 Adjusting shim
11 Balance (Angleich) spring
12 Adjusting shim

6-22 Balance mechanism.

left, impelled by the mainspring. As it does it compresses the *angleich* spring inserted behind the pushrod (Fig. 6-22, No. 9) and allows the rod to move nearer the zero-delivery position. As the engine accelerates, the diaphragm moves to the right. The balance spring mechanism moves with it, away from the stop lever. Once it is clear of the lever, the spring has nothing to react against (except the plungers, which should rotate with a force of a few grams) and is out of the circuit.

Disassembly Referring to Fig. 6-20, remove the governor from the pump at the diaphragm chamber mounting screws, and disconnect the vacuum and air lines and the fuel rack link. Begin disassembly of the governor proper by unscrewing the idler spring assembly (5) from the housing. Remove the pushrod (12), adjusting shim (13), and balance spring (11) from the diaphragm. Next remove the stroke setscrew (24), camshaft cap, and the stop lever (23), along with its mounting hardware. Remove the stop lever (23), and after extracting the cotter pins from either end, remove the shaft (36). Finally, remove the diaphragm and spring.

Clean all parts thoroughly, with special attention to the diaphragm. It is made of leather and can become stiff in service. Holding it by the outer ring, pull upward on the center section. It should collapse of its own weight. If it does not, apply diaphragm oil (available from your distributor) to soften the leather, or in doubtful cases, replace the diaphragm. The spring should be inspected for wear and its free length compared to a new one. For absolute reliability you should check the spring tension against the manufacturer's specs. Examine the top lever shaft for excessive clearance. You might have to replace both the shaft and governor housing.

Assembly Assemble in the reverse order. The *angleich* spring is shimmed as shown in Fig. 6-23 to adjust the stroke of the pushrod. Specifications vary with the application, but for the Chrysler Nissan series they are as follows:

SD22: 1 mm (0.039 in.)

SD33: 0.6 mm (0.0236 in.)

When installing the stroke setscrew (24), make sure that it contacts the end of the stop lever.

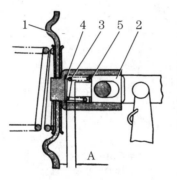

1 Diaphragm
2 Pushrod
3 Balance (Angleich) spring
A Balance (Angleich) spring stroke
4 Shim (stroke adjustment)
5 Shim (spring tension adjustment)

6-23 Pushrod stroke adjustment.

Adjustment Mount the governor pump assembly on the test machine. Set the rack and indicator at zero and connect a vacuum line to the left diaphragm chamber. Run the pump at 500 rpm during the test. (The turning resistance of the plungers varies with speed; static tests are almost meaningless).

Apply a vacuum of 500 mm of water to the diaphragm and verify that the rate of drop averages no more than 2 mm/second for 10 sec. Adjust the stroke setscrew so that the control rack is on specification under full vacuum. If the stroke of the rack is less than prescribed, loosen the screw; if more, tighten it. Check the adjustment of the *angleich* spring again, as shown in Fig. 6-23. Some discretion is required here because the specifications can be voided if the spring resists movement. The best advice is to compare the action of the mechanism with a known-good assembly, and shim for tension as well as stroke.

The mainspring controls the movement of the fuel rack, and adjustment is quite critical. Compare the output with the manufacturer's curve. Increasing the number or thickness of the shims will compress the spring and increase starting pressure.

The idling adjustment has implications over the whole operating range. If the control rack movement does not correspond with the desired curve, you can assume that the idle speed is too high or too low. Tightening the auxiliary idle spring increases the speed. You should use a factory tool when making this adjustment (available from Chrysler as No. 57915-422).

Venturi The venturi section seldom needs attention. After long use the shaft (Fig. 6-24, No. 16) might wear and need replacement. Disassemble using the drawing as a guide on assembly, shim the shaft so that the flap valve (17) moves its full range without binding against the venturi casting.

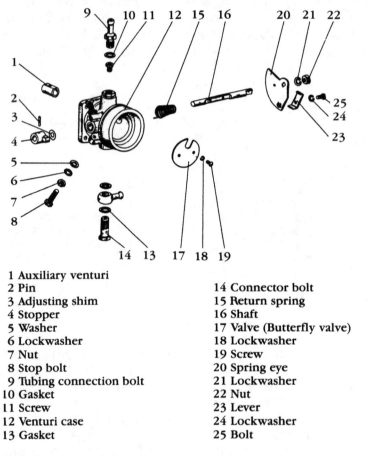

1 Auxiliary venturi
2 Pin
3 Adjusting shim
4 Stopper
5 Washer
6 Lockwasher
7 Nut
8 Stop bolt
9 Tubing connection bolt
10 Gasket
11 Screw
12 Venturi case
13 Gasket

14 Connector bolt
15 Return spring
16 Shaft
17 Valve (Butterfly valve)
18 Lockwasher
19 Screw
20 Spring eye
21 Lockwasher
22 Nut
23 Lever
24 Lockwasher
25 Bolt

6-24 Venturi in exploded view.

Simms GP pneumatic governor

The Simms unit is similar to the MZ, less the *angleich* spring (see Fig. 6-25). The throttle valve A is connected to the throttle lever or pedal and is the only means of speed control. The governor mounts on the back of the injection pump and is spring-loaded to hold the rack open in the absence of vacuum. The diaphragm is similar to the one used on the MV design, consisting of a leather cup secured at the edges by the housing castings. The link assembly transfers diaphragm movement to

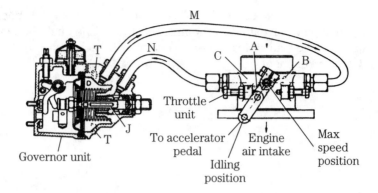

6-25 Simms GP pneumatic governor.

the fuel rack. The damping valve, shown to the right of the governor unit, gives smooth operation by damping diaphragm surge. It is secured by an external locknut.

The main spring is sandwiched between the right side of the diaphragm and the inner wall of the low-pressure chamber. A stop lever moves the rack to the zero-fuel position, overriding the governor. An excess-fuel device is fitted to most models for cold starting.

Adjustment is by turning the valve guide J to vary the area of the inlet port G. Air at or near atmospheric pressure (minus the drop across the filter element) enters through the tube N. The valve is, in effect, a buffer spring.

Disassembly Separate the housing halves by removing the Phillips screws. The diaphragm can be removed by unhooking the link at the diaphragm rod. The diaphragm is delicate and should be handled with care. You will find a tapered lockscrew on the underside of the governor. It secures the link to the control rod. Next, remove the splined stop lever by opening the pinch bolt. If you wish to dismantle the unit further, withdraw the excess-fuel shaft by removing the lock pin and the bearing, which is screwed into place.

Examine the diaphragm for tears and surface cracks. It should be pliable. To replace it remove the diaphragm from its shaft (it is held by two nuts) and install a new one that has been soaked for at least one half-hour in Shell grade C calibration fluid. The filter element should be soaked in solvent and wetted with heavy oil.

Assembly Assembly is straightforward and presents no serious problems. The governor mounting screws should be tightened to 5 ft.-lb. Specifications are not given for the Phillips screws, but they should be drawn up in a crisscross manner to prevent the housing from warping. The damping valve should be coated with a colloidal graphite lubricant. Oildag is recommended—but you might have to go to England to get it.

Check the unit for leaks by compressing the spring and holding your fingers over the air tube fittings. The diaphragm should move and then stop as air pressure equals spring force. Continued movement means an air leak.

Adjustment First, calibrate the pump on a test bench. Then mount the assembly on the engine; after a few minutes of warmup, set the maximum no-load speed at the throttle stop. Next, set the idle stop screw to obtain the specified rpm and adjust the damping valve guide J for smoothest idle. The guide is secured by a locknut, which can allow air to leak into the chamber if backed out too far. Keep it fingertight during adjustment.

7
CHAPTER

Cylinder heads & valves

A vast amount of developmental work was done in the 1920s and 1930s on the optimal shape of the combustion chamber. The work continues—with new emphasis on control of emissions—but the main outlines of the problem and the alternatives appear to be firmly established.

This chapter is divided into two parts. Part 1 approaches the cylinder head assembly from a topographic (or functional) point of view, while Part 2 details service procedures.

Part 1: Combustion chambers

Combustion chamber shapes fall into two groups. Direct injection (DI) chambers, also called open chambers, resemble those used in carbureted engines. A DI chamber consists of an undivided cavity, formed by the piston head (Fig. 7-1). These chambers are commonly used on large trucks, and other applications put a premium on fuel efficiency.

The central problem of diesel combustion chamber design is to provide a mechanism for mixing fuel with the air. The fuel spray must be scattered about the cylinder so that all available oxygen can take part in the combustion process. As fuel droplets move away from the nozzle, the chances of complete combustion are progressively lessened. There is less oxygen available, because some of the fuel has already ignited. In addition, droplets shrink as they travel and successive layers of light hydrocarbons boil off.

Most DI chambers mix by energizing the air. Two mechanisms are involved: swirl and squish. The exit angle of the intake seat imparts a spinning motion, or swirl, to the incoming air stream. Squish is a secondary effect, achieved by making the edges of the piston crown parallel to the chamber roof. As the piston approaches tdc, air trapped between these two faces "squishes" inward, toward the center of the chamber.

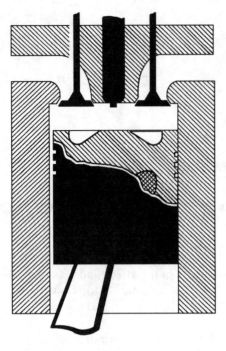

7-1
Open, or direct injection, chamber.

Another approach, limited to engines with ultrahigh pressure unit injectors, is to make the fuel the active element in the mixing process. Injection pressures of 20,000-plus psi are required.

Indirect injection or (IOI) divided chambers come in several styles. The *precombustion chamber* (Fig. 7-2) was the first and actually predates the diesel engine. The low-compression Hornsby-Ackroyd oil engine used such an antechamber. As shown in Fig. 7-2, the smaller chamber, accounting for 25–40% of the main chamber in volume, is connected to the main chamber by a narrow passage. During the compression stroke, air is introduced into the precombustion chamber.

The fuel charge injects into the precombustion chamber. There is not enough oxygen in the chamber for complete combustion, but enough is present to raise chamber pressures. The unburned charge rushes out of the chamber and mixes with air in the main space over the piston. The fuel is dispersed by its own heat energy, rather than by virtue of high pump pressures and fine injector orifices. Consequently, injection gear is simplified and requires less maintenance. Additional benefits include better, or at least more consistent, fuel economy and the ability to burn a wide range of fuels. Because most of these chambers are uncooled, they are referred to as *hot bulb* chambers.

The *turbulence chamber* (Fig. 7-3) is similar in appearance to the precombustion chamber, but its function is different. Both the piston crown and chamber roof are flat, and they form a giant "swish" area. The chamber is usually spherical and is connected to the clearance volume (space above the piston) by a tangential passageway. As the piston approaches tdc, the crown partially masks the entrance to the passageway. Consequently, the velocity of the air entering the chamber is in-

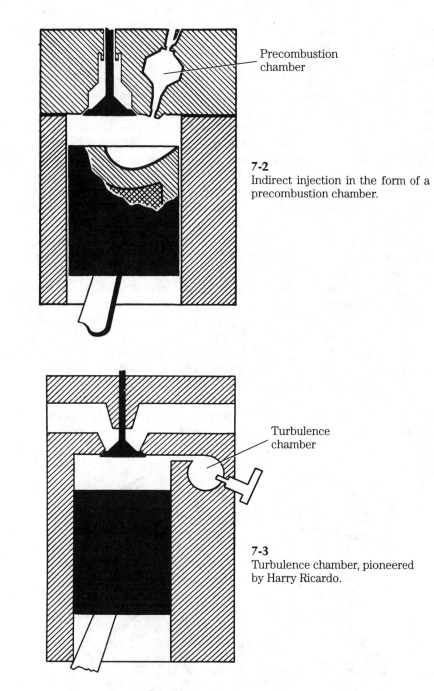

7-2
Indirect injection in the form of a
precombustion chamber.

Precombustion
chamber

Turbulence
chamber

7-3
Turbulence chamber, pioneered
by Harry Ricardo.

creased. This turbulence speed is approximately 50 times crankshaft speed. Fuel injection is timed to occur at the velocity peak. The incoming fuel spray is exposed to
fresh air for the duration of injection. The angle of the passageway and the shape of
the turbulence chamber produce a swirl to the air and flaming fuel particles. As the

piston passes tdc, the direction of movement reverses, and fuel and air swirl out the precombustion chamber and into the cylinder proper. Expansion takes place against the piston.

The precombustion chamber has between a quarter and half of the clearance volume. Its function is to precondition the fuel for combustion, most of which takes place in the main chamber. The turbulence chamber accounts for almost all the volume above the piston. Combustion occurs mainly in the chamber and not above the piston. The area above the piston does contain some oxygen, and combustion does continue in this space; but it might be best thought of as an arena for air/fuel mixing. The precombustion chamber is subject to a simple in-and-out movement of compressed air and expelled fuel. Movement in the turbulence chamber has a definite swirl, first in one direction and then, as combustion commences, in the other.

A third type of chamber is known under several names. The trade name is *Lanova chamber*, but it is also called a *divided chamber* or an *energy cell*. This chamber, used formerly in automotive and tractor engines, has advantages that justify the somewhat elaborate foundry work.

Figure 7-4 shows the figure-8 configuration of the chamber and the opposed injector and antechamber. As in the open chamber, the main volume of air remains above the piston. Most combustion takes place in this space. As in the turbulence chamber, the design generates a high degree of turbulence. But this system does not entail pumping losses (energy is required to force air through the narrow passage leading to the turbulence chamber, which could otherwise be used to turn the crankshaft). Nor is turbulence a function of piston speed in the Lanova design. Turbulence is the result of thermal expansion and is independent of rpm.

The antechamber, or energy cell, consists of an inner and outer chamber. The inner chamber, which is the smaller of the two, opens to a narrow throat, situated between the lobes of the main combustion chamber. If you are familiar with SI engines

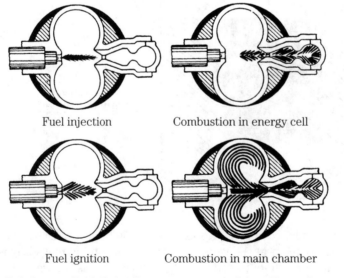

Fuel injection   Combustion in energy cell

Fuel ignition   Combustion in main chamber

7-4 Lanova divided chamber, seen on some Caterpillar engines.

you will recognize the funnel-shaped throat as a venturi not unlike those employed in carburetors. The larger chamber communicates with the inner chamber through a second venturi.

During the compression stroke about 10% of the air volume passes into the energy cell, while the remainder stays above the piston. Fuel is injected in a solid stream, which passes through the small axis of the combustion chamber and into the energy cell. Some small percentage of fuel shears off from the main stream and is ignited in the two lobes of the main chamber, but this is incidental. The bulk of the fuel is trapped in the small inner chamber of the energy cell. Some manages to traverse the second venturi and enter the outer chamber. There it encounters a mass of superheated air and explodes. Pressure rises violently, and the outer cell empties back through the venturi into the second cell and drives the unburned fuel back into the main chamber above the piston. The kidney-shaped walls of the main chamber impart a swirl to the fuel charge so that it mixes thoroughly with the air.

In addition to providing turbulence that is constant regardless of load or speed, the divided chamber allows the use of a simple, single-orifice injector nozzle. Early studies by Caterpillar indicate that this chamber type produces lower emissions than the others, probably because of more consistent fuel mixing and the control exercised by the venturis on the rate of combustion.

Besides forming at least part of the combustion space, the cylinder head forms a major part of the cooling system, and it must be sturdy enough to contain the pressures generated in the chambers. The great majority of heads are cast iron, although a few air-cooled engines employ aluminum heads. Aluminum has about four times the conductivity of cast iron, which is an advantage both for dissipating the heat developed in combustion and for promoting a uniform heat gradient across the head.

The water jacket should be cast with large passages that will resist clogging. But merely pumping coolant through the head will not ensure that local hot spots are cooled. Some designs employ replaceable diverters to direct the stream at the valve seats and other critical areas. Generally the head forms an integral circuit with the block. There is a recent tendency to do away with interconnected water passages and keep the block and head circuits separate. The advantage of this approach is that head gasket failure does not contaminate the oil with water. Contamination becomes quite serious when the water is mixed with ethylene glycol (antifreeze).

For the sake of savings in materials if not weight, the water jacket should form part of the head structure; it should bear part of the loads. The fasteners—either capscrews or, preferably, studs—should be arranged symmetrically insofar as the rocker pedestals, compression release, and other subordinate gear allow.

From these remarks you can appreciate that cylinder head design is one of the most demanding aspects of engine building. The general rule is to stick with old, proven designs and look with jaundiced eye at anything new, even if the computer says it will work.

Valves & valve gear

While a few piston-ported two-cycles survive, most of these engines employ exhaust valves. A few use valves in the inlet side, although this sort of complexity is usually limited to large engines. Ports serve as well and have the advantage of no additional moving parts. Because of the limited amount of time available for two-cycle functions—the complete operating cycle takes place in one revolution of the crank-shaft—two-cycle valves are often paired in the cylinder to increase their area. The Detroit Diesel models represent an example of this practice. Four-cycle engines must, of course, employ intake as well as exhaust valves. Some employ paired valves because of space restrictions in the head and to reduce the inertia of each valve for better control. The type of valve used (Fig. 7-5) is called a *poppet, mushroom,* or *tulip* valve.

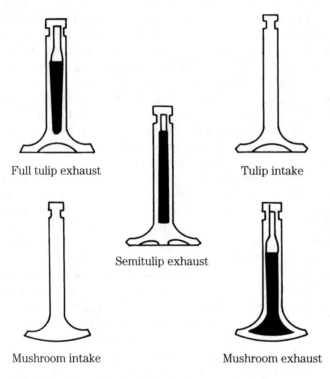

Full tulip exhaust   Tulip intake

Semitulip exhaust

Mushroom intake   Mushroom exhaust

7-5 Valve styles. The dark shading represents sodium.

Because of the temperatures involved, intake valves are made of chromium-nickel alloy and the exhaust valves of silicon and chromium. The intake valve has the benefit of cooling air and is less critical than the exhaust, which is "scrubbed" by the exhaust gases. Some exhaust valves are filled with sodium. When heated, sodium becomes a liquid and transfers heat to the valve guide. Sodium combines explosively with water, and these valves should be treated with respect. When discarded, they should be clearly identified as containing sodium.

The valve inserts, or seats, are a shrink interference fit in the cylinder head. In addition to forming a gas seal with the valve face, the seats provide the main path for heat transfer from the valve head. Consequently, the seats are adjacent to fins in air-cooled designs, and adjacent to the water jacket in the more popular, liquid-cooled types. The bevel is self-centering and can correct some degree of valve guide wear. Normally, the intake seat is at an angle of 30 degrees and the exhaust is at 45 degrees. The 45-degree bevel gives a 20% higher seating load than the more shallow angle, but it exacts a penalty in the form of reduced flow. As a rule of thumb, a 45-degree valve must be lifted 20% more to obtain the same gas flow as a 30-degree valve with the same area. For this reason the intake is normally at 30 degrees, and the exhaust, which operates in more severe environment, is ground to 45 degrees. Usually the cam lift is the same for both, but exhaust restrictions are less meaningful because most of the gases escape during blow-down, at pressures considerably higher than atmospheric.

While the cam forces the valves open, only spring pressure closes them (Fig. 7-6). In some designs the cam has little effect on the upper limit of lift because acceleration is so abrupt that the inertia of the cam follower, pushrod, valve, and part of the rocker arm drives the valve ahead of the cam. This condition is hardly desirable and can be cured by careful attention to the cam profile and to limiting reciprocating masses. Another problem associated with reciprocating motion occurs during seating. Many valves bounce several times before coming to rest. The severity of bounce increases with engine speed.

The valve guides are usually made of chilled iron and pressed into place. They are second in importance to the seats as heat sinks. The springs are contained between two caps and secured by split collars on the valve stem. The valve depicted in Fig. 7-7 features a rotator that turns the valve a few degrees during each lift. Rotating valves is not absolutely essential; new processes, such as spraying the valve faces with aluminum, have almost made it redundant. But rotation does prolong seat life by averaging the wear over the whole surface.

Valve lash is the total of clearances between the cam and the cam follower (tappet), both ends of the push rod, and the rocker arm and valve stem. Some running clearance is required to ensure that the valve will seat. Should it not make solid contact with the seat, compression will be lost and the valve will quickly overheat. Because of expansion most engine makers specify that clearances be adjusted when the engine is hot.

Part 2: Cylinder head service

The remainder of this chapter is an overview of cylinder head service procedures. Some operations, such as cylinder head removal and installation, and repairs to the valve-actuating mechanism, are routinely accomplished in the field. Other operations, including casting repairs, valve work more extensive than replacing the springs, or (on most engines) installing injector sleeves, are the province of the machinist. Although most likely, you will not be actively involved in machine work, it is important to have a general understanding of the procedures involved. One needs to be able to communicate with the machinist and should have some appreciation of the compromises implicit in repairs of this kind.

In general, there are no shortcuts in diesel repair. You must go by the book, measuring every critical surface and replacing or refinishing components as necessary to restore original tolerances. In some cases, for example, when dealing with certain SI-derivative engines or when component failures are endemic, it might be necessary to go an extra mile and build to better-than-new specifications.

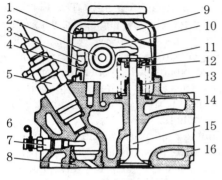

1 Locknut
2 Rocker shaft
3 Adjusting screw
4 Pushrod
5 Nozzle assembly
6 Glow plug cable
7 Glow plug
8 Combustion chamber
9 Rocker cover
10 Valve rocker
11 Spring seat
12 Split collar
13 Stem seal
14 Valve spring
15 Valve
16 Valve seat

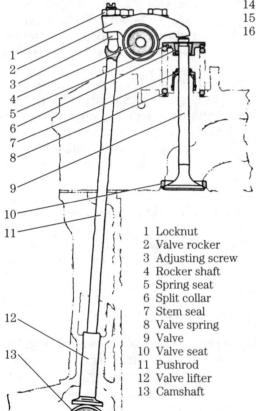

1 Locknut
2 Valve rocker
3 Adjusting screw
4 Rocker shaft
5 Spring seat
6 Split collar
7 Stem seal
8 Valve spring
9 Valve
10 Valve seat
11 Pushrod
12 Valve lifter
13 Camshaft

7-6
Most modern engines employ pushrod-operated valves, located above the piston and articulated through shaft-mounted rocker arms. Marine Engine Div., Chrysler Corp.

Replacement parts can be purchased from the OEM (original equipment manufacturer) or, for the more popular engines, from the aftermarket. Some aftermarket parts are equal to those supplied by the OEM and might, in fact, be the same parts in a different box. Others are not so good in ways that might not be apparent until the engine is in service.

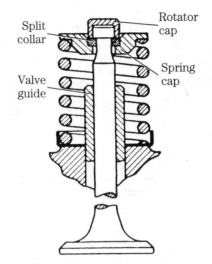

Split collar

Rotator cap

Spring cap

Valve guide

7-7
The GM Bedford valve assembly includes rotator cap (Rotocap), split collar keepers, spring cap, and spring.

It should also be noted that counterfeit parts, usually disguised as OEM parts, have been detected by Caterpillar and other manufacturers. Be suspicious of a part that costs considerably less than it should or that differs from the original in finish, code number, or packaging.

Lifting the cylinder head is a fairly serious operation, involving replacement parts and, in many cases, machine work. Typically, one removes a cylinder head in order to:

- replace a leaking head gasket or, as luck might have it, to repair or replace the head casting.
- reseat the valves.
- decarbonize the combustion chambers.
- replace a defective injector sleeve.
- begin the engine overhaul process.

Diagnosis

Head-related problems usually involve loss of compression in one or two cylinders, a condition that is signaled by a ragged idle, by increased fuel consumption, and in some instances, by exhaust smoke. An exhaust temperature gauge, with switch-controlled thermocouples on each header, will give early warning. Exhaust from weak cylinders will be cooler than the norm. You can cross-check by disabling one injector at a time while the engine ticks over at idle. A cylinder that causes less of an rpm drop than the others does not carry its share of the load. (See chapter 4 for additional information about this procedure.)

At this point, you can check the injectors (the usual suspects) or else go to the heart of the matter with a cranking compression test (Fig. 7-8). As detailed in the diagnostics chapter, we are looking for cylinders with dramatically (at least 20%) lower compression than the average of the others. If the weak cylinder is flanked by healthy cylinders, the problem is either valve- or head-gasket related; or very low compression in an adjacent cylinder points to gasket failure. Abnormally high read-

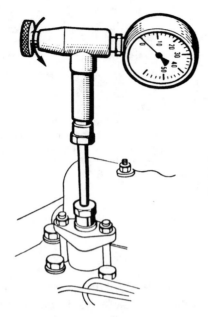

7-8
Cylinder compressor gauge and
adapter. Peugeot

ings on all cylinders indicate heavy carbon accumulations, a condition that might be
accompanied by high pressures and noise. The next step is to make a cylinder leakdown test, which will distinguish between valve and gasket failure.

While most compression leaks bleed into adjacent cylinders or across the fire
deck to the atmosphere, it is possible for a leak path to open into water jacket. The
engine might seem healthy enough, but overheat within a few minutes of start-up.
Coolant in the header tank might appear agitated and might spew violently with the
cap removed. A cooling system pressure test will verify the existence of a leak, which
can be localized with a cylinder leak-down test. However, the leakdown test cannot
distinguish between cracks in the casting and a blown gasket.

Fortunately, it is rare for an engine that has not suffered catastrophic overheating to leak coolant into the oil sump, where it can be detected visually or, in lesser
amounts, by a spectrographic analysis. Likely sources are casting cracks, cracked
(wet-type) cylinder liners, and liner-base gasket leaks.

The cylinder head casting, like the fluid end of a high-pressure pump, will eventually fail. After a large, but finite, number of pressure cycles, the metal crystallizes
and breaks. Owners of obsolete engines for which parts are no longer available would
do well to keep a spare head casting on hand. Even so, most cylinder heads fail early,
long before design life has been realized, because of abnormally high combustion
pressure and temperature.

Combustion pressure and heat can be controlled by routine injector service
(dribbling injectors load the cylinders with fuel), attention to timing, and conservative pump settings. Cooling system maintenance usually stops when the temperature gauge needle remains on the right of center. But local overheating is as critical
as radiator temperature and is rarely addressed. According to engineers at Detroit
Diesel, ¼ inch of scale in the water jacket is the thermal equivalent of four inches of

cast iron. Eroded coolant deflectors and rounded-off water pump impeller blades can also produce local overheating, which will not register as a rise in header-tank temperature.

Local overheating on air-cooled engines can usually be traced to dirty, grease-clogged fins or to loose shrouding.

Gasket life can be extended by doing what can be reasonably done to minimize potential leak paths. As explained below, some imperfection of the head and fire deck seating surfaces must, as a practical matter, be tolerated. Fasteners should be torqued down to specifications and in the suggested torque sequence, which varies between engine makes and models. Asbestos gaskets, without wire or elastomer reinforcements, "take a set" and must be periodically retightened. Newer, reinforced gaskets hold initial torque, but proper diesel maintenance entails retightening head bolts periodically.

Disassembly The drill varies between makes and models, and is described in the manufacturer's manual. Here, I merely wish to add some general information, which might not be included in the factory literature.

It is good and sometimes necessary practice to align the timing marks before the head is dismantled. Bar crankshaft over—in its normal direction of travel—to tdc on No. 1 cylinder compression stroke. Tdc will be referenced on the harmonic balancer or flywheel; the compression stroke will be signaled by closed intake and exhaust valves on No. 1 cylinder. In most cases, the logic of timing marks on overhead cam and unit injector engines will be obvious; when it is not, as for example, when camshaft timing indexes a particular link of the drive chain, make careful notes. A timing error on assembly can cost a set of valves.

Experienced mechanics do not disassemble more than is necessary. Normally you will remove both manifolds, unit injector rocker mechanisms, and whatever hardware blocks access to the head bolts. Try to remove components in large bites, as assemblies, by lifting the intake manifold with turbocharger intact, removing the shaft-type rocker arms at the shaft hold-down bolts, and so on.

Miscellaneous hardware should remain attached, unless the head will be sent out for machine work. In this case, it should be stripped down to the valves and injector tubes. Otherwise, the head might be returned with parts and fasteners missing.

Note: Injector tubes on some DI heads extend beyond the head parting surface and, unless removed, might be damaged in handling.

Examine each part and fastener as it comes off. If disassembly is extensive, time will be saved by storing the components and associated fasteners in an orderly fashion. You might wish to use plastic baggies, labeled with a Sharpie pen or Marks-A-Lot for this purpose. Keep the old gaskets for comparison with the replacements.

Rocker assemblies Two rocker arm configurations are used on ohv engines. Commercial and automotive engines developed from industrial engines generally pivot the rockers on single or double shafts, as illustrated several times in this chapter, including Fig. 7-6 and, as applied to ohc engines, in Figs. 7-9 and 7-10. The earlier drawing is the most typical, because a single rocker drives each intake and exhaust valve. Figures 7-9 and 7-10 illustrate Detroit Diesel solutions to the problem of driving two valves from the same rocker arm. Regardless of the rocker configuration, the hollow pivot shaft doubles as an oil gallery, distributing oil to the rocker bushings through

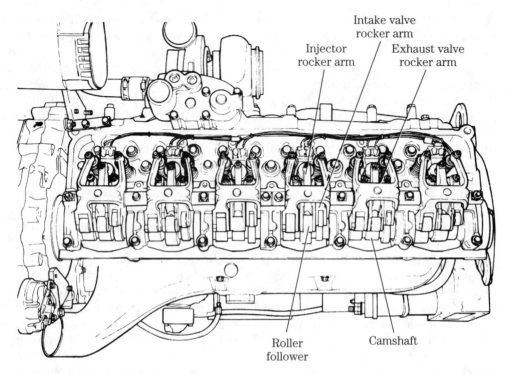

Intake valve
rocker arm

Injector
rocker arm

Exhaust valve
rocker arm

Roller
follower

Camshaft

7-9 Detroit Diesel Series 60 valves and injectors actuate from a single overhead camshaft.

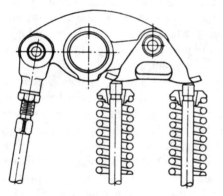

7-10
Detroit Diesel dual valve
actuating mechanism, used on
the firm's two-cycle engines.

radially drilled ports. Rockers are steel forgings, case-hardened on the valve end, and generally include provision for valve lash adjustment.

Shaft-type rockers are detached from the head as an assembly with the shaft, while the pushrods are still engaged. Note the lay of the shaft and rockers, and if no identification marks are present, tag the forward end of the shaft as an assembly reference. Loosen the hold-down bolts slowly, a half-turn or so at a time, following the factory-recommended breakout sequence. If no information is provided on this point, work from the center bolt outward. This procedure will distribute valve spring forces over the length of the shaft.

Set the rocker arm assembly aside and remove the pushrods, racking them in the order of removal with the cam-ends down. A length of four-by-four, drilled to accept the pushrods and with the front of the engine clearly marked, makes an inexpensive rack.

Although the task is formidable on a large engine, the rocker arms, together with spacers, locating springs, and wave washers, should be completely disassembled for cleaning and inspection. Critical areas are:

- *Adjusting screws*—check the thread fit and screw tips. Screws are case-hardened: once the carburized "skin" is penetrated, it will be impossible to keep the valves adjusted.
- *Rocker tips*—with proper equipment, worn tips can usually be recontoured, quieting the engine.
- *Rocker flanks*—check for cracks radiating out from the fulcrum. Cracks tend to develop on the undersides of the rockers at the fillets.
- *Bushings*—wear concentrates on the engine side of the bushing and, when severe, is accompanied with severe scoring, which almost always involves the shaft. Clearance between the bushing and an unworn part of the shaft should be on the order of 0.002 in. Replacement bushings are generally available from the OEM or aftermarket. The old bushing is driven out with a suitable punch, and the replacement pressed into place and reamed to finish size. Aligning the bushing oil port with the rocker arm port is, of course, critical.
- *Shaft*—inspect the surface finish, mike bearing diameters, and carefully clean the shaft ID, clearing the oil ports with a drill bit. Do the same for the oil supply circuit. Rocker arms are remote from the pump, and lubrication is problematic.

Small engines in general and automotive plants derived from SI engines use pedestal-type rockers, of the type illustrated in Fig. 7-11. This technology, pioneered by Chevrolet in 1955, represents a considerable cost saving because the pivots compensate for dimensional inaccuracies and the rockers are steel stampings. Most examples lubricate through hollow pushrods. If rockers are removed, it is vital that they be assembled as originally found, together with fulcrum pieces and hold-down hardware. Rack the pushrods as described in the previous paragraph.

The locknuts that secure the rockers to their studs should be renewed whenever the head is serviced. Other critical items are:

- *Studs*—check for thread wear, nicks, distortion, and separation from the head. As far as I am aware, all diesel pedestal-type rocker studs thread into the cylinder head.
- *Rocker pivots*—the rocker pivots on a ball or a cylindrical bearing, secured by the stud nut, and known as the fulcrum seat (Fig. 7-12A). Reject the rocker, if either part is discolored, scored, or heat checked. How much wear is permissible on the rocker pivot is a judgment call.
- *Fasteners*—replace locknuts if nut threads show low resistance to turning, or for the considerable insurance value. Replace stud nuts if faces exhibit fractures (Fig. 7-12B).
- *Rocker tips*—look for evidence of impact damage that could point to a failed hydraulic lifter and possible valve tip, valve guide, or pushrod damage. For want of a better rule, replace the rocker when tip wear is severe enough to hang a fingernail.

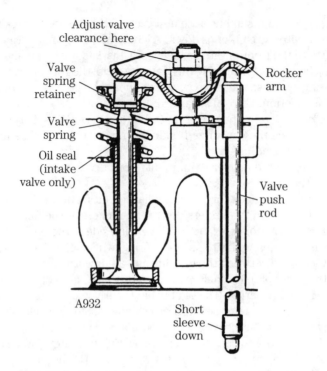

A932

7-11 Pedestal-type rocker arm, pivots on an adjustable fulcrum for lash adjustment. Onan

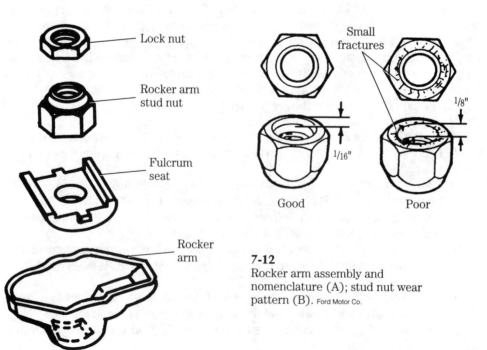

Lock nut

Rocker arm stud nut

Fulcrum seat

Rocker arm

Small fractures

Good

Poor

7-12
Rocker arm assembly and nomenclature (A); stud nut wear pattern (B). Ford Motor Co.

Pushrods Inspect the push rod for wear on the tips and for bends. The best way to determine trueness is to roll the rods on a machined surface or a piece of optically flat plate glass.

Valve lifters Valve lifters, or tappets, are serviced at this time when the lifters are driven from an overhead camshaft or when battered rocker arm tips indicate the need. It should be mentioned that GM 350 hydraulic lifters must be bled down before the cylinder head is reinstalled. Factory manuals recommend that lifters be collapsed twice, the second time 45 minutes after the first. According to technicians familiar with these engines, the factory is not kidding.

Figure 7-13 illustrates the roller tappet used on GM two-cycle engines. This part is subject to severe forces and, in the typical applications, experiences fairly high wear rates. Inspect the roller for scuffing, flat spots, and ease of rotation. Damage to the roller OD almost always is mirrored on the camshaft.

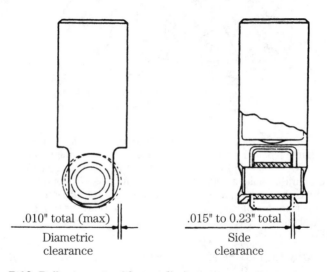

7-13 Roller tappet, with wear limits. General Motors Corp.

Roller and pin replacement offers no challenge, but lubrication is critical. During the first few seconds of operation, the only lubrications the follower receives is what you provide. Engine-supplied lube oil is slow to find its way between the roller and pin.

Ensure proper lubrication by removing the preservative from new parts with Cindol 1705; clean used parts with the same product. Just before installation, soak the followers in a bath of warm (100–125°E) Cindol. Turn the rollers to release trapped air.

Install with the oil port at the bottom of the follower pointed away from the valves. There should be 0.005-in. clearance between the follower legs and guide. The easiest way to make this adjustment is to loosen the bolts slightly and tap the ends of the guide with a brass drift. Bolts should be torqued to 12–15 ft.-lb.

Overhead camshafts Disengage the drive chain or belt (Fig. 7-14). Camshaft mounting provisions vary; some ride in split bearings and are lifted vertically, others slip into full-circle bearings and might require a special tool to open

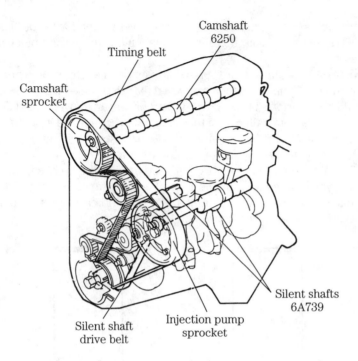

Camshaft
6250

Timing belt

Camshaft
sprocket

Silent shafts
6A739

Silent shaft
drive belt

Injection pump
sprocket

7-14 Ford 2.3L turbo is a new design that employs a cogged (or Gilmer) belt for camshaft drive. Gilmer belts might fail without obvious warning and should be replaced on a rigid engine-hours or mileage schedule. Belts must be tensioned as described by the engine maker and should not be subjected to reverse rotation. Baring the engine backwards can shear or severely damage the teeth.

the valves temporarily for lobe clearance during camshaft withdrawal. Split bearing journals must be assembled exactly as originally found. Make certain that bearing caps are clearly marked for number and orientation. Loosen the caps one at a time, working progressively from the center cap out to the ends of the shaft. (Center, first cap right of center, first cap left of center, second cap right of center, and so on.)

Head bolts Head bolts should now be accessible, but not always visible. Olds 350 engines hide three of the bolts under pipe plugs (Fig. 7-15); some Japanese engines secure the timing cover to the head with small-diameter bolts that, more often than not, are submerged in a pool of oil.

The practice of using an impact wrench on head bolts should be discouraged. A far better procedure and one that must be used on aluminum engines is to loosen the bolts by hand in three stages and in the pattern suggested by the manufacturer. Make careful note of variations in bolt length and be alert for the presence of sealant on the threads. Sealant means that the bolt bottoms into the water jacket, a weight-saving technique inherited from SI engines.

Note: To prevent warpage, the head must be cold before the bolts are loosened.

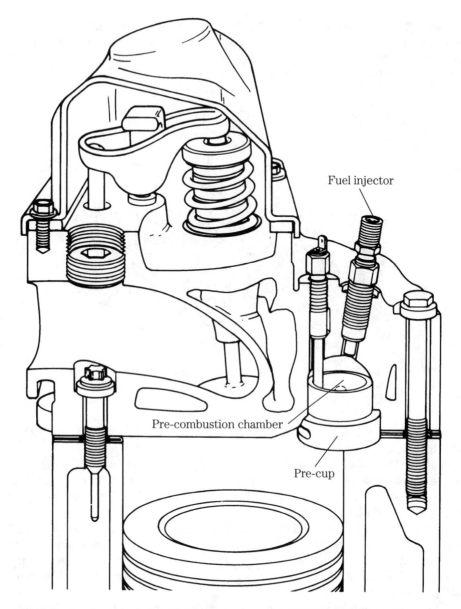

Fuel injector

Pre-combustion chamber

Pre-cup

7-15 Peek-a-boo head bolts hide under pipe plugs on GM 350 engines. Upon assembly, plug threads should be coated with sealant. Fel-Pro, Inc.

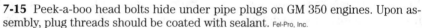

Clean the bolts and examine carefully for pulled threads, cracks (usually under the heads), bends, and signs of bottoming. Engines have come off the line with short bolt holes.

GM and a few other manufacturers use torque-to-yield bolts, most of which are throwaway items. When this is the case, new bolts should be included as part of the gasket set.

Lifting Large cylinder heads require a lifting tackle and proper attachment hardware (Fig. 7-16); heads small enough to be manhandled might need a sharp blow with a rubber mallet to break the gasket seal. Lift vertically to clear the alignment pins, which are almost always present.

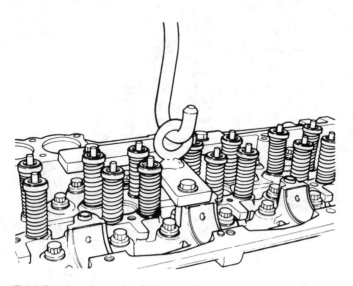

7-16 Lift brackets should be at the approximate center of gravity of the casting, so that the head lifts vertically off its alignment pins. Detroit Diesel

Mount the head in a holding fixture (Fig. 7-17), or lacking that, support it on wood blocks.

Valves Valves are removed by compressing the springs just far enough to disengage the keepers, or valve locks (Fig. 7-18). It is good practice to replace valve rotators, shown in Fig. 7-7. Seals (Fig. 7-19) should also be renewed. Specify high temperature Viton for the seal material.

If all is right, the valves will drop out of their guides from their own weight. The usual cause of valve bind is a mushroomed tip, which can be dressed smooth with a stone and which means that the engine has been operating with excessive lash. The associated camshaft lobe might be damaged. A bent valve suggests guide seizure or piston collision.

Cleaning Cleaning techniques depend on the available facilities. In the field, cleanup usually consists of washing the parts in kerosene or diesel fuel. Gasket fragments can be scraped off with a dull knife (a linoleum knife with the blade ground square to the handle is an ideal tool). The work can be speeded up by using one of the aerosol preparations that promise to dissolve gaskets. In general, it is not a good practice to use a wire brush on head and cooling system gaskets that, with few exceptions, contain asbestos. Sealant, sometimes used in lieu of conventional valve cover and cooling system gaskets, can be removed with four-inch 3M Scotch-Brite Surface Conditioning Discs, mounted on a high-speed die grinder. Use the coarse pad (3M 07450) on steel surfaces, the medium (3M 07451) on aluminum.

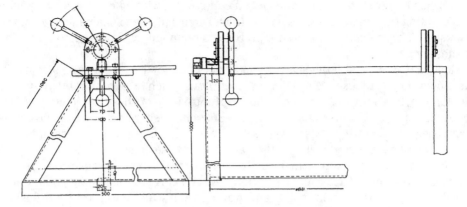

7-17 Cylinder head holding fixture. Marine Engine Div., Chrysler Corp.

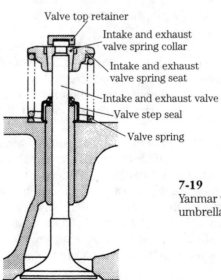

7-18
Lever-type valve spring
compressor.

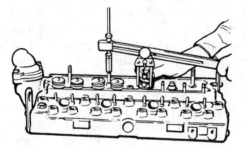

Valve top retainer

Intake and exhaust
valve spring collar

Intake and exhaust
valve spring seat

Intake and exhaust valve

Valve step seal

Valve spring

7-19
Yanmar valves employ
umbrella-type seals.

Carbon responds to a dull knife and an end-cutting wire wheel. Clean the piston tops, rotating the crank as necessary. Overhead camshaft drive chains will foul if the crankshaft is turned, and one cannot pretend to do much by way of piston cleaning on these engines.

Machine shops and large-scale repair depots employ less labor-intensive methods. Some shops still use chlorinated hydrocarbons (such as perchloroethylene and trichloroethylene) for degreasing, although the toxicity of these products has limited their application. A peculiar side effect of trichloroethylene (TCE) exposure is "degreaser's flush." After several weeks of contact with the solvent, consumption of alcohol will raise large red welts on the hapless degreaser's face.

Once the head (and other major castings) are degreased, ferrous parts are traditionally "hot-tanked" in a caustic solution, heated nearly to the boiling point. Caustic will remove most carbon, paint, and water-jacket scale. Parts are then flushed with fresh water and dried. In the past, some field mechanics soaked iron heads in a mild solution of oxalic acid to remove scale and corrosion from the coolant passages.

Caustic and other chemical cleaners pose environmental hazards and generate an open-ended liability problem. The Environmental Protection Agency holds the producer of waste responsible for its ultimate disposition. This responsibility cannot be circumvented by contract; if a shop contracts for caustic to be transported to a hazardous waste site, and the material ends up on a country road somewhere, the shop is liable.

Consequently, other cleaning technologies have been developed. Pollution Control Products is perhaps the best-known manufacturer of cleaning furnaces and claims to have more than 1500 units in service. These devices burn natural gas, propane, or No. 2 fuel oil at rates of up to 300,000 Btu/hour to produce temperatures of up to 800°E. (Somewhat lower temperatures are recommended for cylinder heads and blocks.) An afterburner consumes the smoke effectively enough to meet EPA emission, OSHA workplace safety, and most local fire codes.

Such furnaces significantly reduce the liability associated with handling and disposal of hazardous materials, but are not, in themselves, the complete answer to parts stripping. Most scale flakes off, but some carbon, calcified gasket material, and paint might remain after cleaning.

Final cleanup requires a shot blaster such as the Walker Peenimpac machine illustrated in Fig. 7-20. Parts to be cleaned are placed on a turntable inside the machine and bombarded with high-velocity shot. Shot size and composition determine the surface finish; small diameter steel shot gives aluminum castings a mat finish, larger diameter steel shot dresses iron castings to an as-poured finish. Delicate parts are cleaned with glass beads.

Head casting Make a careful examination of the parting surfaces on both the head and block, looking for fret marks, highly polished areas, erosion around water-jacket ports, and scores that would compromise the gasket seal.

The next step is to determine the degree of head distortion, using a machinist's straightedge and feeler gauges, as illustrated in Fig. 7-21. Distortion limits vary with the application and range from as little as 0.003 in. to 0.008 in. or so. In theory, the block deck has the same importance as the head deck; in practice, a warped block would not be welcome news when head work was all that was contemplated, and most mechanics do not look.

7-20 Walker Peenmatic shot blaster adds an "as-new" finish used casting.

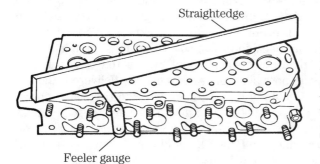

Straightedge

Feeler gauge

7-21
Head warp should be checked in three planes: diagonally as shown, longitudinally, and transversely with particular attention to areas between combustion chambers.
Ford Motor Co.

However, head bolt holes deserve special attention. Be alert for:
* *Stripped threads*—strip-outs can be repaired with Heli-Coil inserts. It is rare to find more than one bolt hole stripped; if the problem is endemic, check with the manufacturer's technical representative on the advisability of using multiple inserts. At least one manufacturer restricts the number of Heli-Coils per engine and per cylinder.
* *Pulled threads*—this condition, characterized by slight eruption in the block metal adjacent to the bolt holes, can be cured by chamfering the uppermost threads with an oversized drill bit or countersink. Limit the depth of cut to one or two threads.

• *Dirty threads*—chase head bolt and other critical threads with the appropriate tap. In theory, one should use a bottoming tap, recognized by its straight profile and squared tip. In practice, one will be lucky to find any tap for certain metric head bolt threads, which are pitched differently than the run of ISO (International Standards Organization) fasteners. Also realize that "bargain" taps can, because of dimensional inaccuracies, do more harm than good. Blow out any coolant that has spilled into the bolt holes with compressed air, protecting your eyes from the debris.

Head resurfacing Minor surface flaws and moderate distortion can usually be corrected by resurfacing, or "milling." However, there are limits to how much metal can be safely removed from either the head or the block. These limits are imposed by the need to maintain piston-to-valve clearance and, on overhead cam engines, restraints imposed by the valve actuating gear. The last point needs some amplification. Reducing the thickness of the head retards valve timing when the camshaft receives power through a chain or belt. Retarded valve timing shifts the torque curve higher on the rpm band. The power will still be there, but it will be later in coming. The effect of lowering a gear-driven camshaft is less ambiguous; the gears converge and ultimately jam.

Fire deck spacer plates are available for some engines to minimize these effects, and ferrous heads have been salvaged by metal spraying. These options are worth investigating, but one is usually better off following factory recommendations for head resurfacing, as in other matters.

The minimum head thickness specification, expressed either as a direct measurement between the fire deck and some prominent feature on the top of the casting, or as the amount of material that can be safely removed from the head and block. Detroit Diesel allows 0.020 in. on four-cycle cylinder heads and a total of 0.030 in. on both the heads and block. Other manufacturers are not so generous, especially on light- and medium-duty engines. For example, the head thickness on Navistar 6.9L engines (measured between the fire deck and valve cover rail) must be maintained at between 4.795 and 4.805 in., a figure that, when manufacturing tolerances are factored in, practically eliminates the possibility of resurfacing. As delivered, the exhaust valve might come within 0.009 in. of the piston crown, and the piston crown clears the roof of the combustion chamber by 0.025 in. Admittedly, these are minimum specifications, but it is difficult to believe that any 6.9L can afford to lose much fire deck metal. Navistar's 9.0L engine, the Ford 2.2L, and the Volkswagen 1.6L engines simply cannot be resurfaced and remain within factory guidelines. Thicker-than-stock head gaskets are available for the VW, but are intended merely to compensate for variations in piston protrusion above the fire deck.

The cardinal rule of this and other machining operations is to remove as little metal as possible, while staying within factory limits. Minor imperfections (gasket frets, corrosion on the edges of water jacket holes, etc.) should be brazed slightly overflush before the head is milled.

Precombustion chambers, or precups, are pressed into the head work on Caterpillar engines, threaded in. In some cases, precups can remain installed during resurfacing; others must be removed and (usually) machined in a separate operation (Fig. 7-22) When it is necessary to replace injector tubes, the work is done after the head has been resurfaced and is always followed by a pressure test.

Note: Ceramic precups are integral components, serviced by replacing the head.

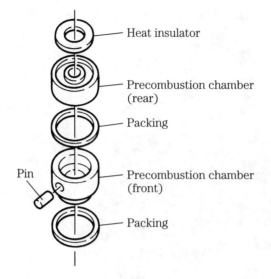

Heat insulator

Precombustion chamber
(rear)

Packing

Pin

Precombustion chamber
(front)

Packing

7-22
Precombustion chamber and
gaskets. An installed pre-cup is
shown in Fig. 7-15. Yanmar Diesel Engine
Co., Ltd.

Most shops rely on a blanchard grinder, such as the one shown in Fig. 7-23 for routine resurfacing. Heavier cuts of 0.015 in. or more call for a rotary broach, known in the trade as a "mill." When set up correctly, a mill will give better accuracy than is obtainable with a grinder. Some shops, especially those in production work, use a movable belt grinder. These machines are relatively inexpensive, require zero set-up time, and produce an unsurpassed finish. However, current belt grinders, which support the workpiece on a rubber platen, are less accurate than blanchard grinders.

Typically, iron heads like a dead smooth surface finish in the range of 60 and 75 rms (root mean square). Some machinists believe that a rougher finish provides the requisite "tooth" for gasket purchase, although there is little evidence to support the contention.

Crack detection Cylinder heads should be crack tested before and after resurfacing. The apparatus used for ferrous parts generates a powerful magnetic field that passes through the part under test. Cracks and other discontinuities at right angles to the field become polarized and reveal their presence by attracting iron filings. The magnetic particle test, known generically by the trade name Magnaflux, is useful within its limits. It cannot detect subsurface flaws, nor does it work on nonferrous metals. But fatigue and thermal cracks always start at the surface, and will be seen.

Nonferrous parts are tested with a penetrant dye (Fig. 7-24). A special dye is sprayed on the part, the excess is wiped off, and the part is treated with a developer that draws the dye to the surface, outlining the cracks. In general, penetrant dye is considered less accurate than Magnaflux, but, short of X ray, it remains the best method available for detecting flaws in aluminum and other nonmagnetic parts.

Neither of these detection methods discriminates between critical and superficial cracks. The cylinder head might be fractured in a dozen places and still be serviceable. But cracks that extend across a pressure regime—that is, from the combustion chamber, cylinder bore, water jacket, or oil circuits—require attention.

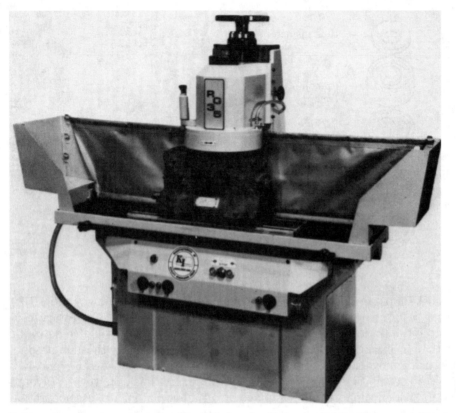

7-23 A blanchard grinder is used for light head milling, manifold milling, and other jobs where some loss of precision can be tolerated.

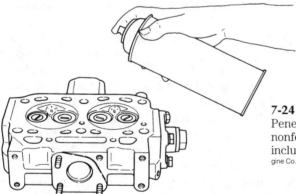

7-24
Penetrant dye detects cracks in nonferrous parts, a category that includes valve heads. Yanmar Diesel Engine Co., Ltd.

Mack and a handful of other manufacturers build surface discontinuities into the roofs of the combustion chambers, which appear as cracks under Magnafluxing, but which are intended to stop crack propagation. These built-in "flaws" run true and straight, in contrast to the meandering paths followed by the genuine article. In general, cracks that are less than a half-inch long and do not extend into the valve seat area might be less serious than they appear. Short cracks radiating out from pre-combustion chamber orifices can also be disregarded.

Crack repairs Assuming that both ends of the crack are visible, it is normally possible to salvage an iron head by gas welding. For best results, the casting should be preheated to 1200°F. Skilled TIG practitioners can do the same for aluminum heads.

Some shops prefer to use one of several patented "cold stitching" processes on iron castings. The technician drills holes at each end of the crack (to block further propagation) and hammers soft iron plugs into the void, which are then ground flush. The plugs must not be allowed to obstruct coolant passages.

Metal spraying is another alternative, demonstrated to have utility, but rarely practiced.

Pressure testing Figure 7-25 illustrates a Detroit Diesel cylinder head, partially dressed out for pressure testing. At this point, most shops would introduce high-pressure water into the jacket and look for leaks, which might take the form of barely perceptible seepage. Detroit Diesel suggests that air be used as the working fluid. Their approach calls for pressurizing the head to 30 psi and immersing the casting in hot (200°F) water for 30 minutes. Leaks register as bubbles.

Test strips

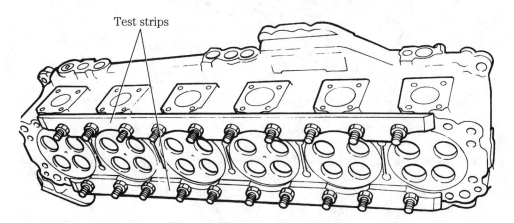

7-25 The water jacket should be sealed as shown, pressurized to about 30psi, and immersed in hot water. Bubbles indicate the presence of cracks.

Straightening Aluminum heads usually carry the camshaft and are often mounted to a cast-iron block. The marriage is barely compatible. Aluminum has a thermal coefficient of expansion four times greater than that of iron. Even at normal temperatures, the head casting "creeps," with most of the movement occurring in the long axis (Fig. 7-26). Pinned at its ends by bolts, the head bows upward—sometimes as much as 0.080 in. The camshaft cannot tolerate misalignments of such magnitude and, if it does not bend, binds in the center bearings.

Such heads are best straightened by stress relieving. The process can be summarized as follows: the head is bolted to a heavy steel plate, which has been drilled and tapped to accommodate the center head bolts. Shims, approximately half the thickness

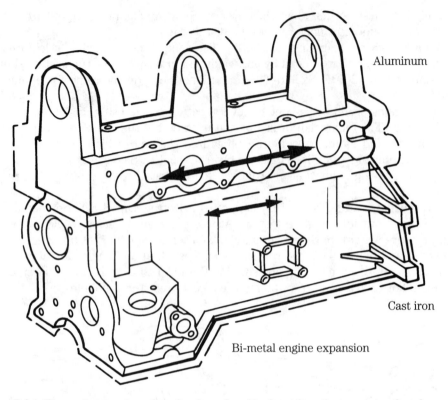

Aluminum

Cast iron

Bi-metal engine expansion

7-26 Thermal expansion of an aluminum head is about four times greater than for a cast iron block. Something must give. Fel-Pro, Inc.

of the bow, are placed under the ends of the head, and the center bolts are lightly run down. Four or five hours of heat soak, followed by a slow cool down, usually restores the deck to within 0.010 in. of true. Camshaft bearings are less amenable to this treatment, and will require line boring or honing.

Corrosion can be a serious problem for aluminum heads, transforming the water jacket into something resembling papier-mâché. Upon investigation, one often finds that the grounding strap—the pleated ribbon cable connecting the head to the firewall—was not installed.

Valve guides Rocker-arm geometry generates a side force, tilting the valves outward and wearing away the upper and lower ends of the guides. The loss of a sharp edge at the lower end of the guides encourages carbon buildup and accelerates stem wear; the bellmouth at the upper end catches oil, which then enters the cylinder. Figure 7-27 illustrates a split ball gauge used to determine guide ID.

Nearly all engines employ replaceable guides or, if lacking that, have enough "meat" in the casting to accept replaceable guides (Fig. 7-28). A Pep replacement guide for 0.375-in. valve stem measures 0.502 in. on the OD. The BMW 2.4L engine is one of the few for which replacement guides are not available. However, the integral (i.e., block metal) guides can be reamed to accept valves with oversized stems.

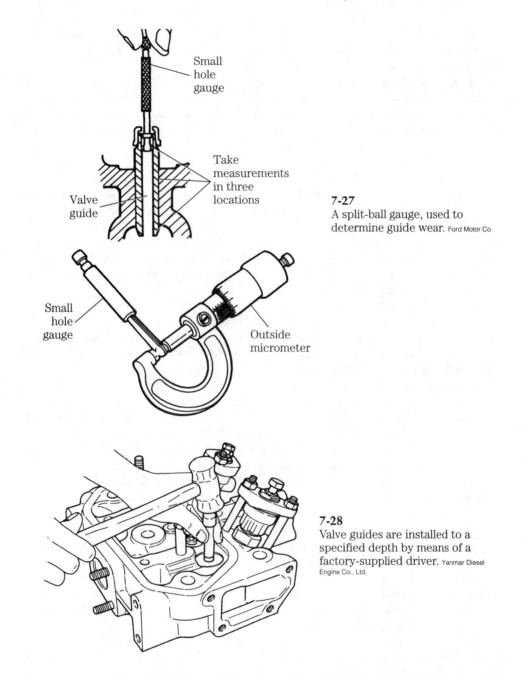

7-27
A split-ball gauge, used to determine guide wear. Ford Motor Co

7-28
Valve guides are installed to a specified depth by means of a factory-supplied driver. Yanmar Diesel Engine Co., Ltd.

The GM 350 offers the option of replaceable guides, plus oversized stems. The latter are apparently a manufacturing convenience and such parts are difficult to come by.

Old guides drive out with a punch, and new guides install with a driver sized to pilot on the guide ID (Fig. 7-28). Cast-iron heads can be worked cold, but a careful technician will heat aluminum heads so that the guide bores do not gall. Engine manufacturers seem to prefer perlitic cast-iron or iron-alloy guides; many machinists claim phosphor bronze has better wearing qualities. Whatever the material, replacement guides rarely are concentric with the valve seat, and some corrective machine work is almost always in order. Stem-to-guide clearances vary with engine type and service; light- and medium-duty engines will remain oil-tight longer with a 0.0015-in. clearance. Heavy-duty engines, which run for long periods at full rated power, need to be set up looser—as much as 0.005 in. when sodium-cooled valves are fitted.

Valve service

Stem and guide problems can be the result of wear or carbon and gum accumulations that hold the valve open against spring pressure. These deposits are caused by the wrong type of lubricant, ethylene glycol leakage into the sump, and low coolant temperatures (below 160°F). Low temperatures are usually the result of long periods at idle and aggravate any mismatch between fuel characteristics and engine demands. Fuel that burns cleanly at normal temperatures can gum the guide and carbon over the valve heads when the engine runs consistently cool. Sticking can also be caused by bent stems.

Premature valve burning, which in extreme cases torches out segments of valve face and seat, has one cause: excessive heat. This can be the result of abnormal combustion chamber temperatures or of a failure of the cooling system. Faulty injector timing and failure of the EGR (exhaust gas recirculation), and sustained high-speed/high-load operation system will create high cylinder temperatures. The thermal path from the valve face to the water jacket can be blocked by insufficient lash (which holds the valve off its relatively cool seat), local water-jacket corrosion, or a malfunction associated with flow directors. Also known as diverters, flow directors are inserted into the water jacket to channel coolant to the valve seats and injector sleeves. These parts can corrode or vibrate loose. Of course, a generalized failure of the cooling system, affecting all cylinders, is possible.

Valve breakage usually takes the form of fatigue failure from repeated shock loads or bending forces. Fatigue failure leaves a series of rings, not unlike tree growth rings, on the parting surfaces. Shock loads are caused by excessive valve lash, which slams the valve face against the seat, or weak valve springs. Bending forces are generated if the seat and/or guide is not concentric with the valve face.

The most dramatic form of valve failure comes about because of collision with the piston. As mentioned previously, Navistar allows as little as 0.009 in. between the exhaust valve and the piston crown at convergence, which occurs 4.¼ degrees btdc. Admittedly, the 6.9 is a "tight" engine, but no diesel will tolerate excessive valve lash, weak valve springs, or overspeeding.

Valve service involves two operations, both of which are done by machine. The face is reground to the specified angle, which is usually (but not always) 30 degrees on the intake and 45 degrees on the exhaust, less a small angle, usually about 1.5 degrees. Thus, the 30-degree intake seats specified for Bedford engines (Fig. 7-29) would be cut to 28.5 degrees. A valve face grinder adjusts to any angle. Seats are cut true to specification, which creates a slight mismatch, known as the interference angle. Once the engine starts, valve faces pound into conformity with the seats. The interference angle ensures a gastight seal on initial start-up and eliminates the need for lapping.

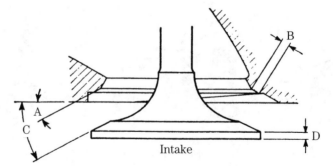

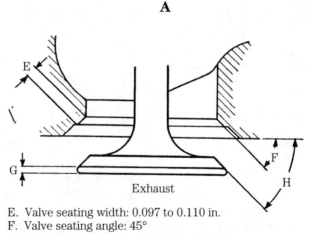

A. Valve seating angle: 30°
B. Valve seating width: 0.055 to 0.069 in.
C. Valve seat angle: 29°
D. Valve head minimum thickness: 0.035 in. valve head
 depth in relation to cylinder head face (minimum
 permissisble): 0.023 in.

A

E. Valve seating width: 0.097 to 0.110 in.
F. Valve seating angle: 45°
G. Valve head minimum thickness: 0.035 in.
 Valve head depth in relation to cylinder head face
 (minimum permissible): 0.041

B

7-29 Typical valve specifications, Bedford engines.

Seats are typically ground in three angles: *entry* (which ranges from about 60 to 70 degrees), seat, and exit (which is usually 15 degrees). Thus seat width can be controlled by undercutting either of the flanking angles (Fig. 7-30). The seat should center on the valve face.

Valve protrusion or, as the case might be, recession, must be held to tight limits on these engines. A valve that extends too far into the chamber might be struck by the piston; one that is sunk into its seat hinders the combustion process, sometimes critically. Figure 7-31 illustrates how the measurement is made. Recession can be

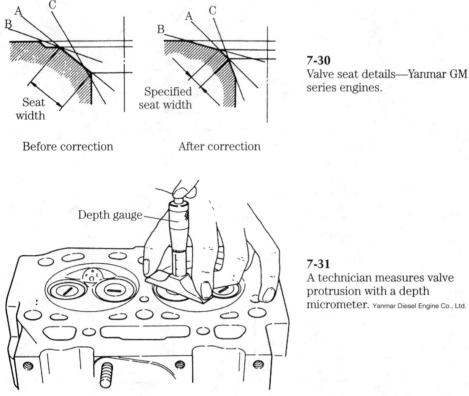

7-30
Valve seat details—Yanmar GM series engines.

Before correction   After correction

Depth gauge

7-31
A technician measures valve protrusion with a depth micrometer. Yanmar Diesel Engine Co., Ltd.

cured by replacing the valve and/or seat. Protrusion is a more difficult problem, encountered when the head has been resurfaced; a deep valve grind might help, provided one has the requisite seat thickness and can tolerate the increase in installed valve height (see immediately below). A handful of OEMs supply thinner-than-stock seat inserts. These inserts require corresponding thicker valve spring seats or shims, in order to maintain valve spring tension. Ultimately, one might be forced to replace the cylinder head.

Installed valve height, measured from the valve seat to the underside of the spring retainer, or cap, has become increasingly significant (Fig. 7-32). Springs on newer engines come perilously close to coil bind with as little as 0.005 in. separating the coils in the full open position.

Finally, the machinist dresses the valve stem tip square to restore rocker-tip to increase available contact area and to center hydraulic-lifter pistons. The average lifter has a stroke of 0.150 in. If the plunger is centered, available travel is 0.075 in., before it bottoms and holds the valve off its seat. But it is rare to find new engines with precisely centered lifter plungers, and available stroke might be considerably less than indicated. Deep valve grinding, head and block resurfacing, and camshaft grinding will, unless compensated for, collapse the lifters. The machinist can remove about 0.030 in. from the tip without penetrating the case hardening. Another limiting factor is rocker/spring-cap contact.

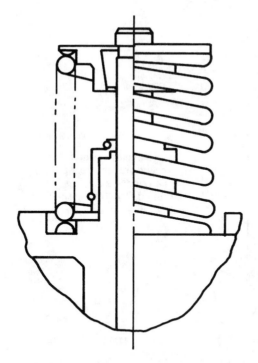

7-32
Installed valve height is the distance between the spring seat and the underside of the spring retainer. Often the factory neglects to provide the specification, and it must be determined by direct measurement during disassembly.

Seat replacement Valve seat inserts are pressed into recesses machined into the head. (Some very early engines used spigoted inserts with mixed results.) Seats must be replaced when severely burned, cracked, loose, or as a means of obtaining the correct valve protrusion. Many shops routinely replace seats during an overhaul for the insurance value.

Seats in iron heads are customarily driven out with a punch inserted through the ports, although more elegant tools are available (Fig. 7-33); seats in aluminum heads should be cut out to prevent damage to the recess (Fig. 7-34).

While new seats can be installed in the original counterbores, it is good practice to machine the bores to the next oversize. Replacement seats for most engines are available in 0.010-, 0.015-, 0.020-, and 0.030-inch oversizes. Material determines the fit: iron seats in iron heads require about 0.005-in. interference fit; Stellite expands less with heat, and seats made of this material should be set up a little tighter; seats in aluminum require something on the order of 0.008-in. interference.

Seat concentricity should be checked with a dial indicator mounted in a fixture that pilots on the valve guide. Often the technician finds it necessary to restore concentricity by lightly grinding the seat.

Springs Valve spring tension is all that keeps the valves from hitting the pistons. A "swallowed" valve is the mechanic's equivalent of the great Lisbon earthquake or the gas blowout at King Christian Island, which illuminated the Arctic night for eight months and could be seen from the moon. Thus, I suggest that valve springs be replaced (regardless of apparent condition) during upper engine overhauls.

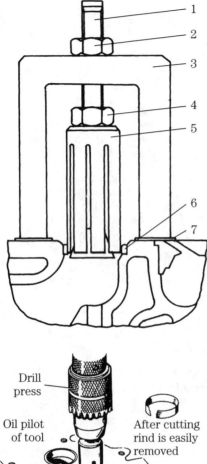

7-33
Chrysler-Nissan supplies this valve seat puller. The split jaws (5) can open as the nut (4) is tightened. Tightening the upper nut (2) lifts the seat.

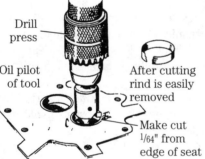

Drill press

Oil pilot of tool

After cutting rind is easily removed

Make cut 1/64" from edge of seat

7-34
Seats in aluminum heads should be machined out, rather than forced out. Onan

If springs are to be used, inspect as follows:
- Carefully examine the springs for pitting, flaking, and flattened ends.
- Measure spring freestanding height and compare with the factory wear limit.
- Stand the springs on their ends and, using a feeler gauge and machinist's square, determine the offset of the uppermost coil (Fig. 7-35). Compare with the factory specification (in angular terms, maximum allowable tilt rarely exceeds 2 degrees). Most keeper failures arise from unequal loading.
- Verify that spring tension falls within factory-recommended norms. Figure 7-36 illustrates the tool generally used to make this determination.

Valve spring shims have appropriate uses, chiefly to restore the spring preload lost when heads and valve seats are refurbished. But shims should not be used as a tonic for tired springs, because the fix is temporary and can result in coil bind.

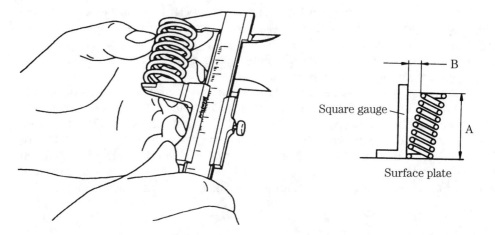

7-35 Most engine makers provide a linear valve spring tilt specification; Yanmar is more sophisticated. First spring free length is determined as shown in the left-hand drawing. The amount of offset is measured with a machinist's square. Offset (B) divided by free length (A) gives the specification, which for one engine series must not exceed 0.0035 inches.

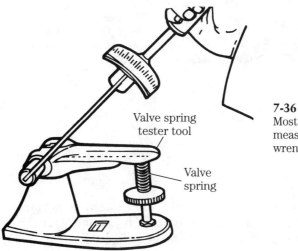

7-36
Most valve spring testers measure force with a torque wrench. Ford Motor Co.

Top clearance Top clearance, or the piston-to-head clearance at tdc, is critical. Unfortunately, the position of piston crown varies somewhat between cylinders because of the stacked tolerances at the crankshaft, rod, piston pin bosses, and deck (which might be tilted relative to the crankshaft centerline). No two pistons have the same spatial relationship to the upper deck. Resurfacing, or decking, the block lowers the fire deck 0.010 in. or more with no better accuracy than obtained by the factory.

Most manufacturers arrive at top clearance indirectly by means of a piston deck height specification. Either of these measurements must be made when:

- the block is resurfaced
- the manufacturer supplies replacement head gaskets in varying thickness to compensate for production variations.

The geometry of some engines (flat pistons and access to the piston top with head in place) invites direct measurement of top clearance. The cylinder head is installed and torqued, using a new gasket of indeterminate thickness. The mechanic then removes No. 1 cylinder glow plug and inserts the end of a piece of soft wire, known as fuse wire, into the chamber. Timing the flywheel through tdc flattens the wire between the piston crown and cylinder head. Wire thickness equals top clearance, or piston deck height plus compressed gasket thickness (Fig. 7-37).

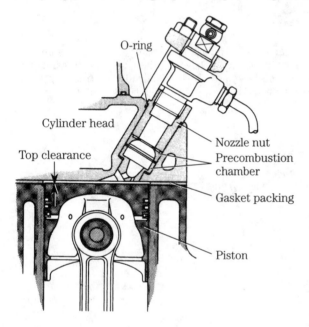

O-ring

Cylinder head

Top clearance

Nozzle nut

Precombustion chamber

Gasket packing

Piston

7-37
Top clearance equals piston deck height plus the thickness of the compressed head gasket. Yanmar Diesel Engine Co. Ltd.

Scrupulous engine builders sometimes make the same determination using modeling clay as the medium. Upon disassembly the clay is removed and carefully miked. This method applies equally well to flat and domed pistons.

The more usual approach is to measure piston deck height with a dial indicator. The procedure involves three measurements, detailed as follows:

1. Zero the dial indicator on the fire deck, with the piston down (left-hand portion of Fig. 7-38).
2. Position the indicator over a designated part of the piston crown, shown as A in Fig. 7-38.
3. Turn the crankshaft in the normal direction of rotation through tdc. Note the highest indicator reading.
4. Repeat Step 3, taking the measurement at B.
5. Average measurements A and B.

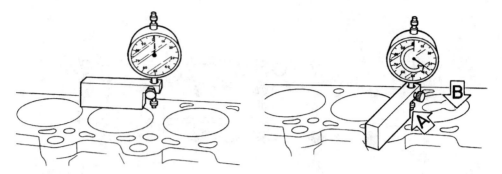

7-38 The first step when measuring piston deck height is to zero the dial indicator on the fire deck. The second step in the measurement process is to determine piston protrusion at two points, the average of which gives piston protrusion. Ford Motor Co.

6. Repeat the process for each piston. Use the piston with the highest average deck height to determine the thickness of the replacement head gasket.

Assembly

The manufacturer's manual provides detailed assembly instructions but includes little about the things that can go wrong. Most assembly errors can be categorized as follows:

- *Insufficient lubrication.* Heavily oil sliding and reciprocating parts, lightly oil head bolts and other fasteners, except those that penetrate into the water jacket. These fasteners should be sealed with Permatex No. 2 or the high-tech equivalent.
- *Reversed orientation.* Most head gaskets, many head bolt washers, and all thermostats are asymmetrical.
- *Mechanical damage.* Run fasteners down in approved torque sequences and in three steps—1/2, 2/3, and 1/1 torque (Fig. 7-39). Exceptions are torque-to-yield head bolts and rocker arm shaft fasteners. The former are torqued as indicated by the manufacturer, whose instructions will be quite explicit. The latter—rocker shaft fasteners—should be brought down in very small increments, working from the center bolts out.

Gaskets, especially head gaskets, might also be damaged during assembly. Lower the head on a pair of guide pins lightly threaded into the block. Pins can be fashioned from discarded head bolts by cutting the heads off. If pins are too short to extend through the head casting, slot the ends for screwdriver purchase.

Set initial valve lash adjustments, bleed the fuel system, start the engine. Final lash adjustments are usually made hot, after the engine has run for 20 minutes or so on the initial settings (Fig. 7-40).

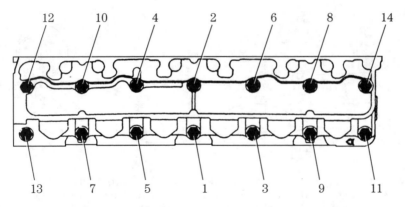

7-39 Cylinder head torque sequence. GM Bedford Diesel

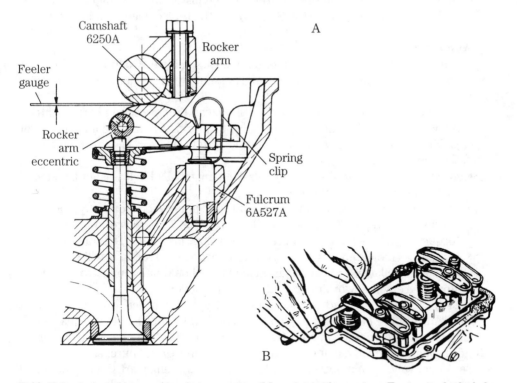

7-40 Valve lash can be measured at any accessible point in the system. For example, lash for the Ford-supplied 2.4L ohc is read as cam-to-rocker clearance (A); Onan measures the clearance between rocker arm tip and valve stem (B).

8

CHAPTER

Engine mechanics

A mechanic needs to be a part-time electrician, semipro fuel system specialist, self-taught millwright, amateur machinist, and back-bench welder. But what he or she is supposed to do is to rebuild engines, the subject of this chapter.

Scope of work

Block assemblies can be *repaired*, *overhauled*, or *rebuilt* (Fig. 8-1). Spot repairs are either triggered by local failure (e.g., a sticking oil pressure relief valve or a noisy valve train bearing) or by a need to extract a few more hours from a worn-out engine. Many a poor mechanic has replaced an oil pump more out of hope than conviction.

While an overhaul is also an exercise in parts replacement, the scope is wider and usually occasioned by moderate cylinder and crankshaft-bearing wear. At the minimum, an overhaul entails grinding the valves and replacing piston rings and bearing inserts and whatever gaskets have been disturbed. The effort might extend to a new oil pump, timing and accessory drive parts, oil seals, cylinder liners (when easily accessible), together with new piston, ring, and wrist pin sets. Because the block and crankshaft remain in place, machine work is necessarily limited to the cylinder head.

In the classic sense, rebuilding an engine means the restoration of every frictional surface to its original dimension, alignment, and finish. The engine should theoretically be as good as new, or even better than new in the sense that used castings tend to hold dimension better than "green" parts. (Repeated heating and cooling cycles relieve stresses introduced during the casting process.) In addition, an older engine might benefit from late-production parts.

Although some mechanics would disagree, the rebuilding process cannot repeal the law of entropy. A competently rebuilt engine will be durable over the long run and will be reasonably reliable in the short term, but it will not quite match the factory norms. Subsurface flaws will not be detected. Metal lost to water jacket corrosion is irretrievably lost. Nor can original deck height, timing gear mesh, main bearing cap height, and camshaft geometry be achieved in any commercially practical sense. And the potential for error, on the part of both the machinist and the as-

1 Cam follower
2 Camshaft
3 Cylinder block
4 Crankshaft
5 Crankshaft
   counterweight
6 Oil pan
7 Connecting rod
8 Liner packing rings
9 Cylinder liner
10 Piston
11 Piston pin
12 Piston rings
13 Oil pump

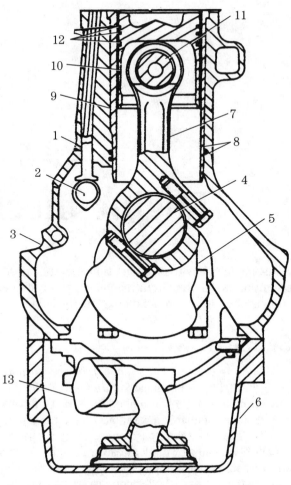

8-1 Sectional view of Deere 6076 block assembly and nomenclature.

sembler, affects reliability. More often than not, a freshly rebuilt engine will experience "teething" difficulties.

On the other hand, the cost should not exceed half of the replacement cost, and engine life will be nearly doubled.

Traditionally, the work is divided among operator mechanics, who remove the engine from its mounts, dismantle it, and consign the components to a machinist for inspection and refurbishing. The machinist might supply some or all of the replacement parts, which, together with the reworked parts, are returned to the mechanics for final assembly.

This approach organizes the work around specialists' skills, keeps the critical business of assembly in-house (where it probably belongs), and minimizes out-of-pocket expenses. One working mechanic—not the shop foreman—should have undiluted responsibility for the job, a responsibility that includes new and refurbished parts quality control (QC), assembly, installation, and start-up.

Engine machine work is an art like gunsmithing or watch repair in the sense that proficiency comes slowly, through years of patient application. In my anachronistic opinion, the best work comes out of small shops, where Model T crankshafts stand in racks, waiting for customers that never come, and the coffee pot hasn't been cleaned since 1940. These shops, in short, are places where a mill means a Bridgeport, a grinder is a Landis, and the lathes were made in South Bend.

Diagnosis

Before you begin you should have a good idea—or at least a plausible theory— about the nature of the problem. The diagnostic techniques described in chapter 4 indicate whether or not major work is in order and, when supplemented by oil analysis, will suggest which class of parts—rings, gears, soft metal bearings, and so on— are wearing rapidly.

Test/analysis data, combined with a detailed operating history, should fairly well pinpoint the failure site (cylinder bore, crankshaft bearings, accessory drive, and the like). But analysis should not stop with merely verbal formulations. For example, it is hardly meaningful to say that a bearing or a piston ring set has "worn out" or "overheated." One should try to identify what associated failure or special operating condition selected those parts to fail. City-bus wheel bearings are a good example of the selection process; experience shows that the right front bearings tend to fail more often than those on the left. Traces of red oxide in the lubricant suggest that failure comes about because of moisture contamination, (i.e., water splash), which is more likely to occur on the curb side of the vehicle.

Once the mechanic understands the failure mechanism, it might be possible to correct matters by performing more frequent maintenance, upgrading parts quality, or modifying operating conditions.

Rigging

Figure 8-2 illustrates the proper lifting tackle for a large engine. Note how the spreader bar and adjustable crossheads keep the chains vertical and in tension to the hook loads.

The engine shown incorporates lift brackets; less serious engines do not, and the mechanic is left to his own devising. In general, attachment points should straddle the center of gravity of the engine in two planes, so that it lifts horizontally without tilting. Chains must clear vulnerable parts, such as rocker covers and fuel lines. Lift brackets made of ⅜-in. flame-cut steel plate are the ideal, but *forged* eyebolts (available from fastener supply houses) are often more practical. A spreader bar will eliminate most bending loads, but there are times when chain angles of less than 90 degrees cannot be entirely avoided. Reduce the bending load seen by the eyebolt with a short length of pipe and heavy washer, as shown in Fig. 8-3.

In no case should chains be bolted directly to the block without the intermediary of a bracket or eyebolt. Nor is multistrand steel cable appropriate for this kind of knock-about service.

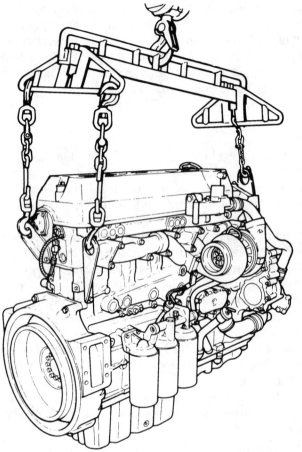

8-2
A proper engine sling is a
necessity. Detroit Diesel

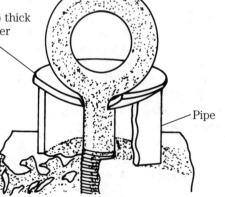

1/8 in. (3 mm) thick
flat washer

Pipe

8-3
Hardened eyebolts occasionally
must be used in lieu of lift
brackets. Bolts should thread into
a minimum depth of three times
diameter and should be
reinforced as shown to limit
bending forces.

Figure 8-4 illustrates a minimal engine stand, suitable for engines in the 600-lb. class. Better stands usually attach at the side of the block (as opposed to the fly-wheel flange) and can be raised and lowered.

Note: A mechanic can get into trouble with one of these revolving-head stands. Inverting an assembled engine with the turbocharger intact can dump oil from the turbocharger sump into one or more cylinders. Subsequent attempts to start the flooded engine might result in bent connecting rods or worse.

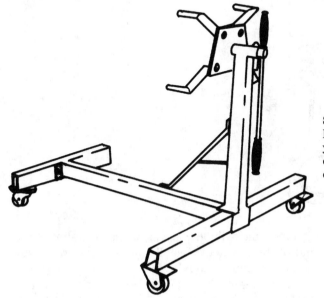

8-4
Purchase the best engine stand you can afford, with a weight capacity that provides a comfortable safety margin.

Special considerations

Mechanics who are knowledgeable about gasoline engine repair will find themselves doing familiar things, but to more demanding standards. Diesel engines are characterized by:

Close tolerances Close tolerances impose severe requirements in terms of inspection, cleanliness, and torque limits. Tolerance stack—unacceptable variations in dimension of assemblies made up of components that fall on the high or low ends of the tolerance range—becomes a factor to contend with. As delivered from the factory, some engines employ a selective assembly of bearing inserts and pistons.

High levels of stress Hard-used industrial engines are subject to structural failures, a fact that underscores the need for careful inspection of crankshafts, main-bearing webs and caps, connecting rods, pistons, cylinder bore flanges, harmonic balancers, and all critical fasteners.

Inflexible refinishing and assembly norms Stress and close tolerances give little latitude for quick fixes and, unless contradicted by hard experience, factory recommendations should be followed to the letter.

Special features While nothing in four-cycle diesel crankcases can be considered uniquely diesel, some features of these engines might be unfamiliar to most gasoline-engine mechanics. Nearly all engines employ oil-cooled pistons. This is accomplished with rifle-drilled connecting rods or, as is more often the case today, by means of oil spray tubes, or jets, which direct a stream of oil to the piston undersides (Fig. 8-5). Jets that bolt or press into place, must be removed for cleaning, and usually require alignment.

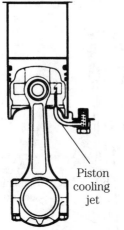

8-5
Spray jet aiming is critical.
Ford Motor Co.

Piston
cooling
jet

Diesel engines often employ removable cylinder sleeves, pressed or slipped into counterbores and standing proud of the fire deck. Installation is quite critical and special honing techniques are usually recommended.

Another difference is the apparent complexity of power transmission, which on some engines can have an almost baroque ornamentation. Figure 8-6 illustrates the bull gear/idler gear and accessory drive gear constellation on the DDA Series 60 engine. Belt drives might be hardly less imposing, as the drawing in Fig. 8-7 indicates.

In actual fact, this complexity is more illusory than real. One merely comes to terms with timing marks, deals with one component at a time, and builds the power train brick by brick. Of course, the work goes more slowly than it would on a simpler engine, and the parts cost can be daunting, especially when gears need to be replaced.

The power train can include one or more balance shafts, a technology that is rarely seen on spark ignition engines (the Mitsubishi/Chrysler 2.6L is one of the few exceptions).

Figure 8-8 illustrates the balance shaft configuration for a four-cylinder in-line engine. Two contra-rotating shafts, labeled *Silent shafts* in the drawing, run at twice crankshaft speed to generate forces that counter the "natural" vibration of the engine.

Engines of this type employ single-plane, two-throw crankshafts. Pistons 1 and 4 move in concert, as do pistons 2 and 3. When pistons 1 and 4 are down, 2 and 3 are up. Consequently, vertical forces generated by pistons 1 and 4 (indicated by the dotted line in Fig. 8-9) oppose the forces generated by the center pair of pistons (represented by the fine line). These vertical forces, known as primary shaking forces, almost cancel and can be ignored.

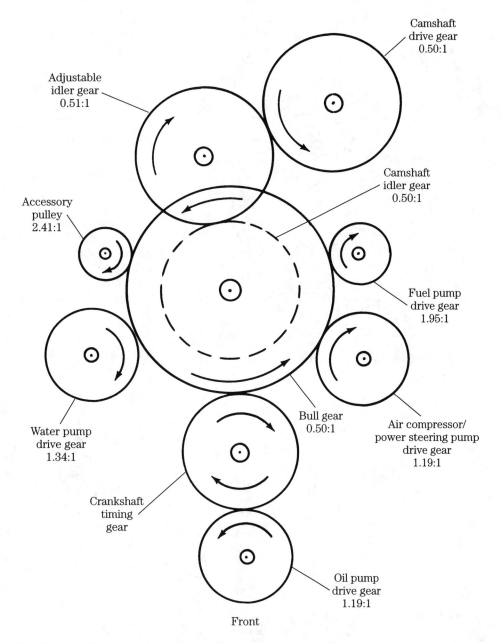

8-6 Detroit Diesel engines are known for their sophisticated gear trains.

Secondary forces pose a more serious problem. Created by connecting-rod angularity and by piston acceleration during the expansion stroke, these forces tend both to rotate the engine around the crankshaft centerline and to shake the engine vertically. Magnitude increases geometrically with rpm, to produce the dull rattle characteristic of in-line four-cylinder engines at speed.

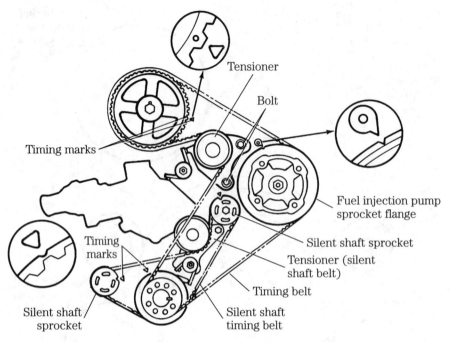

8-7 Ford 2.3L turbocharged diesel makes extensive use of toothed belts.

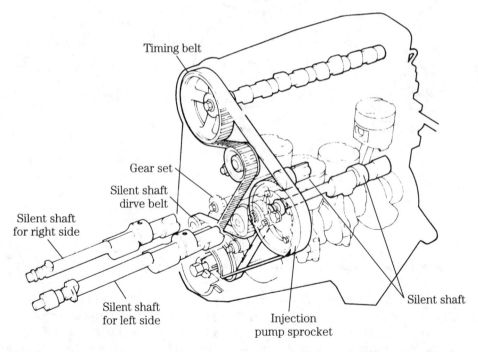

8-8 Balancing secondary forces requires two counterweighted shafts running at twice crankshaft speed. Ford Motor Co.

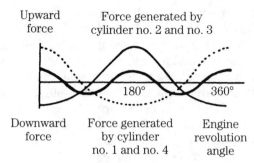

Upward force

Force generated by cylinder no. 2 and no. 3

180°   360°

Downward force

Force generated by cylinder no. 1 and no. 4

Engine revolution angle

8-9
Primary forces in an in-line four-cylinder engine very nearly balance. In an opposed four, with two crankshaft throws in the same plane 180° apart, balance is nearly perfect. A slight rocking couple, imposed by the offset between paired connecting rods on each crankshaft throw, does, however, exist. Ford Motor Co.

Secondary vertical and rolling forces are neutralized by deliberately induced imbalances in the balance shafts. Figure 8-10 diagrams the sequence of countervailing forces through full crankshaft revolution (two balance shaft revolutions).

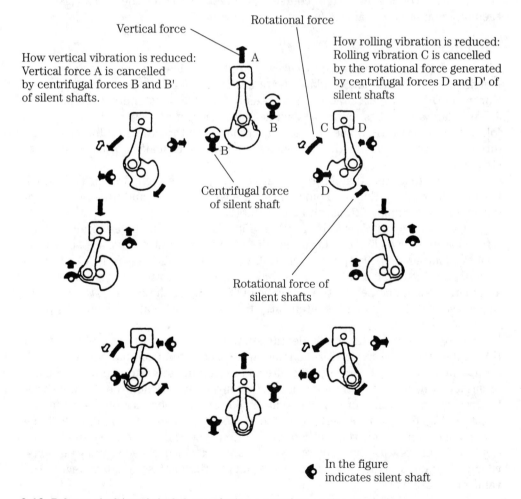

Vertical force

Rotational force

How vertical vibration is reduced: Vertical force A is cancelled by centrifugal forces B and B' of silent shafts.

How rolling vibration is reduced: Rolling vibration C is cancelled by the rotational force generated by centrifugal forces D and D' of silent shafts

Centrifugal force of silent shaft

Rotational force of silent shafts

In the figure indicates silent shaft

8-10 Balance shaft/crankshaft forces during a complete engine revolution. Ford Motor Co.

Detroit Diesel approaches the question of balance differently on its two-cycle engines. Here the concern is to balance the rocking couple created by crankpin offset. Such couples exist unless all pistons share the same crankpin, as for example, in a radial engine with one connecting rod articulated from a central master rod. DDA practice is to use a counterweight can and balance shafts driven at crankshaft speed through counterweighted gears. In other words, each shaft has two sources of imbalance, one integral with the drive gear and the other in the form of a bob weight on the free end of the shaft. Shaft counterweights and shaft gear weights are disposed radially to create a countervailing couple, which acts in opposition to the crankshaft-induced couple. No attempt is made to balance secondary forces.

From a mechanic's point of view, the critical aspects of this technology are the shaft bearings, which are subject to severe radial loads and catastrophic failure. In some cases, bearing bosses must be sleeved before new bearings can be installed. Give the oiling circuit close scrutiny; endemic bearing can justify modifications to increase the rate of oil flow. And of course it is necessary to time the shafts relative to each other and to the crankshaft.

Fasteners

Contemporary foreign and, to a great extent, American engines are built to the metric ISO (International Standards Organization) standards, developed from the European DIN. For most practical purposes DIN and ISO fasteners interchange. A JIS standard also exists, but most Japanese fasteners made since the early 1970s follow the ISO pattern. Some JIS bolts interchange (although head dimension can differ) with those built to the current standard; others make up just enough to strip out.

Few American manufacturers remain wedded to the inch standard, although leaving is hard to do. Engines come off the line with both ISO and fractional fasteners, inch-standard pipe fittings and metric fuel systems.

ISO fasteners are classed by nominal bolt diameter in millimeters and thread pitch measured as millimeters between adjacent thread crests. Thus, a specification might call for a M8 × 1.0 cap screw or stud. Wrench size markings reflect bolt diameter, not the flat-to-flat distance across the screw head. Yield strength is indicated by a numeric code embossed on the screw head. The higher the number, the stronger the cap screw. Metric hex nuts often carry the same numerical code, but this practice is not universal.

Figures 8-11 and 8-12 supply identification data and suggested torque limits for U.S. and metric cap screws. Torque limits were calculated from bolt yield strength ratings, and do take into account the effects of clamping forces on vulnerable parts or gaskets. Consequently, these values should not be used when the engine manufacturer provides a different torque limit or tightening procedure for a specific application. Tighten plastic insert- or crimped steel-type locknuts to about half the amount shown in the charts; toothed- or serrated-type locknuts receive full torque. Replace fasteners with the same or higher grade, except in the case of shear bolts, which are grade specific. When substituting a better-grade fastener, torque it to the value of the original.

SAE Grade	Head Markings	SAE Grade	Nut Markings	SAE Grade	Head Markings	SAE Grade	SAE Grade	Nut Markings	SAE Grade
SAE GRADE 1 SAE GRADE 2	No Mark	2	No Mark	SAE GRADE 5 SAE GRADE 5.1 SAE GRADE 5.2		5 Nut Markings	SAE GRADE 8 SAE GRADE 8.2		8 Nut Markings

DIA.	WRENCH SIZE	SAE GRADE 1		*SAE GRADE 2		SAE GRADE 5		SAE GRADE 8	
		OIL N•m(lb-in)	DRY N•m(lb-in)	OIL N•m(lb-in)	DRY N•m(lb-in)	OIL N•m(lb-in)	DRY N•m(lb-in)	OIL N•m(lb-in)	DRY N•m(lb-in)
#6		0.5(4.5)	0.7(6)			1.4(12)	1.7(15)		
#8		0.9(8)	1.2(11)			2.4(21)	3.2(28)		
#10		1.4(12)	1.8(16)			3.4(30)	4.6(41)		
#12		2(19)	2.8(25)			5.4(48)	7.3(65)		
		N•m(lb-ft)	N•m(lb-ft)	N•m(lb-ft)	N•m(lb-ft)	N•m(lb-ft)	N•m(lb-ft)	N•m(lb-ft)	N•m(lb-ft)
1/4	7/16	3.5(2.5)	4(3)	5(4)	7(5)	8(6)	11(8)	12(8.5)	16(12)
5/16	1/2	7(5)	9(6.5)	10(7.5)	14(10)	16(12)	23(17)	24(18)	33(24)
3/8	9/16	12(8.5)	16(12)	19(14)	24(18)	30(22)	41(30)	41(30)	54(40)
7/16	5/8	19(14)	26(19)	30(22)	41(30)	47(35)	68(50)	68(50)	95(70)
1/2	3/4	24(21)	41(30)	47(35)	61(45)	75(55)	102(75)	102(75)	142(105)
9/16	13/16	41(30)	54(40)	68(50)	88(65)	108(80)	142(105)	149(110)	203(150)
5/8	15/16	54(40)	75(55)	88(65)	122(90)	149(110)	197(145)	203(150)	278(205)
3/4	1-1/8	102(75)	136(100)	163(120)	217(160)	258(190)	353(260)	366(270)	495(365)
7/8	1-5/16	163(120)	244(165)	163(120)	224(165)	414(305)	563(415)	590(435)	800(590)
1	1-1/2	244(180)	332(245)	244(180)	332(245)	624(460)	848(625)	881(650)	1193(880)
1-1/8	1-11/16	346(255)	468(345)	346(255)	468(345)	780(575)	1058(780)	1248(920)	1695(1250)
1-1/4	1-7/8	488(360)	664(490)	488(360)	665(490)	1098(810)	1492(1100)	1763(1300)	2393(1765)
1-3/8	2-1/16	637(470)	868(640)	637(470)	868(640)	1438(1061)	1953(1440)	2312(1705)	3140(2315)
1-1/2	2-1/4	848(625)	1153(850)	848(625)	1153(850)	1912(1410)	2590(1910)	3065(2260)	4163(3070)

8-11 Inch cap screw torque values.

Cleaning

Ford Motor and other manufacturers say that dirt is the chief cause of callbacks after major work. The direct effect is to contaminate the oil supply; the indirect effect is to create an environment that makes craftsmanship difficult or impossible.

The need for almost septic standards of cleanliness argues against the practice of opening the engine for less than comprehensive repairs. Of course, it happens that such repairs must be made, regardless of the long-term consequences. Nor is it possible to maintain reasonable standards of cleanliness during in-frame overhauls, although the damage can be minimized by moving in quickly, cleaning only those friction surfaces that are opened for inspection, and getting out. Dirt accumulations on internal parts of the engine cannot be removed from below, while parts are still assembled, and attempts to do so will only release more solids into the oil stream.

		4.6		4.8		8.8 or 9.8		10.9		12.9	
DIA.	WRENCH SIZE	OIL	DRY	OIL	DRY	OIL	DRY	OIL	DRY	OIL	DRY
		N·m(lb-ft)	N·m(lb-ft)	N·m(lb-ft)	N·m(lb-ft)	N·m(lb-ft)	N·m(lb-ft)	N·m(lb-ft)	N·m(lb-ft)	N·m(lb-ft)	N·m(lb-ft)
M3	5.5mm	0.4(0.2)	0.5(0.3)	0.5(0.4)	0.7(0.5)	1(0.8)	1.3(1)	1.5(1)	2(1.5)	1.5(1)	2(1.5)
M4	7mm	0.9(0.6)	1.1(0.8)	1(0.9)	1.5(1)	2.5(1.5)	3(2)	3.5(2.5)	4.5(3)	4(3)	5(4)
M5	8mm	1.5(1)	2.5(1.5)	2.5(1.5)	3(2)	4.5(3.5)	6(4.5)	6.5(4.5)	9(6.5)	7.5(5.5)	10(7.5)
M6	10mm	3(2)	4(3)	4(3)	5.5(4)	7.5(5.5)	10(7.5)	11(8)	15(11)	13(9.5)	18(13)
M8	13mm	7(5)	9.5(7)	10(7.5)	13(10)	18(13)	25(18)	25(18)	35(26)	30(22)	45(33)
M10	16mm	14(10)	19(14)	20(15)	25(18)	35(26)	50(37)	55(41)	75(55)	65(48)	85(63)
M12	18mm	25(18)	35(26)	35(26)	45(33)	65(48)	85(63)	95(70)	130(97)	110(81)	150(111)
M14	21mm	40(30)	50(37)	55(41)	75(55)	100(74)	140(103)	150(111)	205(151)	175(129)	240(177)
M16	24mm	60(44)	80(59)	85(63)	115(85)	160(118)	215(159)	235(173)	315(232)	275(203)	370(273)
M18	27mm	80(59)	110(81)	115(85)	160(118)	225(166)	305(225)	320(236)	435(321)	375(277)	510(376)
M20	30mm	115(85)	160(118)	165(122)	225(166)	320(236)	435(321)	455(356)	620(457)	535(395)	725(535)
M22	33mm	160(118)	215(159)	225(167)	305(225)	435(321)	590(435)	620(457)	840(620)	725(535)	985(726)
M24	36mm	200(148)	275(203)	285(210)	390(288)	555(409)	750(553)	790(583)	1070(789)	925(682)	1255(926)
M27	41mm	295(218)	400(295)	415(306)	565(417)	810(597)	1100(811)	1155(852)	1565(1154)	1350(996)	1835(1353)
M30	46mm	400(295)	545(402)	565(417)	770(568)	1100(811)	1495(1103)	1570(1158)	2130(1571)	1835(1353)	2490(1837)
M33	51mm	545(402)	740(546)	770(568)	1050(774)	1500(1106)	2035(1500)	2135(1575)	2900(2139)	2500(1844)	3390(2500)
M36	55mm	700(516)	950(700)	990(730)	1345(992)	1925(1420)	2610(1925)	2740(2021)	3720(2744)	3205(2364)	4355(3212)

8-12 Metric cap screw torque values.

When, on the other hand, the engine is rebuilt, the block, cylinder head, pan, and other steel stampings are sent out for thermal or chemical cleaning (see chapter 7). These processes also remove the paint, which is all to the good. Crankshaft, piston assemblies, and other major internal parts receive a preliminary wash-down for inspection by the mechanic in charge of the job. These parts are then forwarded to the machine shop for evaluation. When pistons are reused, the machinist will chemically clean the grooves and piston undersides—chores that save hours of labor. About all that remains for the shop mechanic is to degrease fasteners and accessories that have been detached from the block and remove the rust preventative from new parts.

Teardown

Drain the oil and coolant and degrease the outer surfaces of the engine. Disconnect the battery, wiring harness (make a sketch of the connections if the harness is not keyed for proper assembly), and exhaust system. Attach the sling and undo the drive line connection and the motor mounts. With the engine secured in a stand, detach the manifolds, cylinder heads, and oil sump. The block should be stripped if you contemplate machine work or chemical cleaning of the jackets.

Lubrication system

The first order of business should be the lubrication system. To check it out you must have a reasonably good notion of the oiling circuits. Figure 8-13 is a drawing of the Onan DJ series lubrication system. The crankcase breather is included because it has much to do with oil control. Should it clog, the engine will leak at every pore. Oil passes from the screened pickup tube (suspended in the pan) to the pump, which sends it through the filter. From there the filtered oil is distributed to the camshaft, the main bearings (and through rifle-drilled holes in the crankshaft to the rod bearings and wrist pin), and to the valve gear. Valve and rocker arm lubrication is done through a typically Onan "showerhead" tube. Tiny holes are drilled in the line

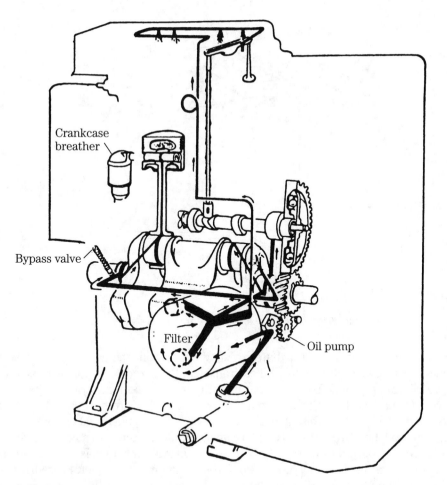

8-13 Lubrication system. Onan

and deliver an oil spray at about 25 psi. On its return to the sump, the oil dribbles down the push rods to lubricate the cam lobes and tappets.

The system in Fig. 8-14 is employed in six-cylinder engines. From the bottom of the drawing, oil enters the pickup tube, then goes to the Gerotor pump. Unlike most oil pumps, this one is mounted on the end of the crankshaft and turns at engine speed. The front engine cover incorporates inlet and pump discharge ports. Oil is sent through a remote cooler (6), then directed back to the block, where it exits again to the filter bank (11). Normally oil passes through these filters. However, if the filters clog or if the oil is thick from cold, a pressure differential type of bypass valve (12) opens and allows unfiltered flow.

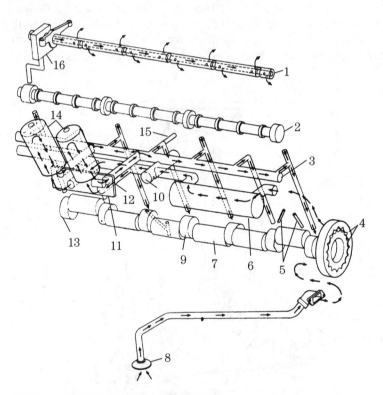

8-14 A more complex lubrication system. International Harvester

The main oil gallery (3) distributes the flow throughout the engine. Some goes to the main bearings and through the drilled crankshaft to the rods. The camshaft bearings receive oil from the same passages that feed the mains. The rearmost cam journal is grooved. Oil passes along this groove and up to the rocker arms through the hollow rocker shaft. On its return this oil lubricates the valve stems, pushrod ball sockets, tappets, and cam lobes. Other makes employ similar circuits, often with a geared pump.

Figure 8-15 describes the Ford 2.3L lubrication system, which is surprisingly sophisticated for a small and, by diesel standards, inexpensive product. Pressurized oil goes first to the filter, described in the following section.

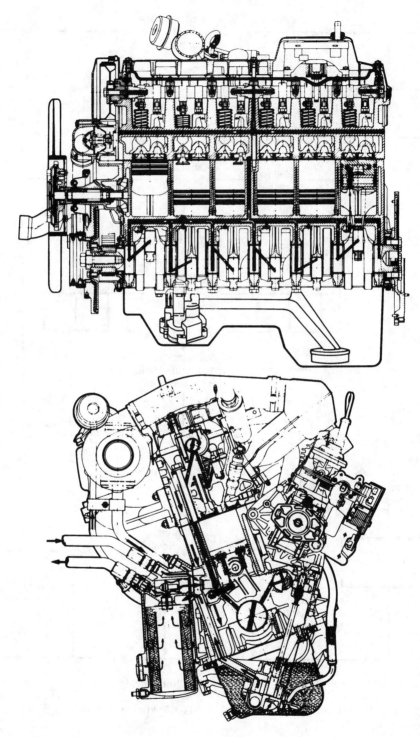

8-15 Ford 2.4L lubrication system.

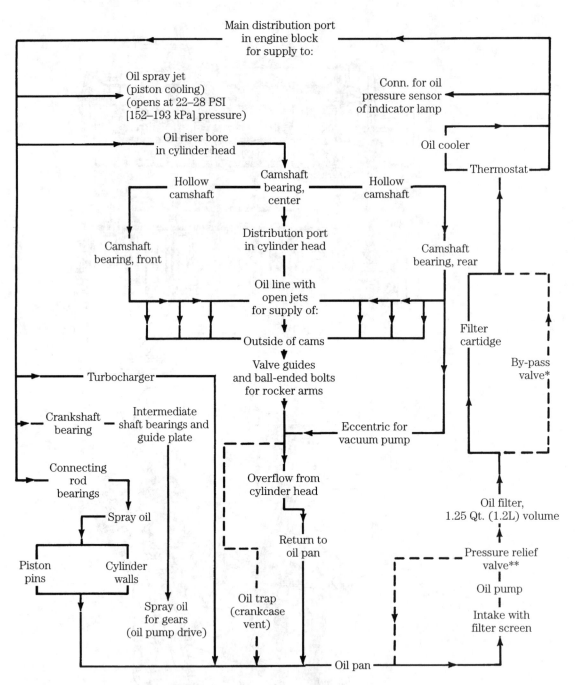

Main distribution port
in engine block
for supply to:

Oil spray jet
(piston cooling)
(opens at 22–28 PSI
[152–193 kPa] pressure)

Conn. for oil
pressure sensor
of indicator lamp

Oil cooler

Oil riser bore
in cylinder head

Thermostat

Hollow
camshaft

Camshaft
bearing,
center

Hollow
camshaft

Camshaft
bearing, front

Distribution port
in cylinder head

Camshaft
bearing, rear

Oil line with
open jets
for supply of:

Filter
cartridge

By-pass
valve*

Outside of cams

Valve guides
and ball-ended bolts
for rocker arms

Turbocharger

Eccentric for
vacuum pump

Crankshaft
bearing

Intermediate
shaft bearings and
guide plate

Connecting
rod
bearings

Overflow from
cylinder head

Spray oil

Oil filter,
1.25 Qt. (1.2L) volume

Piston
pins

Cylinder
walls

Return to
oil pan

Pressure relief
valve**

Oil pump

Spray oil
for gears
(oil pump drive)

Oil trap
(crankcase
vent)

Intake with
filter screen

Oil pan

*(Opens at 36 PSI [241 kPa]) oil supply guaranteed if filter cartridge is plugged

**(Oil pump) with cold oil (opens with 79–92 PSI [544–648 kPa] pressure)

8-15 Continued.

Depending on oil temperature, output from the filter goes either directly to the main distribution gallery or to the gallery by way of an oil cooler (Fig. 8-16). A thermostatic valve, located in the filter housing, directs the flow (Fig. 8-17). Cooling the oil during cold starts is counterproductive, and the valve remains closed; as oil temperature increases, the valve extends to split the flow between the gallery and cooler. At approximately 94°C, all flow is diverted to the cooler. As a safety measure, the bypass valve opens if cooler pressure drop exceeds 14 psi. Thus, a stoppage will not shut down oil flow, although long-term survivability is compromised.

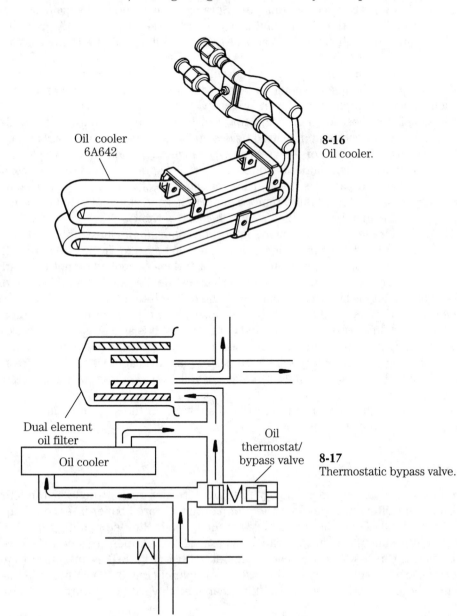

Oil cooler
6A642

8-16
Oil cooler.

Dual element
oil filter

Oil cooler

Oil
thermostat/
bypass valve

8-17
Thermostatic bypass valve.

Camshaft lubrication, a weak point on many ohc engines, shows evidence of careful attention. A large-diameter riser, feeding from the main gallery, supplies oil to the center camshaft bearing. At that point the flow splits, part of it entering the hollow camshaft and part of it going to an overhead spray bar. The camshaft acts as an oil gallery for the remaining two shaft bearings; the spray bar provides lubrication for camshaft lobes, rockers, and valve tips.

The main bearings are lubricated through rifle-drilled ports connecting the webs and main gallery. As a common practice, diagonal ports drilled in the crankshaft convey oil from the webs to adjacent crankpins. Spray jets, again fed by the main gallery, cool the undersides of the pistons and provide oil for the piston pins. A gear and crescent pump—the lightest, most compact type available—supplies the necessary pressure.

This cursory examination of the oil circuitry on three very different engines should underscore the need to come to terms with these systems. No other system has such immediate implications for the integrity of the mechanic's work.

Clogging is the most frequent complaint, caused by failure to change oil and filters at proscribed intervals, and abetted by design flaws. Expect to find total or partial blockage wherever the oil flow abruptly changes direction or loses velocity. Prime candidates are the cylinder head/block interface, spray jets, and the junctions between cross-drilled passages. Neither immersion-type chemicals, heat, nor compressed air can be depended on to remove metal chips and sludge. Drilled passages must be cleared by hand, using riflebore brushes and solvent.

Thorough cleaning is never more important than after catastrophic failure, which releases a flood of metallic debris into the oiling system. In this case, the oil cooler can present a special problem. Modern, high-density coolers do not respond well to chemical cleaners and frustrate even the most flexible swabs. A new or used part—whose history is known—is sometimes the only solution.

Leaks can develop at the welch (expansion) plugs or, less often, at the pipe plugs that blank off cross-drilled holes and core cavities. It is also possible for cracks to open around lifter bores and other thin-section areas. A careful mechanic will discard welch plugs (after determining that spares are available!) and apply fresh sealant to pipe plug threads during a rebuild. Most oil-wetted cracks can be detected visually.

Internal leaks that have been missed will show when the system is filled with pressurized oil before start-up.

Filters

All contemporary engines employ full-flow filtration, using single or tandem paper-element filters in series with the pump outlet. Pleated paper filters can trap particles as small as one micron, but are handicapped by limited holding capacity. Consequently, such filters incorporate a bypass valve that opens to shunt the element when the pressure drop reaches about 15 psi (Fig. 8-18), an action that should trigger a warning lamp on the instrument panel (Fig. 8-19). Otherwise the operator should occasionally feel the filter canister to verify that warm oil is circulating.

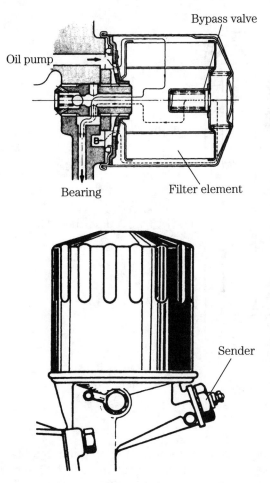

8-18
Yanmar cartridge-type filter mounts downstream of the pressure-regulating valve and includes a bypass valve.

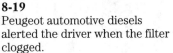

8-19
Peugeot automotive diesels alerted the driver when the filter clogged.

The Ford 2.3L filter consists of two concentric elements, one in series with the oiling circuit, the other in parallel (Fig. 8-20). So long as the bypass valve, shown at the top of the drawing, remains seated, pump output passes through the outer, series-connected element and into the lubrication circuit. Pump delivery rates are greater than the system requires and the surplus migrates through the inner filter and returns to the sump through the 9-mm orifice at the base of the assembly. I have been unable to obtain precise data for this engine, which was designed in Japan. However, standard design practice is to calculate diesel lube oil requirements as 0.003 times the ratio of oil to piston displacement times rpm. Pumps are then sized to meet twice these requirements. If this holds, the parallel filter makes a major contribution.

The most important single maintenance activity is to change the oil and filter on the engine-maker's schedule, which is more demanding than suggested for equivalent SI engines. A turbocharger increases bearing loads and blowby, further shortening oil and filter life.

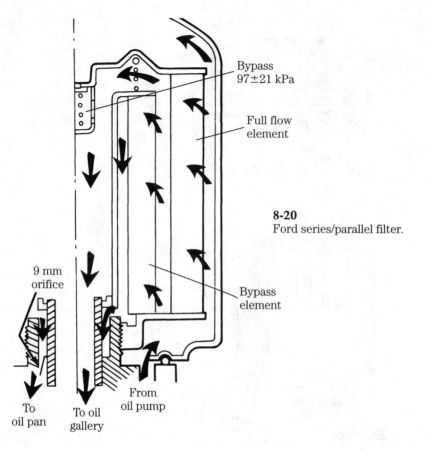

Bypass
97±21 kPa

Full flow
element

8-20
Ford series/parallel filter.

9 mm
orifice

Bypass
element

To
oil pan

To oil
gallery

From
oil pump

The final steps in an overhaul or rebuild are to prime the oil pump by rotating the shaft with the impeller submerged in a container of clean motor oil, fill the turbocharger cavity and feed line, and pressurize the oiling circuit. The latter is accomplished by connecting a source of oil pressure to the main distribution cavity, usually by way of the oil-pressure sender tap. This step is sometimes omitted and the bearings—even though preoiled during assembly—suffer for it. The system *must* be pressurized after a new camshaft is installed.

Any repairs involving the turbocharger should be followed by disconnecting the oil return line and cranking until flow is established.

Before start-up after routine oil and filter changes, it is good practice to dry-crank the engine until the panel gauge indicates that oil pressure has been restored.

Lube oil

Oil collects moisture from condensation in the sump and is the repository for the liquid by-products of combustion. These by-products include several acid families that, even in dilute form, attack bearings and friction surfaces. No commercially practical filter can take out these contaminants. In addition oil, in a sense, wears out. The petroleum base does not change, but the additives become exhausted and no longer suppress foam, retard rust, and keep particles suspended. Heavy sludge in the filter is a sure indication that the change interval should be shortened, because the detergents in the oil have been exhausted. In heavy concentrations water emulsifies to produce a white, mayonnaise-like gel that has almost no lubrication qualities.

Oil that has overheated oxidizes and turns black. (This change should not be confused with the normal discoloration of detergent oil.) If fresh oil is added, a reaction might be set up that causes the formation of hard granular particles in the sump, known as "coffee grounds."

Oil change intervals are a matter of specification—usually at every 100 hours—and sooner if the oil shows evidence of deterioration. Most manufacturers are quite specific about the type and brand to be used. Multigrade oils (e.g., 10W-30) are not recommended for some engines, because it is believed they do not offer the protection of single-weight types. Other manufacturers specify SF grades. As a practical matter this specification means that multigrade oils can be used. Brand names are important in diesel service. Compliance with the standards jointly developed by the American Petroleum Institute, the Society of Automotive Engineers, and the American Society for Testing and Materials is voluntary. You have no guarantee that brand X is the equivalent of brand Y, even though the oil might be labeled as meeting the same API-SAE-ASTM standards. GM suggests that you discuss your lubrication needs with your supplier and use an oil that has been successful in diesel engines and meets the pertinent military standards. MIL-L-2104B is the standard most often quoted, although some oils have been "doped" with additives to meet the API standards. These additives cause problems with some engines often in the form of deposits around the ring belt. For GM two-cycles zinc should be held to between 0.07% and 0.10% by weight, and sulfated ash to 1.0%. If the lubricant contains only barium detergent-dispersants, the sulfated-ash content can be increased by 0.05%. High-sulfur fuels might call for *low-ash series* 3 oils, which do not necessarily meet military low-temperature performance standards. Such oils might not function as well as MIL-L-2104B lubricants in winter operation.

Oil pumps

The Gerotor pump shown in Fig. 8-21 is gear-driven; a more common practice is to drive pumps of this type directly from the crankshaft. Like the Wankel engine, which employs a similar trochoidal geometry, operation is somewhat difficult to visualize on paper and painfully obvious in the hardware. The inner rotor, which has one lobe less than the outer rotor, "walks" as it turns, imparting motion to the outer rotor and simultaneously varying the volume of the working cavity (shaded in the drawing). Volume change translates as a pressure head.

The clearance between the rotors and the end cover is best checked with *Plasti-Gage*. Plasti-Gage consists of rectangular-section plastic wire in various thicknesses. A length of the wire is inserted between the rotors and cover. The working clearance is a function of how much the wire is compressed under assembly torque. The package has a scale printed on it to convert this width to thousandths of an inch. Plasti-Gage is extremely accurate and is fast. The only precautions in its use are that the parts must be dry, with no oil or solvent adhering to them, and that the wire must be removed after measurement. Otherwise it might break free and circulate with the oil, where it can lodge in a port. Typical end clearance for a Gerotor pump is on the or-

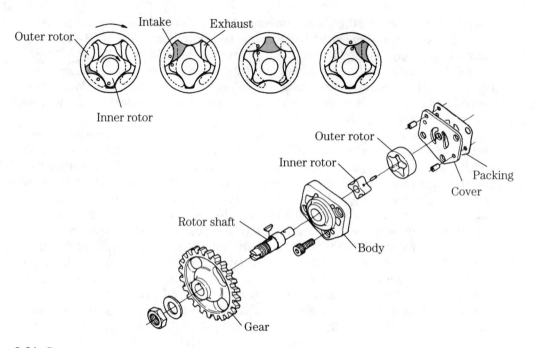

8-21 Gerotor pump. Yanmar Diesel Engine Co., Ltd.

der of 0.003 in. Of course, all rubbing surfaces should be inspected for deep scratches, and the screen should be soaked in solvent to open the pores.

Other critical specifications are the clearance between the outer rotor and the case (Fig. 8-22A) and the approach distance between inner and outer rotors (B).

Camshaft-driven gear-type pumps are the norm (Fig. 8-23). Check the pump for obvious damage—scores, chipped teeth, noisy operation. Then measure the clearance; the closer the better as long as the parts are not in physical contact.

End clearance can be checked by bolting the cover plate up with Plasti-Gage between the cover and gears, or with the aid of a machinist's straightedge. Lay the straightedge on the gear case and measure the clearance between the top of the gear and the case with a feeler gage. When in doubt, replace or resurface the cover. Check the diameter of all shafts and replace as needed. The idler gear shaft is typically pressed into the pump body. Use an arbor press to install, and make certain that the shaft is precisely centered in its boss. Allowing the shaft to cant will cause interference and early failure. Total wear between the drive shaft and bushing can be determined with a dial indicator as shown in Fig. 8-24. Move the shaft up and down as you turn it. Check the backlash with a piece of solder or Plasti-Gage between the gears. For most pumps this clearance should not exceed 0.018 in.

In general, it is wise not to tamper with the pressure relief valve, which might be integral with the pump or at some distance from it. If you must open the valve, because of either excessive lube oil pressure or low pressure, observe that the parts are generally under spring tension and can "explode" with considerable force. Check the spring tension and free length against specs, and replace or lap the valve parts as

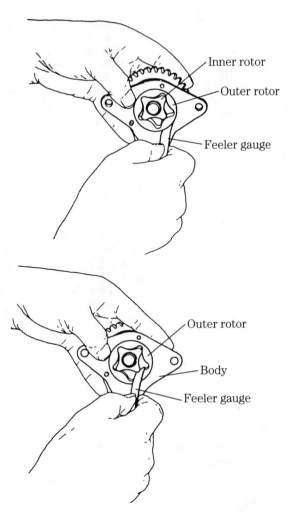

8-22
Use a feeler gauge to determine outer rotor to case clearance (A) and inner to outer rotor clearance. Wear limits for the Yanmar and most other such pumps are 0.15 in.

needed. When this assembly has been disturbed it is imperative that the oil pressure be checked, with an accurate instrument such as the gauge shown in Fig. 8-25.

Oil pressure monitoring devices

Because oil pressure is critical, it makes good sense to invest in some sort of alarm to supplement the usual gauge. Gauge failure is rare (erratic operation of mechanical gauges is usually caused by a slug of hardened oil impacted at the gauge fitting), but it does happen. Failure to shut down when pressure is lost will destroy the engine in seconds.

Stewart-Warner monitor Stewart-Warner makes an audiovisual monitor that lights and buzzes when oil pressure and coolant temperatures exceed safe norms. The indicator is a cold-cathode electron tube, which is much more reliable than the usual incandescent bulb. Designed for panel mounting, the 366-T3 is relatively inexpensive insurance.

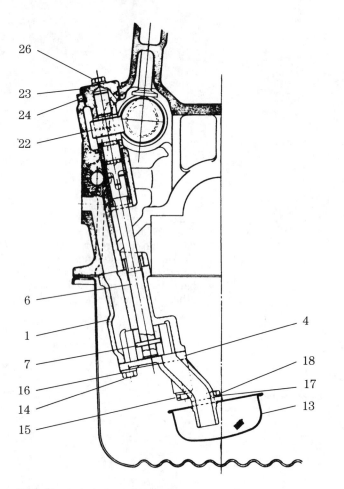

1 Oil pump body
2 Idler shaft
3 Packing (SD22)
4 Oil pump cover
5 Drive gear
6 Drive shaft
7 Pin
8 Driven gear
9 Relief valve
10 Relief valve spring
11 Washer
12 Cotter pin
13 Oil screen
14 Bolt
15 Bolt
16 Washer
17 Lockwasher
18 Bolt
19 Gasket
20 Bolt
21 Lockwasher
22 Driving spindle
23 Driving-spindle support
24 O-ring
25 Bolt
26 Bolt
27 Lockwasher

8-23 Exploded view of typical gear-type oil pump. Marine Engine Div., Chrysler Corp.

Onan monitor Onan engines can be supplied with an automatic shutdown feature. The system employs a *normally on* sensor element and a time delay relay. In conjunction with a 10 W, 1-Ω resistor, this relay allows the engine to be started. When running pressure drops below 13 psi, the sensor opens, causing the fuel rack to pull out.

The sensor (Fig. 8-26) is the most complicated part of the system. It has a set of points that should be inspected periodically, cleaned, and gapped to 0.040 in. When necessary, disassemble the unit to check for wear in the spacer, fiber plunger, and spring-loaded shaft plunger. The spacer must be at least 0.35 in. long. Replace as needed. Check the action of the centrifugal mechanism by moving the weights in their orbits. Binding or other evidence of wear dictates that the weights and cam assembly be replaced. The cam must not be loose on the gear shaft.

Ulanet monitor The George Ulanet Company makes one of the most complete monitoring systems on the market (Fig. 8-27). It incorporates an optional water level

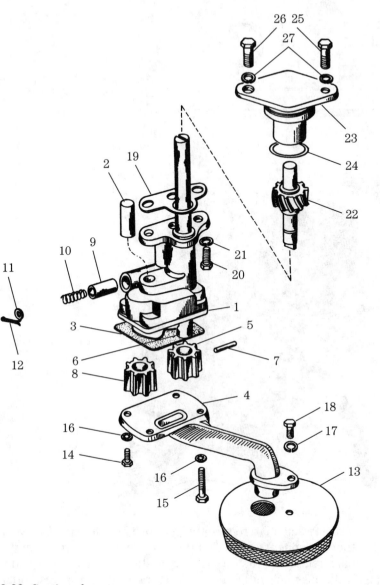

8-23 Continued.

sensor, overheat sensor, fuel-pressure sensor, and low- and high-speed oil sensors. For the GM 3-71 and 4-71 series, these sensors are calibrated at 4 psi and 30 psi, respectively.

Crankcase ventilation

The crankcase must be vented to reduce the concentration of acids and water in the oil. A few cases are at atmospheric pressure; others (employing forced-air scavenging) are at slightly higher than atmospheric; and still other systems run the

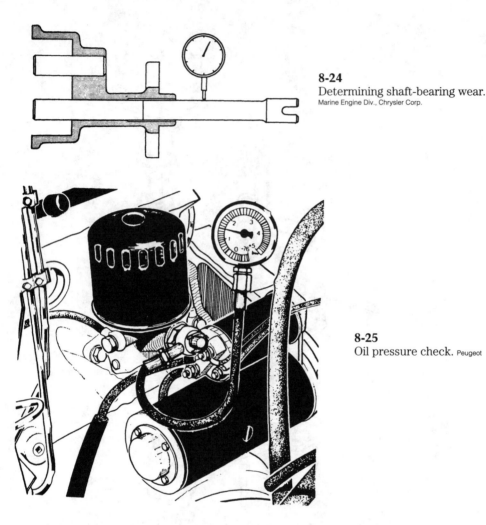

8-24
Determining shaft-bearing wear.
Marine Engine Div., Chrysler Corp.

8-25
Oil pressure check. Peugeot

crankcase at a slight vacuum to reduce the possibility of air leaks. The vapors might be vented to the atmosphere or recycled through the intake ports. In any event, the system requires attention to ensure that it operates properly. A clogged mesh element or pipe will cause unhealthy increases in crankcase pressures, forcing oil out around the gaskets and possibly past the seals, as well as increasing oil consumption. Figure 8-28 shows a breather assembly with check valve to ensure that the case remains at less than atmospheric pressure.

Block casting

Carefully examine the fire deck for pulled head bolt threads, eroded coolant passages, missing or damaged coolant deflectors, and cracks, particularly between adjacent cylinders. In some cases, minor cracks can be ground out and filled. Several dimensional checks also need to be made at this time.

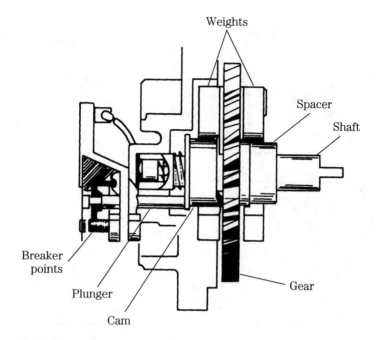

8-26 Onan oil pressure monitor—sensor section.

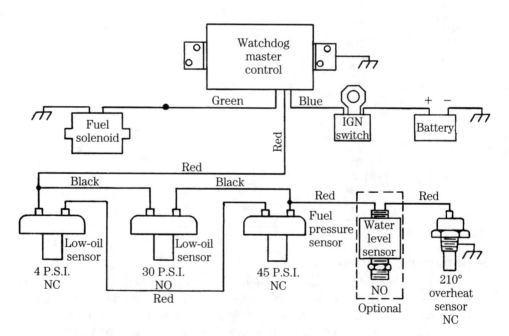

8-27 Ulanet monitor as configured for DD two-cycle engines.

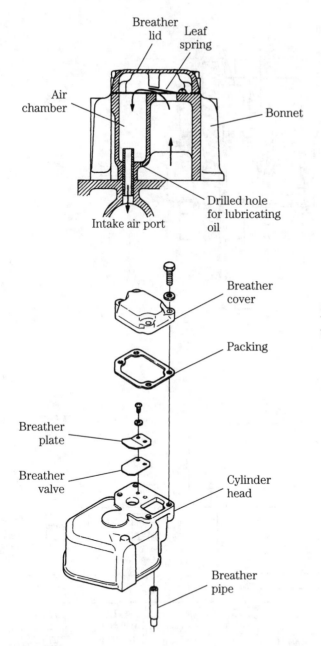

8-28
Yanmar atmospheric crankcase
breather.

Deck flatness

Figure 8-29 illustrates the basic technique, which involves a series of diagonal and transverse measurements along the length and width of the deck, using a machinist's straightedge and feeler gauge. Block deck surfaces can be milled flat (a process called *decking*), although contemporary design practices give little leeway for corrective machining. As mentioned in chapter 7, several modern engines cannot

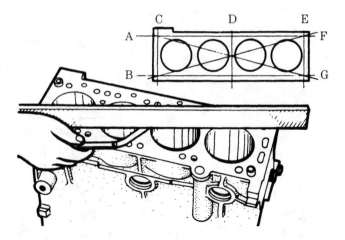

8-29 If the head gasket is to live, the fire deck must be flat and true. Chrysler Corp.

be safely decked, and warped blocks must be replaced. However, when it comes to it, most machinists will push these limits and obtain piston-to-cylinder head clearance by shimming the gasket or by selective assembly, fitting the shortest piston and rod sets that can be found. The owner should be aware of these expedients and agree to them.

Piston height

While this is an assembly dimension, an early check is not out of order. Measure piston protrusion or regression with a dial indicator, as described in the previous chapter.

Liner height

A few engines continue to be built with integral cylinder bores, machined directly into the block metal or in the form of a sleeve permanently installed during the casting process. These engines have flat decks, and what will be said here does not apply.

Most diesel engines employ discrete cylinder bore liners, or sleeves, which can be more or less easily replaced. Dry liners press into block counterbores; wet liners insert with a light force fit and come into direct contact with the coolant. Seals confine the coolant to areas adjacent to ring travel (Fig. 8-30). Wet liners simplify foundry work and give better control of water-jacket dimensions. The interface between dry liners and cylinder bores erects a minor, but real, thermal barrier between combustion and coolant. However, the liner is a structural member and eliminates the possibility of coolant leaks into the crankcase or back into the combustion chamber. Liners of either type stand proud of the deck, a practice that gives additional compression to the head gasket in a critical area and, at the same time prestresses the liner (Fig. 8-31). Any liner that does not meet specification must be replaced.

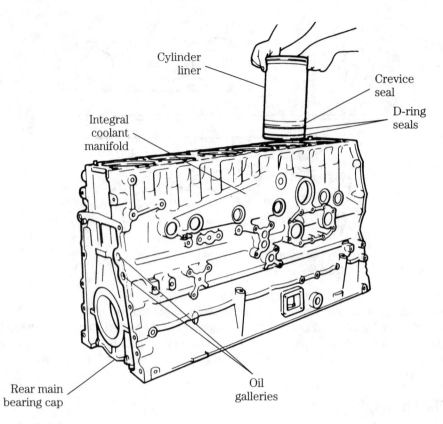

8-30 Cylinder liner—wet type. Detroit Diesel

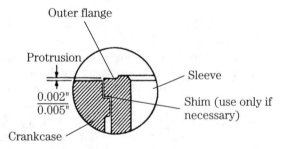

8-31
Liner protrusion measurement is taken at several places around the circumference of the part and averaged. International Harvester

This measurement taken at several points around the circumference of the liner, should be made during initial teardown and after replacement liners are installed. Figure 8-32 illustrates the liner hold-down hardware used in machine shops. Clamps are normally fabricated on site from cold-rolled steel, fitted with hardened washers, and pulled down securely. This technique works for dry liners and wet liners with radial (O-ring type) seals. Liner counterbores—the ledges that establish liner height—can be remachined and shimmed when necessary. The more usual procedure is to shim the liner shoulder, as shown in the illustration.

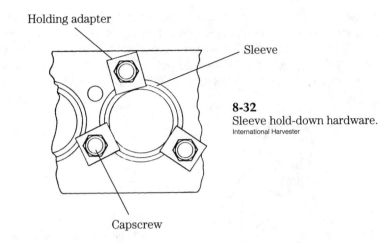

Holding adapter

Sleeve

8-32
Sleeve hold-down hardware.
International Harvester

Capscrew

Some wet liners fit so loosely that seal resiliency can affect liner height when the cylinder head is removed. John Deere and other manufacturers that use this type of liner provide detailed loading instructions. Neither do engines fitted with these liners tolerate moving the crankshaft with the head detached. If this is done, piston-ring friction will raise the liners and possibly damage the O-ring seals.

Boring

Ascertain the amount of bore wear. Most wear occurs near the top of the cylinder at the extreme end of ring travel. This wear is caused by local oil starvation and combustion-related acids. Pronounced wear, sometimes taking the form of scuffing, can occur at right angles to the crankshaft centerline on the bore surface that absorbs piston angular thrust. An engine that turns clockwise when viewed from the front will show more wear on the right side of its cylinders than on the left. In addition, axial thrust forces can generate wear in bore areas adjacent to the crankshaft centerline (Fig. 8-33). These forces account for most of the taper and eccentricity exhibited by worn cylinders. Block distortion accounts for the rest.

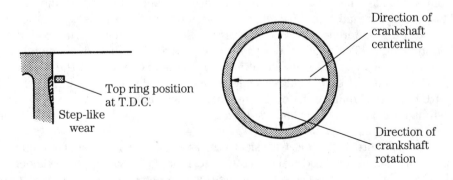

Top ring position at T.D.C.

Step-like wear

Direction of crankshaft centerline

Direction of crankshaft rotation

8-33 Areas of accelerated cylinder wear. Yanmar Diesel Engine Co., Ltd.

Some idea of bore condition can be had by inserting a ring into the cylinder with the flat of a piston. The difference in ring end gap between the upper and lower portions of the cylinder, as determined with a feeler gauge, roughly corresponds to cylinder wear. However, such techniques do not substitute for repeated and averaged measurements with a cylinder bore gauge (Fig. 8-34).

8-34
Using a cylinder gauge. Chrysler Corp.

Study the surface of the bore under a strong light. Deep vertical scratches usually indicate that the air filter has at one time failed. The causes of more serious damage—erosion from contact with fuel spray, galling from lack of lubrication, rips and tears from ring, ring land, or wrist pin lock failure—will be painfully obvious.

Integral or dry-sleeve bores can be overbored and fitted with correspondingly oversized pistons. When the bore limit is reached, replacement sleeves can be fitted to either type, although the work is considerably easier on an engine that was originally sleeved.

The common practice is to use a boring bar for the initial cuts and finish to size with a hone, preferably an automatic, self-lubricating hone such as the Sunnen CV. The finish will approximate that achieved by the OEM.

The accuracy of the job can only be as good as the datum—the reference point from which all dimensions, including the bore-to-crankshaft relationship, are taken. Most boring bars index to the deck, on the assumption that the deck is parallel to the main bearing centerline. This is a large assumption. The better tools index to the main bearing saddles. The cylinder with most wear is bored first; how much metal must be removed to clean up this cylinder determines the bore oversize for the engine.

Oversized piston and ring assemblies are normally supplied in increments of 0.010 in (or 0.25 mm) to the overbore limit. A few manufacturers offer 0.05 in. over pistons for slightly worn cylinders. Occasionally one runs across a 0.015-in. piston and ring set.

The overbore limit varies with sleeve thickness, cylinder spacing (too much overbore compromises the head gasket in the critical "bridge" area between cylinders), and the thickness of the water jacket for engines with integral bores. Jacket thickness depends, in great part, on the quality control exercised by the foundry. Some blocks and whole families of engines have fairly uniform jacket thickness; others are subject to core shift and the unwary machinist can strike water. A final consideration, not of real concern unless the engine is really "hogged out" beyond continence, is obtaining a matching head gasket. The gasket must not be allowed to overhang the bore.

Of course, it is always possible to replace dry liners and to install liners in worn integral bores. The latter operation can be expensive, and most operators would be advised to invest in another block.

Dry sleeves are, by definition, difficult to move; wet sleeves can also stick and some have more propensity for this than others. In short, liner removal and installation tools must be used. Figure 8-35 illustrates a typical combination tool. This or a

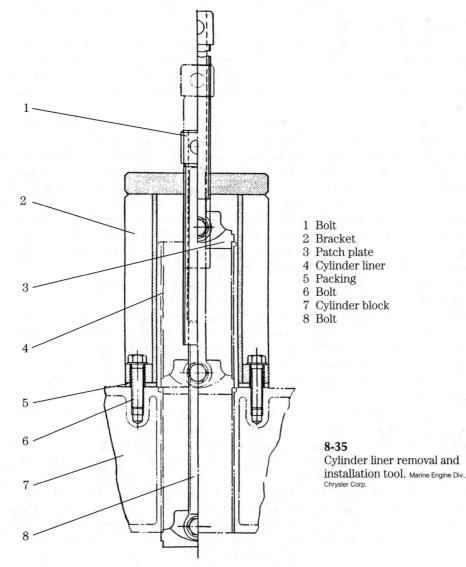

1 Bolt
2 Bracket
3 Patch plate
4 Cylinder liner
5 Packing
6 Bolt
7 Cylinder block
8 Bolt

8-35
Cylinder liner removal and installation tool. Marine Engine Div., Chrysler Corp.

similar tool must be used for extraction, but a press and a stepped pilot—one diameter matching liner ID, the other, liner OD—is the better choice for installation.

Liner bores and counterbores (the ledge upon which the liner seats) require careful measurement and inspection. Out-of-spec counterbores can sometimes be machined and restored to height with shims, Cumming's fashion.

Wet-liner seals must be lubricated just prior to installation to control swelling. Seal contact areas on the water-jacket ID should be cleaned and inspected under a light. Some machinists oil dry sleeves; others argue against the practice.

Some sleeves, wet or dry, tend to crack at the counterbore area after a few hours of operation. The most common causes have to do with chamfers, either the chamfer on the counterbore or the chamfer on the liner installation pilot.

Detroit Diesel reboring

These two-cycle engines require some special instruction. Many Series 71 engines are cast in aluminum. Early production inserts were a slip fit in the counterbore; current standards call for the liner to be pressed in. In any event, the block should be heated to between 160° and 180°F in a water tank. Immerse the block for at least 20 minutes.

Counterbore misalignment can affect any engine, although mechanics generally believe that the aluminum block is particularly susceptible to it. You will be able to detect misalignment by the presence of bright areas on the outside circumference of the old liner. The marks will be in pairs—one on the upper half of the liner and the other diagonally across from it, on the lower half. The counterbore should be miked to check for taper and out-of-roundness.

Small imperfections—but not misalignment between the upper and lower deck—can be cleaned up with a hone. Otherwise, the counterbore will have to be machined. Torque the main-bearing caps. Oversize liners are available from the OEM and from outside suppliers such as Sealed Power Corporation.

Liners on early engines projected 0.002–0.006 in. above the block (Fig. 8-36). These low-block engines used a conventional head gasket and shims under the liner to obtain flange projection. Late-model high-block engines use an insert below the liner. Narrower-than-stock inserts are available to compensate for metal removed from the fire deck, and in various oversized diameters to accommodate larger liners. A 0.002-in. shim is also available for installation under the insert.

Note: Inserts can become damaged in service and can contribute to upper liner breakage.

Series 53 engines employ a wet liner. The upper portion of the liner is surrounded by coolant and sealed with red silicone seals in grooves on the block. Early engines had seals at the top and above the ports. Late-model engines dispense with the lower seal. A second groove is machined at the top of the cylinder to be used in the event of damage to the original.

The seals must be lubricated to allow the liners to pass over them. Do not presoak the seals, because silicone expands when saturated with most lubricants. The swelling tendency is pronounced if petroleum products are used. Lubricate just prior to assembly with silicone spray, animal fat, green soap, or hydraulic brake fluid.

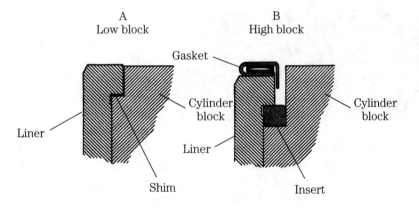

8-36 Detroit Diesel liner arrangements. Sealed Power Corp.

Carefully lower the liners into the counterbores, without twisting the seals or displacing them from their grooves.

The eccentricity (out-of-roundness and taper) must be measured before final assembly. On the 110 series you are allowed 0.0015-in. eccentricity. The 53 and 71 engines will tolerate 0.002 in. Eccentricity can often be corrected by removing the liners and rotating them 90 degrees in the counterbores. Do not move the inserts in this operation.

Honing

Honing is used to bring rebored cylinders to size and to remove small imperfections and glaze in used cylinders. Glaze is the hard surface layer of compacted iron crystals formed by the rubbing action of the rings. Most engine manufacturers recommend that the glaze be broken to aid in ring seating and to remove the ridge that forms at the upper limit of ring travel. The Perfect Circle people suggest that honing can be skipped if the cylinder is in good shape.

The pattern should be diamond-shaped, as shown in Fig. 8-37, with a 22-32 intersection degree at the horizontal centerline. The cut should be uniform in both directions, without torn or folded metal, leaving a surface free of burnish and imbedded stone particles. These requirements are relatively easy to meet if you have access to an automatic honing machine. However, satisfactory work can be done with a fixed-adjustment hone turned by a drill press or portable drill motor.

The hone must be parallel to the bore axis. Liners can be held in scrap cylinder blocks or in wood jigs. The spindle speed must be kept low—a requirement that makes it impossible to use a quarter-inch utility drill motor. Suggested speeds are shown below.

Bore diameter (inches)	Spindle (rpm)	Speed strokes (per minute)
2	380	140
3	260	83
4	190	70
5	155	56

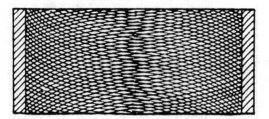

8-37
Preferred crosshatch pattern.

Move the hone up and down the bore in smooth oscillations. Do not let the tool pause at the end of the stroke, but reverse it rapidly. Excessive pressure will load the stone with fragments, dulling it and scratching the bore. Flood the stones with an approved lubricant (such as mineral oil), which meets specification 45 SUV at 100°F.

Stone choice is in part determined by the ring material. Most engines respond best to 220-280 grit silicon carbide with code J or K hardness.

Cleaning the bore is a chore that is seldom done correctly. Never use a solvent on a honed bore. The solvent will float the silicon carbide particles into the iron, where they will remain. Instead, use hot water and detergent. Scrub the bore until the suds remain white. Then rinse and wipe dry with paper towels. The bore can be considered "sanitary" when there is no discoloration of the towel. Oil immediately.

Piston rings

Piston rings are primarily seals to prevent compression, combustion, and exhaust gases from entering the crankcase. The principle employed is a kind of mechanical jujitsu—pressure above the ring is conducted behind it to spread the ring open against the cylinder wall. The greater the pressure above the ring, the more tightly the ring wedges against the wall (Fig. 8-38).

Gas pressure

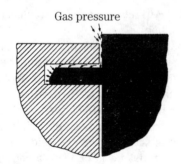

8-38
A compression ring is a dynamic seal, expanding under the effects of gas pressure.

The rings also lubricate the cylinder walls. The oil control ring distributes a film of oil over the walls, providing piston and ring lubrication. One or more scraper rings control the thickness of the film, reducing chamber deposits and oil consumption. In addition to sealing and lubrication, the ring belt is the main heat path from the piston to the relatively cool cylinder.

Rings are almost always cast iron, although steel rings are used in some extreme-pressure situations. Cast iron is one of the very few metals that tolerate rubbing contact with the same material. Until a few years ago rings were finished as cast.

Today almost all compression rings are flashed with a light (0.004 in.) coating of chrome. Besides being extremely hard and thus giving good wear resistance, chrome develops a pattern of microscopic cracks in service. These cracks, typically accounting for 2% of the ring's surface, serve as oil reservoirs and help to prevent scuffing. A newer development is to *fill*, or *channel*, the upper compression ring with molybdenum. The outer diameter of the ring is grooved and the moly sprayed on with a hot-plasma or other bonding process. Besides having a very low coefficient of friction and a very high melting temperature, moly gives a piston ring surface that is 15–30% void. It retains more oil than chrome-faced rings and is, at least in theory, more resistant to scuffing.

Rings traditionally have been divided into three types, according to function. Counting from the top of the piston, the first and second rings are *compression* rings, whose task is to control blowby. The middle ring is the *scraper*, which keeps excess oil from the combustion space. The last ring is the *oil* ring, which is serrated to deliver oil to the bore.

This rather neat classification has become increasingly ambiguous with the development of multipurpose ring profiles and the consequent reduction in the number of rings fitted to a piston. Five- and six-ring pistons have given way to three- and four-ring pistons on many of the smaller engines. The function of the middle rings is split between gas sealing and oil control. The lower rings, while primarily operating as cylinder oilers, have some gas-seating responsibilities. Design has become quite subtle, and it is difficult for the uninitiated to distinguish between *compression* and *scraper* rings.

The drawing in Fig. 8-39 illustrates the ring profiles used on the current series of GM Bedford engines. Note the differences in profile among the three. These profiles are typical, but by no means, universal. The Sealed Power Corporation offers several hundred in stock and will produce others on special order.

What this means to the mechanic is that he or she must be very careful when installing rings. Most have a definite *up* and *down*, which might or might not be indicated on the ring. Usually the top side is stamped with some special letter code. Great care must be exercised not to install the rings in the wrong sequence. New rings are packaged in individual containers or in groups that are clearly marked 1 (for first compression), 2, and so forth. Reusable rings should be taken off the piston and placed on a board in the assembly sequence.

Ring wear

The first sign of ring wear is excessive oil consumption, signaled by blue smoke. But before you blame the rings, you should check the bearing clearances at the main and crankpin journals. Bearings worn to twice normal clearance will throw off five times the normal quantity of oil on the cylinder walls. You can make a direct evaluation of oil spill by pressurizing the lubrication system. If appreciable amounts of oil are getting by the rings, the carbon pattern on the piston will be chipped and washed at the edges of the crown.

Check the rings for sticking in their grooves (this can be done on two-cycles from the air box), breaks, and scuffing. The latter is by far the most common malady, and results from tiny fusion welds between the ring material and cylinder walls. Basically it can be traced to lack of lubrication, but the exact cause might require the

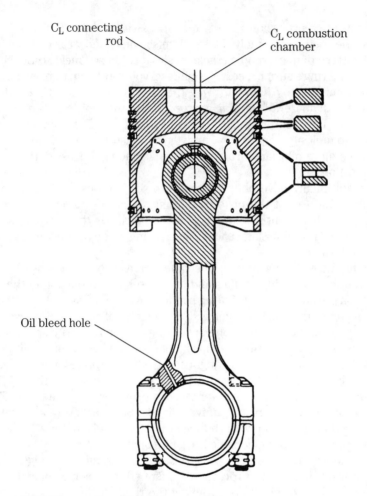

C$_L$ connecting rod

C$_L$ combustion chamber

Oil bleed hole

8-39 Ring configuration and nomenclature. GM Bedford Diesel

deductive talents of a Sherlock Holmes. Engineers at Sealed Power suggest these possibilities:

Symptom	Possible cause
Overheating	Clogged, restricted, sealed cooling system
	Defective thermostat or shutters
	Loss of coolant
	Detonation
Lubrication failure	Worn main bearings
	Oil pump failure
	Engine lugging under load
	Extensive idle
	Fuel wash on upper-cylinder bores
	Water in oil
	Low oil level
	Failure to pressurize oil system after rebuild
Wrong cylinder finish	Low crosshatch finish

Symptom	Possible cause
	Failure to hone after reboring
Insufficient clearance	Inadequate bearing clearance at either end of the rod
	Improper ring size
	Cylinder sleeve distortion

Usually inadequate bearing clearance, complicated by a poor fit in the block counterbore, results in overheating. The fundamental cause is often poor torque procedures, or improperly installed sealing rings on wet-sleeved engines. A rolled or twisted sealing ring can distort the sleeve.

Ring breakage is due to abnormal loading or localized stresses. It can be traced to:

- Ring sticking—this overstresses the free end of the ring.
- Detonation—this is traceable to the overly liberal use of starting fluid, to dribbling injectors, and to out-of-time delivery.
- Overstressing the ring on installation. Usually the ring breaks directly across from the gap.
- Excessively worn grooves, which allow the ring to flex and flutter.
- Ring hitting the ridge at the top of the bore. The mechanic is at fault because this ridge should have been removed.

The last point—involving blame—can be sticky in a shop situation. Mechanics make mistakes the same as everyone else, and the number of mistakes is, in part, a function of the complexity of the repair. Few people can overhaul a machine as complicated as a multicylinder engine without making some small error. Assessing blame, if only to correct the situation, is sometimes complicated by having the mechanic who built the engine tear it down. But careful examination of the parts usually points to the fault. For example, rings that have been fitted upside down show reversed wear patterns. Rings that have broken in service are worn on either side of the break, from contact with the cylinder walls. The fracture will be dulled. Rings that have been broken during removal show sharp crystalline breaks without local wear spots.

Pistons

Large diesel engines often employ cast-iron or steel pistons; smaller, high-speed engines generally use aluminum castings. That such pistons survive in combustion temperatures that can reach 4500°F and cylinder pressures that, in highly supercharged engines, can exceed 2000 psi, is a triumph of engineering over materials. Aluminum has a melting point of 1220°F and rapidly loses strength as this temperature is approached.

Construction

The heavy construction of these pistons—diesel pistons typically weigh half as much again as equivalent SI engine pistons—provides mechanical strength and the heat conductivity necessary to keep the piston crown at about 500°F. The standard practice is to direct a stream of oil to the underside of the crown, usually by means of spray jets. Some designers go a step further, and insulate the crown, which is relatively easy to cool, from the ring belt and skirt. Turn ahead to Fig. 8-48 for an illustration of a heat dam. The cavity above the piston pin slows heat transfer by reducing piston wall thickness. Another approach is to lengthen the thermal path by grooving the area above the ring belt (Fig. 8-40). Yet another approach is to apply a thermal coating to the upper side of the piston crown, thus confining heat to the combustion chamber (Fig. 8-41).

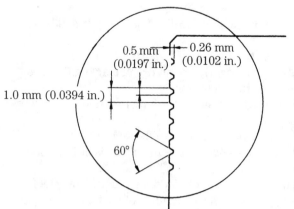

0.5 mm (0.0197 in.)

0.26 mm (0.0102 in.)

1.0 mm (0.0394 in.)

60°

8-40
Fire groves form thermal barriers between the ring band and combustion chamber. Yanmar Diesel Engine Co., Ltd.

The skirts of aluminum pistons run hotter than cast iron and have a coefficient of expansion that is about twice that of iron. Consequently, light-metal pistons are assembled with fairly generous bore clearances to compensate for thermal expansion, and might be noisy upon starting. Semi-exotic alloys, such as Lo-Ex, or cast-in steel struts, help control expansion and knocking. It is interesting that one of the first experimenters with aluminum pistons, Harvey Marmon, found it necessary to sheathe the piston skirt in an iron "sock."

Most alloy pistons are cam-ground; when cold the skirts are ovoid, with the long dimension across the thrust faces. As the piston heats and expands, it becomes circular, filling the bore. Other ways of coping with thermal expansion are progressively to reduce piston diameter above the pin and to relieve, or cut back, the skirts in the pin area (also shown in Fig. 8-41). The heavy struts that support the pin bosses transfer heat from the underside of the crown to the skirts.

Critical wearing points include the thrust faces and the sides of the ring grooves. With the exception of certain two-cycle applications, rings are designed to rotate in their grooves and, according to one researcher, reach speeds of about 100 rpm. Rotation is the primary defense against varnish buildup and consequent ring sticking. But it also wears "steps" into the grooves. Piston thrust faces are also subject to rapid wear.

Some manufacturers run the compression ring against a steel insert, cast integrally with the piston. Another approach is to substitute a long-wearing eutectic alloy for the SAE 334 or 335 usually specified. Eutectic alloys consist of clusters of hard silicon crystals distributed throughout an aluminum matrix. As the aluminum wears, silicon—one of the hardest materials known—emerges as the bearing sur-

Alumite treatment

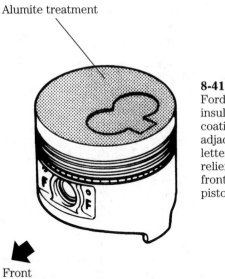

Front

8-41
Ford 2.3L pistons are insulated with a plasma coating and relieved at areas adjacent to the pins. Note the letter "F" embossed on the relief, which should face the front of the engine when the piston is installed.

face. The same mechanism rapidly dulls cutting tools, which is why the cost of eutectic pistons approaches the cost of forgings.

Forged pistons are an aftermarket item, used as a last resort in highly supercharged engines when castings have failed. Forging eliminates voids in the metal and compacts the grain structure at the crown, pin bosses, and ring lands (Fig. 8-42). These pistons have superior hot strength characteristics but require generous running clearances. An engine with forged pistons will be heard from during cold starts.

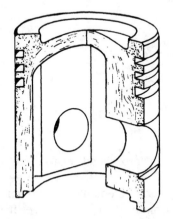

8-42
Forged pistons gain strength from the uniform grain structure.
Sealed Power Corp.

A two-piece piston consists of a piston dome, or ring carrier, element and a skirt element (Fig. 8-43). These parts pivot on the piston pin. Although other manufacturers use this form of the piston, the GM version first appeared on Electromotive railroad engines and, in 1971, replaced conventional trunktype pistons on turbocharged DDA Series 71 engines. It eventually found its way into several other DDA engines, including the Series 60.

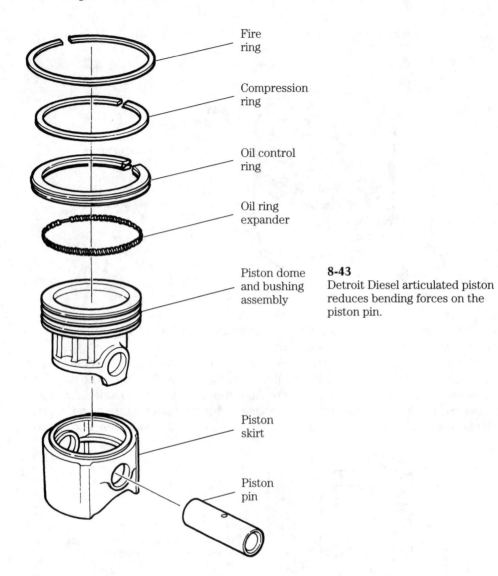

Fire
ring

Compression
ring

Oil control
ring

Oil ring
expander

Piston dome
and bushing
assembly

Piston
skirt

Piston
pin

8-43
Detroit Diesel articulated piston
reduces bending forces on the
piston pin.

Detroit Diesel describes this design as a "crosshead" piston. As the term is usually applied, it refers to a kind of articulated piston used on very large engines. A pivoted extension bar separates the upper piston element and the skirt, which rides
against the engine frame. The crosshead isolates the cylinder bores from crankshaft-
induced side forces. The DDA design does not relieve the bore of side forces, but it
increases the bearing area of the pin and centralizes the load more directly on the
conn rod for a reduction in bending forces. The connecting rod, illustrated in the following section, bolts to the underside of the pin, making the upper half of the pin
available to support piston-dome thrust.

Failure modes

Piston failure is usually quite obvious. Wear should not be a serious consideration in low-hour engines, because the skirt areas are subject to relatively small forces and have the benefit of surplus lubrication. Excessive wear can be traced to dirty or improperly blended lube oil or inadequate air filtration. Poor cylinder finishing might also contribute to it. Piston collapse or shrinkage is usually due to overheating. If the problem shows itself in one or two cylinders, expect water-jacket stoppage or loose liners.

Combustion toughness or *detonation* damage begins by eroding the crown, usually near the edge. The erosion spreads and grows deeper until the piston "holes." Typically the piston will look as if it were struck by a high-velocity projectile. *Scuffing* and *scoring* (a scuff is a light score) might be confined to the thrust side of the piston. If this is the case look for the following:

- Oil pump problems—screen clogged, excessive internal clearances.
- Insufficient rod bearing clearances, which reduce throw-off, robbing the cylinders of oil.
- Lugging.

The probable causes of damage to both sides of the skirt include the ones just mentioned, plus these:

- Low or dirty oil.
- Detonation.
- Overheating caused by cooling system failure.
- Coolant leakage into the cylinder.
- Inadequate piston clearance.

Scuffs or scores fanning out 45 degrees on either side of the pinhole mean one of these conditions:

- Pin fit problems—too tight in the small end of the rod or in the piston bosses.
- Pinhole damage (see below for installation procedures).

Ring land breakage can be caused by the following:

- Excessive use of starting fluid.
- Detonation.
- Improper ring installation during overhaul.
- Excessive side clearance between the ring and groove.
- Water in cylinder.

Free-floating pins sometimes float right past their lock rings and contact the cylinder walls. Several causes (listed below) have been isolated.

- Improper installation: Some mechanics force the lock rings beyond the elastic limit of the material. In a number of cases, it is possible to install lock rings by finger pressure alone.
- Improper piston alignment: This might be caused by a bent rod or inaccuracies at the crankshaft journal. Throws that are tapered or out of parallel with the main journals will give the piston a rocking motion that can dislodge the lock ring. Pounding becomes more serious if the small-end bushing is tight.

- Excessive crankshaft end play: Fore-and-aft play is transmitted to the lock rings and can pound the grooves open. Again, a too-tight fit at the connecting rod's small end will hasten piston failure.

Servicing

Used pistons can give reliable service in an otherwise rebuilt engine, but only after the most exhaustive scrutiny. Scrape and wire-brush carbon accumulations from the crown, but do not brush the piston flanks. Carbon above the compression ring and on the underside of the crown should be removed chemically. A notch, letter, arrow, or other symbol identifies the forward edge of the piston. These marks are usually stamped on the crown, on the pin boss relief (illustrated back in Fig. 8-41), or hidden under the skirt. Make note of the relationship between the leading edge of the piston and the numbered side of the connecting rod.

Lay out the rings in sequence, topsides up, on the bench. Spend some time "reading" the rings—the history of the upper engine is written on them, just as the crank bearings testify to events below. The orientation code—the word Top (T) or *Ober* (O)—will be found on the upper sides of the compression and scraper rings, adjacent to the ring ends.

You might wonder why attention is given to these codes for piston assemblies that, at this stage, are of unknown quality, and for rings that will, in any event, be discarded. The purpose is to become familiar with the concept of orientation as it applies to the particular engine being serviced. New parts might not carry identical codes, but the relationship between coded parts will not change.

While still attached to the rod and before investing any more time in it, inspect the piston for obvious defects. Reject if the piston is fractured, deeply pitted, scored, or if it exhibits ring land damage (Fig. 8-44). Cracks tend to develop where abrupt changes in cross section act as stress risers. Deep pits usually develop at the edges of the piston; scoring is most likely to develop on the thrust faces. Contact with the cylinder bore occurs at two areas, 90 degrees from the piston-pin centerline. The major thrust face lies in the direction of the crankshaft rotation and is normally the first to score. (Viewed from the front of a clockwise-rotation engine, the major thrust face is on the right.) Machine marks, the light cross-hatching left by the cutting tool, should remain visible over most of the contact area. The next section describes the relationship of wear patterns to crankshaft and rod alignment.

If the piston passes this initial examination, make a micrometric measurement of skirt diameter across the thrust faces. Depending on the supplier, the measurement is made at the lower edge of the skirt, at the pin centerline, or at an arbitrary distance between these points (Fig. 8-45). Piston OD subtracted from cylinder bore ID equals running clearance.

All traces of carbon in the ring grooves and oil spill ports must be removed. Whenever possible, this tedious task should be consigned to the machinist and thought about no more. Figure 8-46 illustrates a factory-supplied plug gauge used to measure groove width; the next drawing, Fig. 8-47, shows an alternate method using a *new* ring and a feeler gauge. The latter method is accurate, so long as the gauge bottoms against the back of the groove.

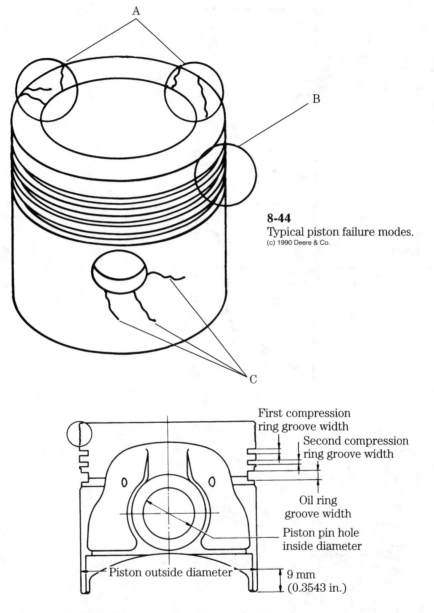

8-44
Typical piston failure modes.
(c) 1990 Deere & Co.

First compression
ring groove width

Second compression
ring groove width

Oil ring
groove width

Piston pin hole
inside diameter

Piston outside diameter

9 mm
(0.3543 in.)

8-45 Piston measurement points. OD measurements may vary
with manufacturers. Yanmar Diesel Engine Co., Ltd.

Most engines of this class employ "full-floating" piston pins that, at running temperatures, are free to oscillate on both the rod and piston (Fig. 8-48). Pins secure laterally with one or another variety of snap rings, some of which are flattened on the inboard (or wearing) sides.

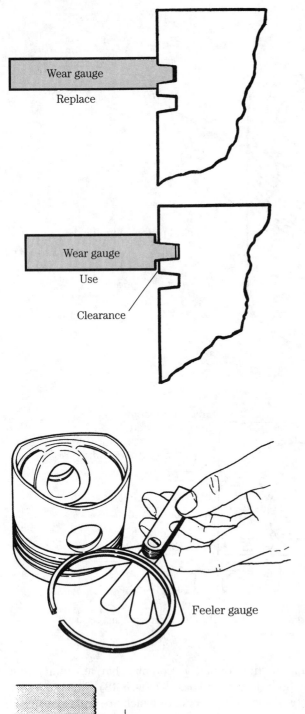

Wear gauge

Replace

Wear gauge

Use

Clearance

8-46
The best way to determine
ring groove wear is with a
factory-supplied go-nogo gauge.
International Harvester

Feeler gauge

8-47
An alternative method of
determining groove wear is to
measure the width
relative to a new ring. Yanmar Diesel
Engine Co., Ltd.

Ring-to-groove clearance

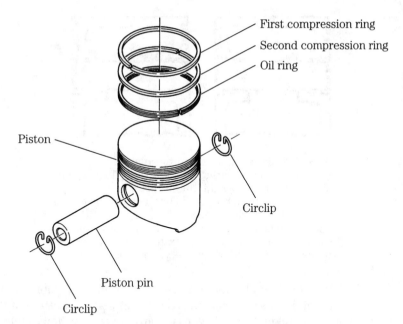

First compression ring
Second compression ring
Oil ring

Piston

Circlip

Piston pin

Circlip

8-48 Like most manufacturers, Yanmar favors full-floating piston pins, pivoting on both the rod and piston, and secured laterally by snap rings ("circlips" in the drawing).

Protecting your eyes with safety glasses, disengage and withdraw the snap rings. Although mechanics generally press out (and sometimes hammer out) piston pins, these practices should be discouraged. Instead, take the time to heat the pistons, either with a heat gun or by immersion in warm (160° F) oil. Pins will almost fall out.

While the piston is still warm, check for bore integrity. Insert the pin from each side. If the pin binds at the center, the bore might be tapered; if the bore is misaligned, the pin will click or bind as it enters the far boss (Fig. 8-49).

Other critical areas are illustrated in Fig. 8-45. Measurements are to be made at room temperature. Damage to retainer rings or ring grooves suggests excessive crankshaft end play, connecting-rod misalignment, or crankpin taper. Slide forces generated by a badly worn crankpin or severely bent rod react against the snap rings. Rod or crankpin misalignment might also appear as localized pin and pin-bearing wear, a subject discussed in the following section.

To install, warm the piston, oil the pin and pin bores, and, with the rod in its original orientation, slip the pin home with light thumb or palm pressure. Use *new* snap rings, compressing them no more than necessary. Verify that snap rings seat around their full diameters in the grooves.

A few engines use pressed-in pins, which make a lock on the piston with an interference fit. Support the piston on a padded V-block and press the pin in two stages: stop at the point of entry to the lower boss, relieve press force to allow the piston to regain shape, and press the pin home. If the pin is installed in one pressing, the lower boss might be shaved.

What remains is to establish the running clearance. Piston-to-bore clearance is fundamentally a matter of specification, but specifications are never so rigid that

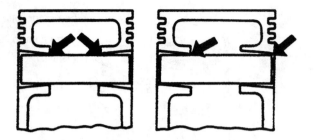

8-49 This drawing, intended by John Deere to show how piston pin bearing checks (free-floating pins should insert with light thumb pressure; if the pin goes in easily from both sides but binds in the center, the problem might be tapered bores (A); if the pin clicks or binds when contacting the far boss, the piston is warped (B), also illustrated the cavity on the underside of the head that serves as a heat dam.

they cannot be bent. For example, one John Deere piston/liner combination requires a running clearance of 0.0034 to 0.0053 in., as measured at the bottom of the piston skirt. A high-volume rebuilder typically goes toward the outside limit, building clearance here and in the crankshaft bearings. A "loose" engine will be more likely to tolerate severe loads as delivered and without the benefit of a break-in period. A custom machinist, who works on one or two engines at a time, often goes in the other direction, aiming at the tightest clearance the factory allows. Besides being aesthetically more satisfying, "tight" engines tend to live long, quiet lives. Of course, such an engine must be carefully run in during the first hours of operation.

In an attempt to stabilize the maintenance process, factory manuals include wear limits for critical components. The concept of permissible wear is a value judgment, made in an engineering office remote from the world of mechanics, back-ordered parts, and budgetary restraints. Permissible piston-to-liner clearance for the Deere engine is 0.0060 in. Suppose that the numming clearance is found to be 0.0055 in.—only 0.0002 in. over the allowable 0.0053 with 0.0030 in. to go before the wear limit is reached. Assuming that the wear is equally distributed over both parts, should the piston or liner or both be replaced? This is only one cylinder of six, none of them worn by identical amounts. Questions like this go beyond mechanics and depend for their answers on the politics of the situation, interpreted in light of the philosophy of maintenance—formal or informal—that characterizes the operation.

Connecting rods

No part is more critical than the connecting rod and none more conservatively designed (Fig. 8-50).

Construction

All diesel engines employ two-piece, H-section rods, heavily faired at the transitions between bearing sections, and forged from medium-carbon steel or from that ubiquitous alloy, SAE 4140. The small end is generally closed and fitted with a re-

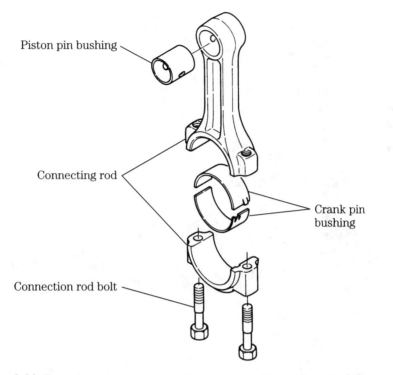

Piston pin bushing

Connecting rod

Crank pin bushing

Connection rod bolt

8-50 Typical connecting rod and bearing assembly. A rifle-drilled oil passage indexes with ports in both the small-end bushing and upper big-end insert. Yanmar Diesel Engine Co., Ltd.

placeable bushing (some automotive OEMs do not catalog replacement bushings, but a component machinist can work around that). The big end carries a two-piece precision insert bearing. A few rods terminate in a slipper, which bolts to the piston pin, as shown in Fig. 8-51. Open construction makes the whole length of the pin available as a bearing surface.

Whenever possible, interface between the shank and cap should be perpendicular to the crankshaft centerline. This configuration reduces side forces on the cap. However, the rod must be split at an angle to allow disassembly through the bore when crankpin diameter is large (as in Onan engines) or to prevent contact with other parts (Deere 6076).

Connecting rods have a definite orientation relative to the piston and to the cap. The former is a function of transverse oil ports, drilled in the rod shank; the latter reflects the way connecting rods are manufactured. The relatively small connecting rods that we are dealing with are forged in one piece. Then the cap is separated (sometimes merely by snapping it off), assembled, and honed to size. Reversing the cap or installing a cap from another rod destroys bearing circularity. Typically, a mismatched assembly overheats and seizes within a few crankshaft revolutions. Rod shanks and caps carry match marks, and both parts are identified by cylinder number (Figs. 8-52 and 8-53). Errors can be eliminated by double- and triple-checking match mark alignment and cylinder numbers. Mechanics do well to observe the rule that no more than one rod cap should be disassembled at a time.

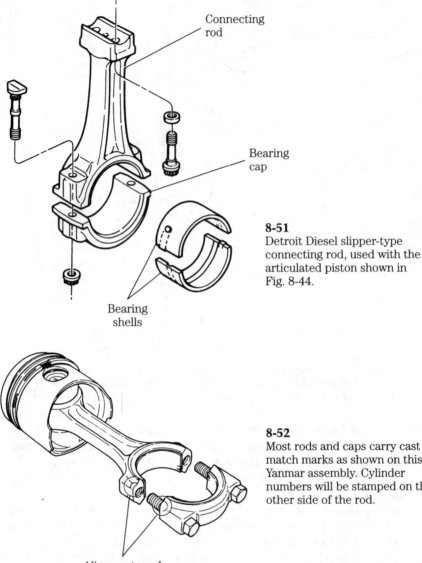

8-51
Detroit Diesel slipper-type connecting rod, used with the articulated piston shown in Fig. 8-44.

Connecting rod

Bearing cap

Bearing shells

8-52
Most rods and caps carry cast match marks as shown on this Yanmar assembly. Cylinder numbers will be stamped on the other side of the rod.

Alignment mark (casting mark)

Service

The number stamped on the rod shank and cap should correspond to the cylinder number. Sometimes these numbers are scrambled or missing, and the mechanic must supply them. Stamp the correct numbers on the pads provided and, to prevent confusion, deface the originals.

Mike rod journals at several places across the diameter, repeating the measurements on both sides to detect taper (Fig. 8-54). Inertial forces tend to stretch the

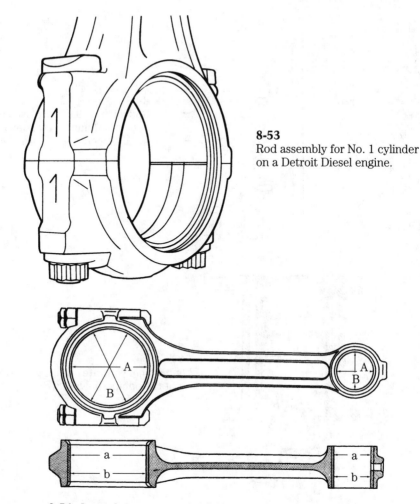

8-53
Rod assembly for No. 1 cylinder
on a Detroit Diesel engine.

8-54 Journal measurement points. Marine Engine Div., Chrysler Corp.

rod cap and pinch the ends together. When this condition is present, the machinist
should mill a few thousandths off the cap interface and, with the cap assembled and
torqued, machine the journal to size. Typically, this is done with a reamer, although
automatic honing machines, such as the Sunnen or the Danish-made AMC, produce
a finer, more consistent surface.

The machinist should check the alignment of each connecting rod with a fixture
similar to the one shown in Fig. 8-55. Even so, these matters should not be entirely left
to the discretion of the machinist. It is always prudent to check the work against the
testimony of the engine. Rod and crankpin misalignment produces telltale wear patterns on piston thrust faces, on pin bores, and on the retainers and retainer grooves.

A bent rod tilts the piston, localizing bearing wear at the points shown in Fig.
8-56. This condition might also be reflected by an hourglass-shaped wear pattern on
the thrust faces, as illustrated in Fig. 8-57. A twisted rod imparts a rocking motion to

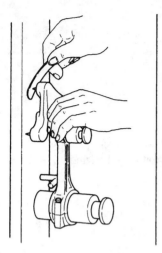
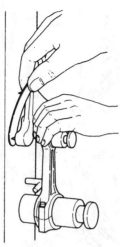

8-55
Rod alignment fixture used to detect bending and torsional misalignment. Yanmar Diesel Engine Co., Ltd.

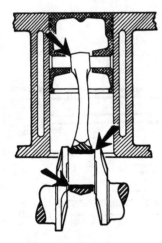
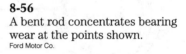

8-56
A bent rod concentrates bearing wear at the points shown.
Ford Motor Co.

8-57
Another effect of loss of rod parallelism is to tilt the piston and, in extreme cases, produce the wear pattern shown on the left. Sealed Power Corp.

the piston, concentrating wear on the ring band and skirt edges (Fig. 8-58). Crankshaft taper generates thrust, which might scuff the skirt and could drive the pin through one or the other retainer.

8-58
A twisted connecting rod causes the piston to rock, a condition that can be signaled by an elliptical wear pattern.
Sealed Power Corp.

Bent or twisted rods generally can be straightened, although some manufacturers warn against the practice.

Most rod bolts can be reused, if visual and magnetic-particle testing fails to reveal any flaws. Some recently developed torque-to-yield bolts can also be reused; earlier types were sacrificial. Rod-bolt nuts should be replaced, regardless of the fastener type.

Used rods should be Magnafluxed. Interpretation of the crack structure thus revealed requires some judgment. Any rod that has seen service will develop cracks. One must distinguish between inconsequential surface flaws and cracks that can lead to structural failure. The drawing in Fig. 8-59 can serve as a guide. In general, longitudinal cracks are not serious enough unless they are $\frac{1}{32}$ in. deep, which can be determined by grinding at the center of the crack. Transverse cracks are causes of concern because they can be the first indication of fatigue. If the cracks do not extend over the edges of the H-section, are no more than $\frac{1}{8}$ in. long, and less than $\frac{1}{64}$ in. deep, they can be ground and feathered. Cracks over the H-section can be removed if 0.005 in. deep or less. Cracks in the small end are cause for rejection.

Small-end bushings are pressed into place with reference to the oil port and are finish-reamed. Loose bushings are a sign that the rod has overheated, and they can cause a major failure by turning and blocking the oil port.

What has been described are standard, industry-wide practices. One can go much further in the quest for a more perfect, less problematic engine. For example, it is good practice to match the weight of piston and rod assemblies to within 10 grams or so—even when parts are not to be sent out for balancing. Surplus weight can be ground from the inner edges of the piston skirts. Some mechanics assemble

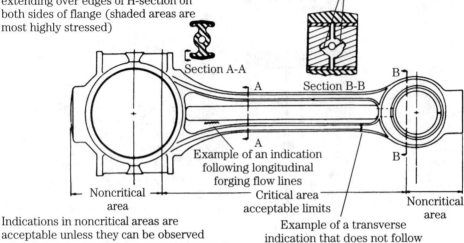

Do not use or attempt to salvage rods with indications over 0.005 in. deep extending over edges of H-section on both sides of flange (shaded areas are most highly stressed)

Start of fatigue crack resulting from overloading (due to hydrostatic lock). Do not attempt to salvage (this type of indication is not visible with bushings in place)

Section A-A

Section B-B

Example of an indication following longitudinal forging flow lines

Critical area acceptable limits

Noncritical area

Noncritical area

Indications in noncritical areas are acceptable unless they can be observed as obvious cracks without magnetic inspection

Example of a transverse indication that does not follow longitudinal forging flow lines can be either a forging lap, heat treat crack, or start of a fatigue crack

Longitudinal indications
Following forged flow lines are usually seams and are not considered harmful if less than 1/32 in. deep

Poor practice   Good practice

Grinding notes
Care should be taken in grinding out indications to assure proper blending of ground area into unground surface so as to form a smooth contour

Transverse indications (across flow lines) Having a minimum length of 1/2 in. which can be removed by grinding no deeper than 164 in. are acceptable after their **complete removal.** An exception to this is a rod having an indication which extends over the edge of H-section and is present on both sides of the flange in this case. Maximum allowable depth is 0.005 in. (see section A-A).

8-59 Interpreting Magnaflux indications. Detroit Diesel

one piston and rod to several crankpins. Differences in deck height between No. 1 and the last cylinder of the bank indicate an alignment problem, either between crankshaft throws or between the crankshaft centerline and deck.

Crankshafts

At this point we have progressed to the heart of the engine, the place where its durability will be finally established.

Construction

Crankshafts for the class of engine under discussion are, for the most part, steel forgings. Materials range from ordinary carbon steels to expensive alloys such as chrome-moly SAE 4140. A number of automotive engines and lightly stressed stationary power plants get by with cast-iron shafts, usually recognized by sharp, well-defined parting lines on the webs and by cored crankpins. Forging blurs the parting lines and mandates solid crankpins, which, however, are drilled for lubrication.

Most crank journals are induction-hardened, a process that leaves a soft, fatigue-resistant core under a hardened "skin." Hardness averages about 55 on the Rockwell scale and extends to a depth of between 0.020 to 0.060 in. in order to accommodate regrinding. But a cautious machinist will test crankshaft hardness after removing any amount of metal.

A few extreme-duty crankshafts are hardened by a proprietary process known as Tufftriding. Wearing quality is comparable to that provided by chrome plating, but unlike chromium, the process does not adversely affect the fatigue life of the shaft.

Service

Remove the crankshaft from the block—a hoist will be required for the heavier shafts—and make this series of preliminary inspections:

- If a main bearing cap or rod has turned blue, discard the shaft, together with the associated cap or rod. The metallurgical changes that have occurred are irreversible. By the same token, be very leery of a crankshaft that is known to have suffered a harmonic balancer failure.
- Check the fit of a new key in the accessory-drive keyway. There should be no perceptible wobble. A competent machinist can rework worn keyways and, if necessary, save the shaft by fabricating an oversized key.
- Check the timing-gear teeth for wear and chipping. Magnetic-particle testing can be of some value when applied to the hub area, but cannot detect the subsurface cracks that signal incipient gear-tooth failure. Timing gear sprockets and related hardware should be replaced as a routine precaution during major repairs.
- Mike the journals and pins as shown in Fig. 8-60. Compare taper, out-of-roundness, and diameter against factory wear limits.
- If corrective machining does not appear necessary, remove with crocus cloth all light scratches and the superficial ridging left by bearing oil grooves. Tear off a strip of crocus cloth long enough to encircle the journal. Wrap the cloth with a leather thong, crossing the ends, to apply force evenly over the whole diameter of the journal. Work the crocus cloth vigorously, stopping at intervals to check progress. An Armstrong grinder works surprisingly fast.
- Using an EZ-Out or hex wrench, remove the plugs capping the oil passages. Back-drilled sections of these passages serve as chip catchers, and must be thoroughly cleaned. Compressed air, shown in Fig. 8-61, helps but is no substitute for rifle-bore brushes, solvent, and elbow grease. Clean the plugs, seal with Loctite, and assemble.
- Check trueness with one or, preferably, tandem dial indicators while rotating the crank in precision V-blocks (Fig. 8-62).

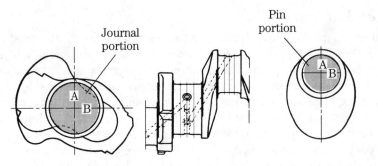

8-60 Crankshaft measurement points. Marine Engine Div., Chrysler Corp.

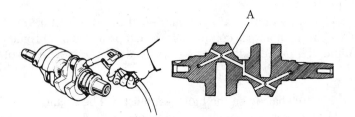

8-61 Blind runs in the crankshaft oiling circuit must be opened, mechanically cleaned, and resealed. Lombardini

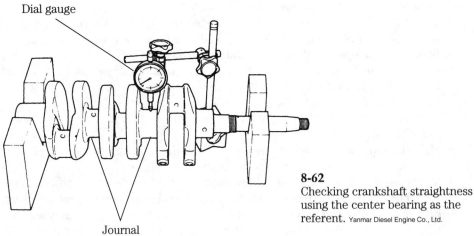

8-62 Checking crankshaft straightness using the center bearing as the referent. Yanmar Diesel Engine Co., Ltd.

Flaw testing

Flaw testing is generally done with the Magnaflux process, although some shops prefer to use the fluorescent-particle method. Both function on the principle that cracks in the surface of the crankshaft take on magnetic polarity when the crank is put in a magnetic field. Iron particles adhere to the edges of these cracks, making them visible. The fluorescent particle method is particularly sensitive because the metal particles fluoresce and glow under black light.

Most cracks are of little concern because the shaft is loaded only at the points indicated in Fig. 8-63. The strength of the shaft is impaired by crack formations which follow these stresses, as shown in Fig. 8-64. These cracks radiate out at 45 degrees to the crank centerline and will eventually result in a complete break.

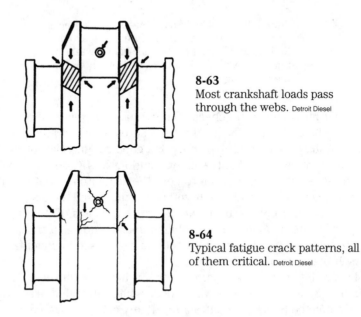

8-63
Most crankshaft loads pass through the webs. Detroit Diesel

8-64
Typical fatigue crack patterns, all of them critical. Detroit Diesel

Abnormal bending forces are generated by main-bearing bore misalignment, improperly fitted bearings, loose main-bearing caps, unbalanced pulleys, or over-tightened belts. Cracks caused by bending start at the crankpin fillet and progress diagonally across.

The distribution of cracks caused by torsional (or twisting) forces is the same as for bending forces. All crankshafts have a natural period of torsional vibration, which is influenced by the length/diameter ratio of the crank, the overlap between crankpins and main journals, and the kind of material used. Engineers are careful to design the crankshaft so that its natural periodicity occurs at a much higher speed than the engine is capable of turning. However, a loose flywheel or vibration damper can cause the crank to wind and unwind like a giant spring. Unusual loads, especially when felt in conjunction with a maladjusted governor, can also cause torsional damage.

Crankshaft grinding

Bearings are available in small oversizes (0.001 and 0.002 in.) to compensate for wear. The first regrind is 0.010 in. Some crankshafts will tolerate as much as 0.040 in., although the heat treatment is endangered at this depth. The crankpin and main-journal fillets deserve special attention. Flat fillets invite trouble, because they act as stress risers. Gently radius the fillets as shown in the left drawing in Fig. 8-65.

All journals and pins should be ground, even if only one has failed. Use plenty of lubricant to reduce the possibility of burning the journal. Radius the oil holes with a stone and check the crankshaft again for flaws with one of the magnetic particle methods.

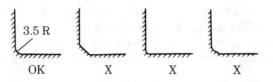

8-65 Fillet profiles. Sharp edges, flat surfaces, and overly wide profiles should be avoided. Rolled, as opposed to ground, fillets extend fatigue life by as much as 60%, but few shops have the necessary equipment. Detroit Diesel

Surface hardness should be checked after machining and before the crank receives final polishing. Perhaps the quickest and surest way to do this is to use Tarasov etch. Clean the shaft with scouring powder or a good commercial solvent. Wash thoroughly and rinse with alcohol. Apply etching solution No. 1 (a solution of 4 parts nitric acid in 96 parts water). It is important to pour the acid into the water, not vice versa.

Rinse with clean water and dry. If you use compressed air, see that the system filter traps are clean. Apply solution No. 2, which consists of 2 parts hydrochloric acid in 98 parts acetone. Acetone is highly flammable and has a sharp odor that can produce dizziness or other unpleasant reactions when used in unventilated areas, so allow yourself plenty of breathing room.

The shaft will go through a color change if it has been burned. Areas that have been hardened by excessive heat will appear white; annealed areas turn black or dark gray. Unaffected areas are neutral gray. If any color other than gray is present, the shaft should be scrapped and the machinist should try again, this time with a softer wheel, a slower feed rate, or a higher work spindle speed. Some experimentation might be necessary to find a combination that works.

Cranks that have been Tufftrided must be treated after grinding to restore full hardness. One test is chemical: a 10% solution of copper ammonium chloride and water applied to the crankshaft reacts almost immediately by turning brown if a traditional heat treatment has been applied. There will be no reaction in 10 seconds if the crankshaft is Tufftrided. Another test is mechanical: Tufftriding is applied to the whole crankshaft, not just to the bearing journals. If a file skates ineffectually over the webs without cutting, one can assume that the shaft is Tufftrided.

Any heavy-duty crankshaft, forged or cast, can benefit from Tufftriding, if the appropriate bearing material is specified. This treatment is especially beneficial when the manufacturer has neglected to treat journal fillets. The abrupt change in hardness acts as a stress riser. Treatment should be preceded by heating the crankshaft for several hours at a temperature above 1060°F. It might be necessary to restraighten the crankshaft.

Camshafts and related parts

Inspect the accessory drive gear train for tooth damage and lash. Crank and camshaft gears are pressed on their shafts and further secured by pins, keys, or bolts.

Ohv cam and balance shafts are supported on bushings, which require a special tool, shown in Fig. 8-66, to extract and install. Normally, the machinist services in-block bushings, removing the old bushings and installing replacements that, depending upon engine type, might require finishing. The OEM practice of leaving bushing bores unfinished and reaming installed bushings to size complicates the rebuilding process. It is also possible for bushings to spin in their bosses, a condition that can be corrected by sleeving.

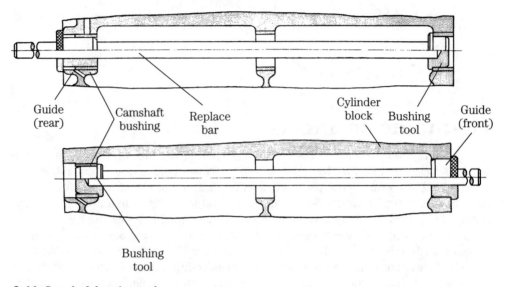

8-66 Camshaft bearing tool. Marine Engine Div., Chrysler Corp.

Camshaft lobes are the most heavily loaded parts of the engine, with unit pressures in excess of 50,000 psi at idle. Valve lifters, whether hydraulic or mechanical, are normally offset relatively to the lobes, so that contact occurs over about half of the cam surface. This offset, together with a barely visible convexity ground on the lifter foot tends to turn the lifter as its reciprocates. As the parts wear, the contact area increases and eventually spreads over the whole width of the lobe. Thus, if you find a camshaft with one or more "widetrack" lobes, scrap the cam, together with the complete lifter set. Although the condition is rare in diesel engines, overspeeding and consequent valve float might batter the flat toes of the lobes into points. Figure 8-67, which illustrates lobe wear measurements, inadvertently illustrates this condition. The same two-point measurements should be made on the journals.

Normally, hydraulic valve lifters are replaced during a rebuild; attempting to clean used lifters is a monumental waste of time. This means that the camshaft must also be replaced, because new lifters will not live with a used camshaft (and vice versa). Assemble with the special cam lubricant provided, smearing the grease over each lobe and lifter foot. Oil can be used on the journals.

Replacement hydraulic lifters normally are charged with oil as received, and ample time must be allowed for bleed when screwing down the rocker arms on ohv engines. Some valves will be open; if the rockers are tightened too quickly, pushrods will bend. As mentioned earlier, the low bleed-down rates of Oldsmobile 350 lifters make rocker assembly an exercise in patience; most other rockers can be assembled in a few minutes of careful wrenching.

Break in the camshaft exactly per maker's instructions. Otherwise, it will score.

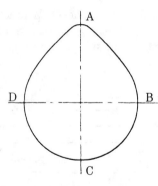

8-67
Cam lobe measurement points.
Navista

Harmonic balancers

The harmonic balancer, or vibration dampener, mounts on the front of the crankshaft where it muffles torsional vibration. Power comes to the crankshaft as a series of impulses that cause the shaft to twist, first in the direction of rotation and then against normal rotation. The shaft winds and unwinds from a node point near the flywheel. This movement can, unless dampened, quickly break the shaft.

Most harmonic balancers consist of an outer, or driven, ring bonded by means of rubber pads to the hub, which keys to the crankshaft. The rubber medium dampens crankshaft accelerations and deceleration, transferring motion to the outer ring at average crankshaft velocity. Some balancers drive through silicon-based fluid that exerts the same braking effect.

It is difficult to test a harmonic balancer in a meaningful way. Rubberized balancers can be stressed in a press and the condition of the rubber observed. Cracks or separation of the bonded joint means that the unit should be replaced. Fluid-filled balancers are checked for external damage and fluid leaks. Some machinists equate noise when the balancer is rolled on edge with failure.

Detroit Diesel's advice is best: replace the balancer whenever the engine is rebuilt.

Crankshaft bearings

All modern engines are fitted with two-piece precision insert bearings at the crankpin, as illustrated in Figs. 8-50 and 8-51. Most employ similar two-piece inserts, or shells, at the main journals, although full-circle bearings are appropriate for barrel-type crankcases. No bearing can be allowed to spin in its carrier.[1] Full-circle bearings pressed into their carriers or locked by pins or cap screws. Two-piece shells secure with a tab and gain additional resistance to spinning from residual tension. Bearing shells are slightly oversized, so that dimension A in Fig. 8-68 is greater than

---

[1] Henry Ford made an exception to the "no-spin" rule; his early V-8s used floating crankpin bearings, babbitted on both sides. This technique cut bearing speed in half, but introduced a complication in the form of big-end connecting rod wear.

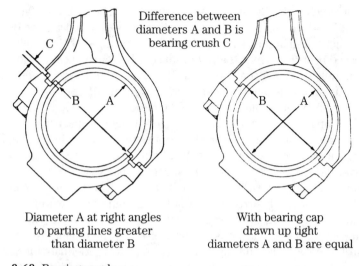

Difference between diameters A and B is bearing crush C

Diameter A at right angles to parting lines greater than diameter B

With bearing cap drawn up tight diameters A and B are equal

8-68 Bearing crush. Navistar

dimension B. The difference is known as the *crush height*. Torquing the cap equalizes the diameters, forcing the inserts hard against their bosses.

Thrust bearings, usually in the form of flanges of main inserts, limit crankshaft fore-and-aft movement (Fig. 8-69). So long as the crankpin is not tapered, connecting-rod thrust loads are insignificant and transfer through rubbing contact between the rod big end and the crankpin flanges.

Insert-type bearings are made up in layer-cake fashion of a steel backing and as many as five tiers of lining material. Although their numbers are dwindling, perhaps half of the high-speed engines sold in this country continue to use copper-lead bearings, the

8-69
Thrust bearing configuration for two Yanmar engines. Other makes use two-piece thrust washers.

best known of which is the Clevite (now Michigan)-77. The facing surface of this bearing consists of 75% copper, 24% lead, and a 1% tin overplate. Daimler-Benz, Perkins, and several other manufacturers have followed the example of Caterpillar and specify aluminum-based bearings, which tolerate acidic oils better than copper-lead types. One material commonly used is Alcoa 750, an alloy of almost 90%-pure aluminum, 6.5% tin, alloyed with zinc and trace amounts of silicon, nickel, and copper.

Normal bearing wear should not exceed 0.0005 in. per 1000 hours of continuous operation. The operative word is "continuous"—frequent startups, cold loads, and chronic overloads can subtract hundreds of hours from the expected life. In practice, main bearings outlast crankpin bearings by about three to one, although No. 1 main, which receives side loads from accessories, can fail early.

According to Michigan Bearing, 43% of failings can be attributed to dirt, 15% to oil starvation, 13% to assembly error, and 10% to misalignment. Overloading, corrosion, and miscellaneous causes account for the remainder. In other words, the mechanic contributes to most failures, by either assembling dirt into the engine, reversing nonsymmetrical bearing shells or bearing caps, or failing to provide adequate lubrication during initial start-up. These matters are discussed in the following section.

Bearing alignment problems come about from improperly seated rod caps or from warped main-bearing saddles. The condition can sometimes be spotted during disassembly as accelerated wear on outboard or center mainbearing shells. The mechanic should verify alignment by spanning the saddles with a precision straightedge (Fig. 8-70). Loss of contact translates as block warp.

Bearing alignment can be restored, but at the cost of raising the crankshaft centerline a few thousandths of an inch. The effect of chain-driven camshafts is to retard valve timing by an almost imperceptible amount; but gear trains are not so forgiving and the machinist might be forced to resort to some fairly exotic (and expensive) techniques to maintain proper tooth contact. The "cleanest" solution is to resize the bearing bosses by a combination of metalizing and honing.

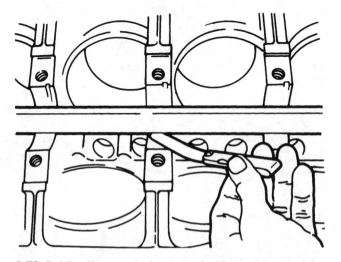

8-70 Saddle alignment is determined with precision straightedge and feeler gauge. Ford Motor Co.

Assembly—major components

Building up the major components—crankshafts, pistons, and rods—is a special kind of activity, that can be neither forced nor hurried.

Although remote from our subject, a certain insight into it can be gained from the experience of U.S. Naval gunners almost a century ago. As the century closed, five warships fired at a hulk from a mile distance for 25 minutes and registered two hits. This was an average performance, achieved by taking approximate aim and waiting for the roll of the ship to bring the sights into alignment with the target. Meanwhile an Englishman by the name of Timms had developed another system of aiming, which involved constant correction, in an attempt to keep the sights on the target however the ship moved. His explanations that the gunner would naturally adjust to ship's motion went unheeded by Navy brass until he got the ear of Teddy Roosevelt, who put him in charge of gunnery training. In 1910, the test described above was repeated, this time using the new method and only one ship. The number of hits doubled.

Timms scored by concentrating directly on the problem and not on some ritualized technique that was supposed to yield an automatic solution. There is a great deal of room in the mechanic's trade for this kind of pragmatic thinking. For example, the whole business of measurement could be reformed, beginning with the establishment of a datum line (such as the crankshaft centerline) from which deck flatness, cylinder bore centerlines, and bearing clearances would be generated. And much thought could be given to substituting a direct measurement of clamping force for the indirect system currently used. When a capscrew is tightened, about 90% of the torque load appears as friction between the underside of the fastener head and the work piece. A 5% miscalculation in friction values represents a 50% clamping force error.

Glaze breaking

Cast-iron bores develop a hard glaze in service, which must be "broken" or roughened, with a hone to facilitate ring sealing. Consequently, worn cylinder bores must be honed whenever piston rings are replaced. Remachined cylinders must also be honed, but this is usually the province of the machine shop.

Most mechanics use a brush hone, sized to cylinder diameter, such as the 120-grit BRM Flex-Hone pictured in Fig. 8-71. Obtaining the desired crosshatch pattern (shown earlier in Fig. 8-37) depends on a four-sided relationship among the ring material, spindle speed, bore diameter, and stroke frequency. For what it is worth, chrome rings running in a 4-in. cylinder need the finish produced by a 280-grit stone, rotated 190 rpm and reciprocated 70 times a minute. Such precision is unobtainable, but a little practice will produce an acceptable surface.

It is important to use plenty of lubricant, such as mineral oil or PE-12, available in aerosol cans from specialty tool houses. Keep the hone moving—it cuts as it rotates—and do not pause at the ends of the strokes. Hone for three or four seconds and stop to inspect the bore. Crosshatched grooves should intersect at angles of 30 to 45 degrees relative to the bore centerline. If the pattern appears flat, reduce the spindle speed or increase the stroke rate. The job should require no more than 15 or 20 seconds to complete. Withdraw the hone while it is still turning.

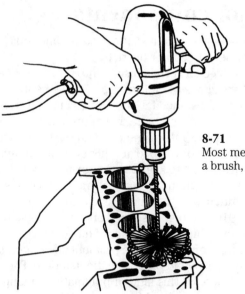

8-71
Most mechanics break glaze with a brush, or ball, hone.

Few mechanics take the time to clean a bore properly. Solvents merely float abrasive hone fragments deeper into the metal. Detergent, hot water, and a scrub brush are the prescription. Scrub until the suds are white, and wipe dry with paper shop towels. Continue to scrub and wipe until the towels no longer discolor. According to TRW, half of early ring failures can be traced to abrasive particles left in the bore.

Laying the crank

This discussion applies to engines with two-piece main-bearing shells; full-circle bearings require a slightly different technique, as outlined in the appropriate shop manual.

1. Although it is late in the day, make one final inspection of the block. Critical areas include saddle alignment for two-piece GM block castings, cracks that are likely to be associated with saddle webbing and liner flanges, and main bearing cap-to-saddle fits. Most caps are buttressed laterally by interference fits with their saddles. Fretting on contact surfaces can, if not corrected, result in crankshaft failure. Some mechanics make it a practice to chase main-bearing bolt threads; if you opt to do this, use the best, most precise tap that money will buy.
2. Verify that main-bearing saddles are clean and dry.
3. Unwrap one set of main-bearing inserts. Bearing size will be stamped on the back of each insert. All machinists can tell stories about mislabeled bearings, but certainly it would give one pause if the crankshaft were turned 0.010 in. and the inserts were marked "Std."
4. Identify the upper insert, which always is ported for crankshaft lubrication. A few lower shells carry a superfluous port; most are blanked off and grooved.

5. Roll the upper insert into place on the web, leading with the locating tab, as shown in Fig. 8-72. Make absolutely certain that the tab indexes with the web recess. When installed correctly, the ends of the insert stand equally proud above the parting face.

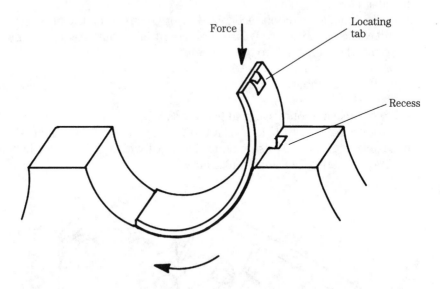

8-72 Main bearing inserts roll into place. Slip the bearing into the cap or saddle, leading with the smooth edge. Then, with thumbs providing the force, push the insert home, seating the locating tab in its recess.

6. Verify that the oil passage drilled in the web indexes with the oil port in the insert. A glance will suffice, but some shops go a step further and run a drill bit through the insert and into the web.
7. Install the remaining upper shells, one of which might be flanged for thrust. Check each insert for indicated size and, once they are installed, verify that oiling passages are open.
8. Where appropriate, install the upper half of the rear crankshaft seal, per manufacturer's instructions. Rope seals must be rolled into place, using a bar sized to match crankshaft journal diameter.
9. Saturate the bearings and seal with clean lube oil. (Some mechanics prefer to use a grease formulated for bearings assembly, such as Lubriplate No. 105. Grease is very appropriate when the engine will not be returned to immediate service.)
10. Wipe down the crankshaft with paper shop towels. Coat journals and crankpins with a generous amount of lubricant.
11. Lower the shaft gently and squarely into position on the webs. Exercise care not to damage thrust flanges.
12. Install lower bearing shells into the caps, lubricating as before.
13. Mount the caps in the correct orientation and sequence. Lightly oil cap bolt threads.

14. Torque the caps, working from the center cap outward. Conventional bolts make up in three steps—1/3, 2/3, and 3/3 torque; torque-to-yield bolts are run down to a prescribed torque limit and rotated past that limit by a set amount. After each cap is pulled down, turn the crankshaft to detect possible binds.

   - Radius ride, a condition recognized by bright edges on the bearing shells and caused by excessively large crankshaft fillets left after regrinding. Return the crankshaft to the machinist.
   - Dirt on the bearing face or OD.
   - Insufficient running clearance, a condition that can be determined with Plasti-Gage (see below).
   - Bent crankshaft or misaligned bearing saddles.

15. Using a pry bar, lever the crankshaft toward the front of the engine, then pull back (Fig. 8-73). Measure crankshaft float with a feeler gauge between the thrust bearing face and crankshaft web.

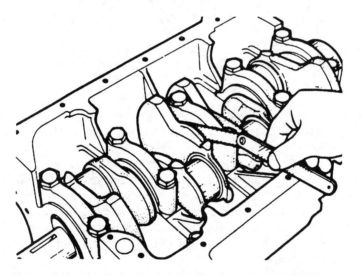

8-73 Lever the crankshaft through its full range of axial movement before measuring thrust bearing clearance. The same measurement can be made with dial indicator, registering off of the crankshaft nose. Chrysler Corp.

Piston and rod installation

It is assumed that piston and rod assemblies have already been made up. If the machinist has not already done so, install the rings on the piston. Rings are packaged with detailed instructions, which supersede those in the factory manual. Here, it is enough to remind you that:

   - The time required to check ring gap is well spent, because rings, like bearing inserts, are sometimes mislabeled. Using the flat of the piston as a pilot, insert each compression ring into the cylinder and compare ring gap with the specification (Fig. 8-74).

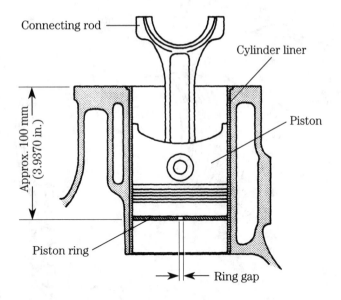

Connecting rod

Cylinder liner

Piston

Approx. 100 mm (3.9370 in.)

Piston ring

Ring gap

8-74 Measure ring gap relative to the lower, and least worn part of the cylinder. Yanmar Diesel Engine Co., Ltd.

- Ring orientation is important; according to one manufacturer, a reversed scraper ring increases oil consumption 500%.
- Using the proper tool, expand the ring just enough to slip over the piston (Fig. 8-75).
- Stagger ring gaps, as detailed by the manufacturer.

At this point, you are ready to install the piston assemblies. Follow this procedure:

1. Remove the bearing cap from No. 1 rod and piston assembly.
2. Mount an upper bearing shell on the rod, indexing oil ports.
3. Coat the entire bearing surface with fresh lube oil. Repeat the process for the cap.
4. Slip lengths of fuel line over the rod bolt ends to protect the crankshaft journals during installation.
5. Turn the crankshaft to bottom dead center on No. 1 crankpin. Saturate the crankpin with oil.
6. Install a ring compressor over the piston. The bottom edge of the compressor should be a half-inch or so below the oil ring. Tighten the compressor bands just enough to overcome ring residual tension.
7. With the block upright and the leading side of the piston toward the front of the engine, place the compressor and captive piston over No. 1 cylinder bore. A helper should be stationed below to guide the rod end over the crankpin. While holding the compressor firmly against the fire deck, press the piston out of the tool and into the bore. As shown in Fig. 8-76, thumb pressure should be sufficient. Stop if the piston binds and reposition the piston in the compressor.

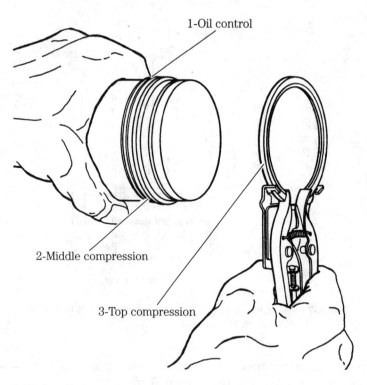

1-Oil control

2-Middle compression

3-Top compression

8-75 Install piston rings in the sequence shown. Kohler of Kohler

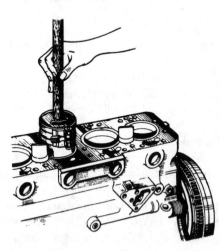

8-76
Piston installation should be a
gentle process, involving no more
than the force exerted by one
hand. Peugeot

8. Install the lower bearing shell in the cap; verify that the locating tab indexes with its groove.
9. Coat the bearing contact surface with lube oil.
10. Working from below, carefully remove the hoses from the bolt ends and pull the rod down over the crankpin. Make up the rod cap, making sure match marks align.

11. Torque the rod bolts to spec (Fig. 8-77).
12. Using a hammer handle or brass knocker, gently tap the sides of the big end journal. The rod should move along the length of the crankpin.
13. Rotate the crankshaft a few revolutions to detect possible binds and the bore scratches that mean a broken ring. Resistance to turning will not be uniform: piston speed increases at mid-stroke, with a corresponding increase in crankshaft drag.
14. Repeat this operation for remaining pistons.
15. Press or bolt on piston cooling jets, aiming them as the factory manual indicates.

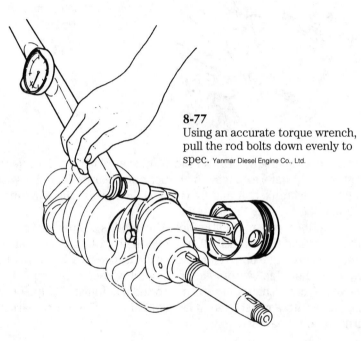

8-77
Using an accurate torque wrench, pull the rod bolts down evenly to spec. Yanmar Diesel Engine Co., Ltd.

Plasti-Gage

Old-time mechanics checked timing gears by rolling a piece of solder between the gear teeth. The width of the solder represented gear lash. (This technique has generally been replaced by direct measurement with a dial indicator.) Plastic gauge wire represents an application of the same principle to journal bearings. Perfect Circle and several other manufacturers supply color-coded gauge wire together with the necessary scales that translate wire width into bearing clearance. Green wire responds to the normal clearance range of 0.0001 to 0.0003 in.; red goes somewhat higher—0.0002 to 0.0006; blue extends to 0.0009 in. Accuracy compares to that obtained with a micrometer.

1. Remove a bearing cap and wipe the lubricant off both the cap and the exposed portion of the journal.
2. Lay a strip of gauge wire longitudinally on the bearing, about ¼ inch off-center, as shown in Fig. 8-78.

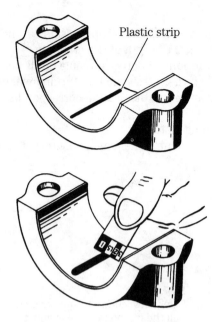

Plastic strip

8-78
Plastic gauge wire is an inexpensive and quite accurate method of determining bearing clearance. Detroit Diesel

Note: Invert the engine or raise the crankshaft with a jack under an adjacent counterweight before gauging main-bearing clearances. Otherwise, the weight of the crankshaft and flywheel will compress the wire and give false readings.

3. Assemble the cap, torquing the hold-down bolts to specifications.
4. Without turning the crankshaft, remove the cap. Using the scale printed on the gauge-wire envelope, read bearing clearance as a function of wire width. Clearance should fall within factory assembly specifications. If it does not, remove the crankshaft and return it to the machinist. Scrape or wipe off the gauge wire with a rag soaked in lacquer thinner. Reoil the bearing and journal before final assembly.

Balancing

Precision balancing is an optional service, difficult to justify in economic terms, but worth having done if the goal is to build the best possible engine. Balanced engines run smoother, should last longer, and sometimes exhibit a small gain in fuel economy.

But do not expect wonders; balancing is a palliative, arrived at by compromise, and effective over a fairly narrow rpm band. No amount of tweaking can take all of the shake, rattle, and roll out of recip engines.

There are three steps in the balancing process. First the technician equalizes piston and pin weights to within one-half a gram (Fig. 8-79). If the technician works out of a large inventory, this can sometimes be achieved by selective assembly. Otherwise, machine work is in order; heavy pistons are lightened by turning the inner edge of the skirts, and when absolutely necessary, aluminum slugs can be pressed into the hollow piston pins to add weight to light assemblies.

Stewart-Warner

8-79 Balancing begins with match-weighing piston assemblies.

The next step is to match-weight the rods using a precision scale and a special rod adapter, such as shown in Fig. 8-80. The Stewart-Warner adapter makes it possible to distinguish between rotating (or lower-end) and reciprocating (upper-end) masses; when such equipment is not available, the technicians assume that the lower half of the rod rotates and that the upper end describes a purely reciprocating path. However arrived at, rod reciprocating and rotating masses are equalized to a one-half gram tolerance by removing metal from the balance pads (shown as the extensions on the ends of the DDA rod in the photograph and more clearly back in Fig. 8-59, where they are labeled "noncritical" areas).

The technician computes the total rotating mass and bolts an equivalent mass to the crankshaft (Fig. 8-81). The shaft is then mounted on the balancer and rotated at the desired speed, which should correspond to actual operating conditions. Balance, or matching counterweight mass to rotating mass, can normally be achieved

8-80 Stewart-Warner fixture splits rod weight into the rotating mass of piston and rod assemblies.

Stewart-Warner

8-81 The crankshaft is dynamically balanced relative to the rotating mass of piston and rod assemblies.

by drilling the counterweights. Most shops work to a 0.5 oz.-in. tolerance; the equipment will support 0.2 oz.-in. accuracy. Once this is done, the process is repeated for the harmonic balancer and clutch assembly.

Oil seals

Oil seals can be a headache, especially if one leaks just after an overhaul. Premature failure is almost always the mechanic's fault and can be traced to improper installation.

Rope or strip seals are still used at the aft end of the crankshaft on some engines. These seals depend on the resilience of the material for wiping action and so must be installed with the proper amount of compression. Remove the old seal from the grooves and install a new one with thumb pressure. (You might want to coat the groove—but not the seal face—with stickum.) The seal should not be twisted or locally bound in the groove.

Now comes the critical part. Obtain a mandrel of bearing boss diameter (journal plus twice the thickness of the inserts) and, with a soft-faced mallet, drive the seal home. Using this tool as a ram, you can cut the ends of the seal flush with the bearing parting line. Without one of these tools you will have to leave some of the seal protruding above the parting line in the elusive hope that the seal will compact as the caps are tightened. Undoubtedly some compaction does take place, but the seal ends turn down and become trapped between the bearing and cap, increasing the running clearance.

Lubricate the seal with a grease containing molybdenum disulfide to assist in break-in.

Synthetic seals, usually made of Neoprene, are used on all full-circle applications and on the power takeoff end of many crankshafts as well. These seals work in a manner analogous to piston rings. They are preloaded to bear against the shaft and designed so that oil pressure on the wet side increases the force of contact.

These seals must be installed with the proper tools. If the seal must pass over a keyway, obtain a seal protector (a thin tube that slides over the shaft) or at least cover the keyway with masking tape. Drive the seal into place with a bar of the correct diameter. The numbered side is the driven side in most applications. The steep side of the lip profile is the wet side. It is good practice to use a nonhardening sealant on the back of metallic seals. Plastic coated seal cams are intended to conform to irregularities in the bore and do not need sealant.

More elaborate seals require special one-of-a-kind factory tools. The better engines often incorporate wear sleeves over the shafts, either as original equipment or as a field option. Figure 8-82 shows the use of a wear sleeve on a Detroit Diesel crankshaft. The sleeve makes an interference fit over the shaft and is further secured with a coating of shellac. Worn sleeves are cut off with a chisel or peened to stretch the metal. The latter method is preferred because there is less chance of damaging the shaft.

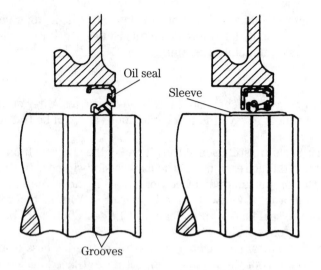

8-82 Wear sleeves extend crankshaft life. When retro-fitted, an oversized seal must be used to compensate for the increase in journal diameter. Detroit Diesel

9
CHAPTER

Air systems

Diesel engines run with open manifolds, restricted only by the pressure drop across the air filter. Consequently, large amounts of air are pumped, typically on the order of 200 cu. ft. per pound of fuel for four cycles and considerably more for two cycles. Most of the air merely cycles through the engine, without taking part in combustion. But it must, for onshore applications, at least, be filtered to remove sand and other abrasives.

Research has established that the most lethal particles measure between 10 and 20 microns in diameter. Larger particles (or particulates) tend to pass out the exhaust without doing much damage. Smaller particulates are less aggressive, although they are by no means harmless.

Air cleaners

The ideal air cleaner would stop all particulates, regardless of size. Unfortunately, such a device does not exist, at least in a practical form. Some of the smaller particulates get through, which is why engines tend to wear rapidly in dusty environments, no matter how carefully maintained.

Construction

Air filters are classified by the type of media, which can be wire mesh, fabric, foam, or fiber. The filter can be dry, like the pleated-paper filter shown in Fig. 9-1, or oil-wetted to improve particulate adhesion. Regardless of manufacturer's claims, I have not seen evidence that one filter media is superior to others. Any well designed, OEM-approved filter should work as long as it is properly maintained.

The oil bath air cleaner is a special case, which combines oil-wetted filtration with inertial separation. The assembly shown in Fig. 9-2 employs two stages of separation. Air enters the top of the unit through the precleaner, or cyclone. Internal vanes cause the air stream to rotate, which tends to separate out the larger and heavier particulates. The air then proceeds through the central tube to the bottom of the canister, where it reverses direction. Some fraction of the remaining particulates fail to make this U-turn and end up in the oil.

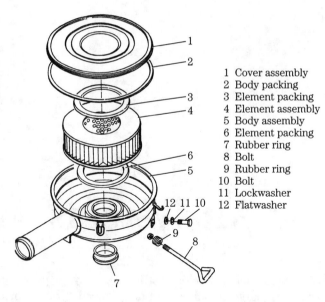

1 Cover assembly
2 Body packing
3 Element packing
4 Element assembly
5 Body assembly
6 Element packing
7 Rubber ring
8 Bolt
9 Rubber ring
10 Bolt
11 Lockwasher
12 Flatwasher

9-1 Automotive-type air cleaner, with disposable paper element. Marine Engine Div., Chrysler Corp.

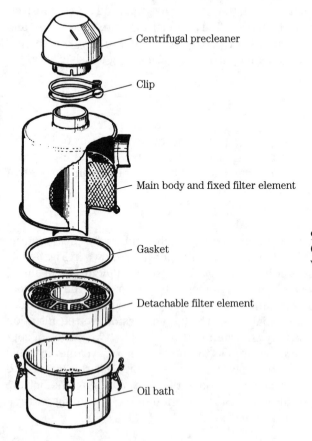

Centrifugal precleaner

Clip

Main body and fixed filter element

Gasket

Detachable filter element

Oil bath

9-2
Oil-bath air cleaner in exploded view.

Oil-bath air cleaners have lost favor because of their relatively high cost, bulk, and fairly intensive maintenance requirements. But, given the necessary maintenance, they appear to work about as well as any.

Maintenance

The operator's manual should provide the necessary maintenance instructions. Polyurethane foam filters require frequent reoiling in normal operation and will effectively go dry in storage as the oil migrates to low spots. Unless oiled, these filters are ineffective. Pleated-paper filters work better when slightly dirty and cannot, in any event, be cleaned. Solvents cause the fibers to swell. Avoid overfilling oil bath air cleaner assemblies—the engine might run away on oil drawn from the reservoir.

Some air cleaners include a pressure-sensing tell-tale that alerts the operator when the filter needs service. If yours does not have this feature, it might be worthwhile to install a manometer on the intake manifold, downstream of the filter. Note the pressure drop across a new filter (usually on the order of 2 in./H_2O at normal engine speeds) and clean or replace the element when the pressure drop exceeds the manufacturer's recommendation.

Turbochargers

A turbocharger is an exhaust-powered supercharger, that unlike conventional superchargers, has no mechanical connection to the engine (Figs. 9-3 and 9-4). The exhaust stream, impinging against the turbine (or "hot") wheel, provides the energy to turn the compressor wheel. For reasons that have to do with the strength of materials, turbo boost is usually limited to 10 or 12 psi. This is enough to increase engine output by 30% to 40%.

Turbocharging represents the easiest, least expensive way to enhance performance. It is also something of a "green" technology, because the energy for compression would otherwise be wasted as exhaust heat and noise (Fig. 9-5). On the other hand, the interface between sophisticated turbo machinery, turning at speeds as great as 140,000 rpm and at temperatures in excess of 1000°F, and the internal combustion engine is not seamless.

Unless steps are taken to counteract the tendency, turbochargers develop maximum boost at high engine speeds and loads. The turbine wheel draws energy from exhaust gas velocity and heat, qualities that increase with piston speed and load. The compressor section behaves like other centrifugal pumps, in that pumping efficiency is a function of impeller speed. At low speeds, the clearance between the rim of the impeller and the housing shunts a large fraction of the output. At very high rotational speeds, air takes on the characteristics of a viscous liquid and pumping efficiency approaches 100%. In its primitive form, a turbocharger acts like the apprentice helper, who loafs most of the day and, when things get busy, becomes too enthusiastic.

Another innate, but not necessarily uncorrectable, characteristic of turbocharged engines is the lag, or flat spot, felt during snap acceleration. Perceptible time is required to overcome the inertia of the rotating mass. By the same token, the wheels continue to coast for a few seconds after the engine stops.

9-3 John Deere 6068T turbocharged engine. One of the appeals of turbocharging is its apparent simplicity. In this instance, a 25-lb turbocharger boosted output to 175 hp for a gain of 45 hp over the naturally aspirated version of the same engine. Torque went from 335 lb-ft to 473 lb-ft.

Background

The exhaust-driven supercharger was first demonstrated in 1915, but remained impractical until the late 1930s, when the U.S. Army, working closely with General Electric, developed a series of liquid-cooled and turbocharged aircraft engines. The expertise in metallurgy and high temperature bearings gained in this project made GE a leader in turbocharging and contributed to its success with jet engines.

As applied to aircraft, turbocharging was a kind of artificial lung that normalized manifold pressure at high altitudes. A wastegate deflected exhaust gases away from the turbine at low altitudes, where the dense air would send manifold pressures and engine power outputs to dangerously high levels. In emergencies, P-51 pilots were instructed to push their throttles full forward, an action that broke a restraining wire, closed the wastegate, richened the mixture, and initiated alcohol injection for a few seconds of turboboost. The broken wire was a telltale sign, signaling the ground crew to rebuild the engine.

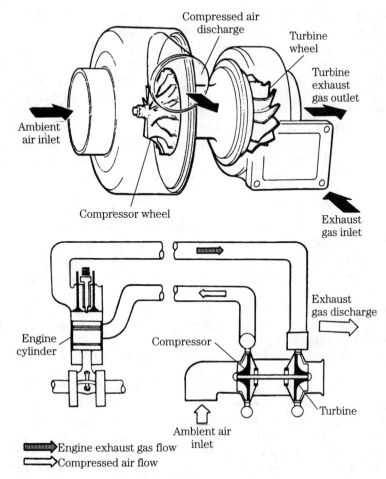

Compressed air discharge

Turbine wheel

Turbine exhaust gas outlet

Ambient air inlet

Compressor wheel

Exhaust gas inlet

Engine cylinder

Compressor

Exhaust gas discharge

Turbine

Ambient air inlet

Engine exhaust gas flow

Compressed air flow

9-4 A turbocharger is an exhaust-driven centrifugal pump, operating at six-figure speeds and typically generating 10 to 12 psi of supercharge.

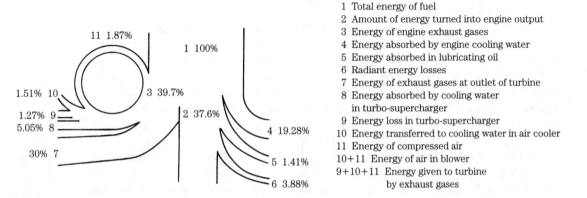

11 1.87%

1 100%

1.51% 10

3 39.7%

1.27% 9

2 37.6%

5.05% 8

4 19.28%

30% 7

5 1.41%

6 3.88%

1 Total energy of fuel
2 Amount of energy turned into engine output
3 Energy of engine exhaust gases
4 Energy absorbed by engine cooling water
5 Energy absorbed in lubricating oil
6 Radiant energy losses
7 Energy of exhaust gases at outlet of turbine
8 Energy absorbed by cooling water
   in turbo-supercharger
9 Energy loss in turbo-supercharger
10 Energy transferred to cooling water in air cooler
11 Energy of compressed air
10+11 Energy of air in blower
9+10+11 Energy given to turbine
      by exhaust gases

9-5 Turbocharger heat balance chart. Note that even with turbocharging 30% of the fuel energy goes out the exhaust.

Turbochargers continue to be used on reciprocating aircraft engines and enjoy some currency in high performance automobiles, thanks to electronic controls that minimize detonation. But diesel engines are the real success story.

Turbocharging addresses the fundamental shortcoming of compression ignition, which is the delay between the onset of injection and ignition. That delay, terminated by the sudden explosion of puddled fuel, results in rapid cylinder pressure rise, rough running and incomplete combustion with attendant emissions. Turbocharging (and supercharging generally) increase the density of the air charge. Ignition temperature develops early during the compression stroke and coincides more closely with the onset of injection. Fuel burns almost as quickly as delivered for a smoother running, cleaner, and more economical engine.

Applications

Depending on how it is accomplished, turbocharging can have three quite distinct effects on performance. If no or little additional fuel is supplied, power output remains static, but emissions go down. Surplus air lowers combustion temperatures and provides internal cooling. Such engines should be more durable than their naturally aspirated equivalents.

The high-boost/low-fuel approach to turbocharging is limited to large stationary and marine plants. Makers of small, high-speed engines are more concerned with maximum power or mid-range torque.

Supplying additional fuel in proportion to boost yields power, which can translate as fuel savings for engines that run under constant load. Some gains in fuel efficiency (calculated on a hp/hour basis) accrue from turbocharging, but the real advantage comes about when high supercharge pressures allow for lower piston speeds. Large marine engines, developing a 1000 hp and more per cylinder, use this approach to achieve thermal efficiencies of 50%. On a more familiar scale, the naturally aspirated International DT-414 truck engine produced 157 hp at 3000 rpm. The addition of a large turbocharger boosted output to 220 hp, for a gain of 40%. Once in the fairly narrow power band, the trucker could save fuel by selecting a higher gear.

The third approach is more characteristic of automotive and light truck engines, which operate under varying loads and rarely, if ever, develop full rated power. What is wanted is torque.

The Navistar 7.3L, developed for Ford pickups and Econoline vans, is perhaps the best demonstration of the way a turbocharger can be throttled for torque production. In its naturally aspirated form, the engine develops 185 hp at 3000 rpm and 360 ft./lb. of torque at 1400 rpm. A Garrett TC43 turbocharger boosts power output marginally to 190 hp; but torque goes up 17.8% to 385 ft./lb. In normal operation the wastegate remains closed, directing exhaust gases to the turbine, until 1400 rpm. At that point, which corresponds to the torque peak, the hydraulically actuated wastegate begins to open, shunting exhaust away from the turbine.

Serious applications of turbocharging, whether for peak power or mid-range torque, impose severe mechanical and thermal loads, which should be anticipated in the design stage. Insofar as they work as advertised, add-on turbocharger kits—which rarely involve more than a rearrangement of the plumbing—are a buyer-beware proposition. The 7.9L is considered a very rugged engine in its naturally aspirated form. But turbocharging called for hundreds of engineering changes, including shot-peened connecting rods, oversized piston pins, Inconel exhaust valves, and special Zollner pistons, with anodized crowns.

Construction

Figure 9-6 illustrates an air-cooled Ishikawajima-Harima turbocharger of the type found on engines in the 100–150-hp range. The inset shows the water-cooled version of the same turbocharger, plumbed into the engine cooling system. All modern small-engine turbochargers follow these general patterns.

Floating bushings, located in the central bearing chamber, support the shaft. These bushings, like the connecting rod bearings specified for the original Ford V-8, float on the ID and OD. Thus, bushing speed is half that of shaft speed, which is to

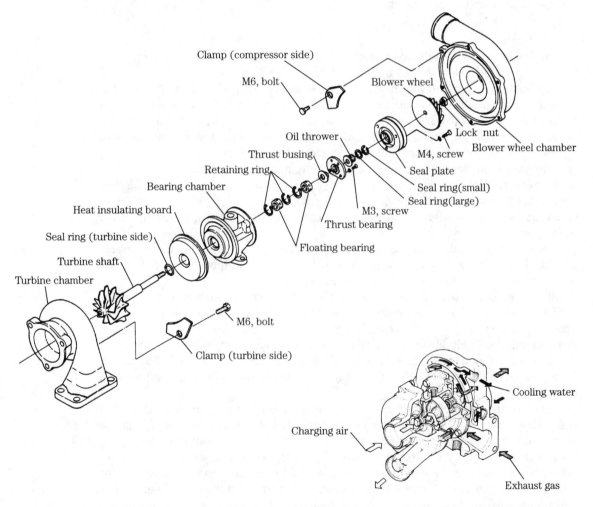

9-6 Air- or water-cooled turbocharger of the type used on small Japanese diesels. Water cooling may prolong bearing life and should reduce the coking experienced during hot shutdowns, when the turbocharger continues to rotate after the engine stops.

say that the bushings can reach speeds of 60,000 or 70,000 rpm. A floating thrust bushing contains axial motion.

Also note the way the impeller wheels cantilever from the bushings, so that the masses of the rotating assembly are concentrated near the ends of the shaft. This "dumb-bell" configuration requires precise wheel balance and extremely accurate shaft alignment.

The bearing section is lubricated and cooled by engine oil, generally routed through external pipes or hoses. Shaft seals keep oil from entering the turbine and compressor sections.

The turbocharger is usually the last component to receive oil pressure and might continue to rotate after the engine stops. To ensure an adequate oil supply to the bearings, operators should idle the engine for at least 30 seconds upon starting and for the same period before shutdown.

Wastegate

All turbocharger installations incorporate some form of boost limitation; otherwise, boost would rise with load until the engine destroyed itself.

Turbocharger geometry, sometimes abetted by inlet restrictions and designed-in exhaust backpressure, limit boost on constant-speed engines. Automotive and light truck engines have surplus turbocharging capability for boost at part throttle. These applications employ a wastegate—a kind of flap valve—that automatically opens at a preset level of boost to shunt exhaust gas around the turbine.

Most wastegates are controlled by a diaphragm, open to the atmosphere on one side and to manifold pressure on the other (Fig. 9-7). The Ford unit shown also incorporates a relief valve. Normally the wastegate opens at 10.7 psi; should it fail to do so, the relief valve opens at 14 psi and, because it is quite noisy, alerts the driver to the overboost condition.

Other wastegates are spring-loaded and usually include an adjustment, appropriately known as the "horsepower screw." As usually configured, tightening the screw increases available boost. Therefore, exercise restraint.

Test wastegate operation by loading the engine while monitoring rpm and manifold pressure. If the installation does not include a boost gauge, connect a 0–20 psi pressure gauge at any point downstream of the compressor. The diaphragm-sensing line (on units so-equipped) serves as a convenient gauge point. Note, however, that the gauge must be connected with a tee fitting to keep the wastegate functional. High gear acceleration from 2500 rpm or so should generate sufficient load to open the gate.

The control philosophy discussed in the preceding paragraphs implies that the turbocharger is used for power enhancement. When mid-range torque is the object, the wastegate opens early, at the engine speed corresponding to peak torque output. Most of these applications employ computer-controlled wastegates, whose response is conditioned by manifold pressure, engine rpm, coolant temperature, and other variables. No generalized test procedure has been developed for these devices.

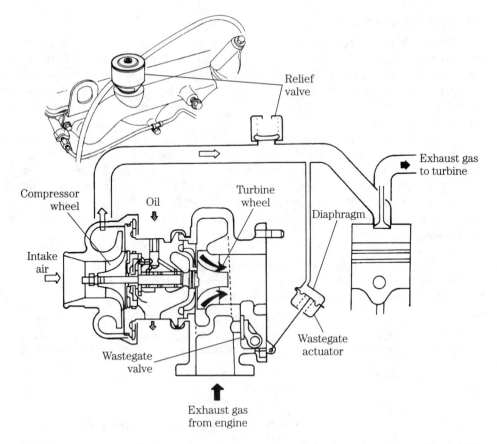

9-7 Diaphragm-controlled wastegate and pressure-relief valve, as used in on Ford 2.3L diesel engines.

Aftercoolers

Compressing air raises its temperature, which reduces change density and tends to defeat the purpose of supercharging. Sophisticated turbocharger installations include a heat exchanger, or aftercooler, between the compressor outlet and intake manifold. The cooling medium can be air, engine coolant, or—for marine applications—water. The plate and tube seawater-fed aftercooler used with the Yanmar 4LH-HTE boosts output to 135 hp, or 30 hp more than an identical engine without aftercooling.

Air-cooled heat exchangers work most efficiently when mounted in front of the radiator. Engine coolant should be taken off at the pump discharge and returned to some point low on the water jacket.

Air-cooled units require little attention, other than an occasional dust off. Liquid-cooled heat exchangers should be cleaned as needed to remove scale and fouling, and periodically tested by blowing low-pressure (25-psi maximum) air through the tubes. Water intrusion into the intake tract can be expensive.

Routine maintenance

The first priority is to obtain actual performance data, particularly with reference turbocharger behavior. The full story would require a dynamometer to extract, but one can gain useful insight by observing the changes in manifold pressure under working loads that, at some point in the test, should be great enough to cause the wastegate to open. The rise in oil temperature and variations in crankcase pressure supply additional parts of the picture.

Do not operate a turbocharged engine unless the air cleaner (or spark arrestor) is in place and the intake-side ducting secure. The compressor acts as a vacuum cleaner, drawing in foreign objects which will severely damage the unit and might cause it to explode. The troubleshooting chart (Table 9-1) makes reference to coast-down speed. If you feel it necessary to observe compressor rotation, cover the turbocharger inlet with a screen to at least keep fingers and other large objects out of the mechanism.

Table 9-1. Turbocharger fault diagnosis

Engine lacks power, black smoke in exhaust

Symptom	Probable causes	Corrective actions
1. Insufficient boost pressure; compressor wheel coasts to a smooth stop when engine is shut down; turns easily by hand.	1. Clogged air filter element.	Clean or replace element.
	2. Restriction in air intake.	Remove restriction.
	3. Air leak downstream of compressor.	Repair leak.
	4. Insufficient exhaust gas energy:	
	• Exhaust leaks upstream of turbocharger.	Repair leak.
	• Exhaust restriction downstream of turbocharger.	Remove restriction.
2. Insufficient boost pressure; compressor wheel does not turn or judders during coastdown; drags or binds when turned by hand.	1. Carbon accumulations on turbine-shaft oil seals.	Disassemble, clean turbocharger and change oil and filter.
	2. Bearing failure, traceable to:	
	• Normal wear.	Rebuild turbo.
	• Insufficient oil.	Rebuild turbo. Inspect oil supply/return circuit and repair as necessary to restore full flow. Change engine oil and filter.
	• Excessive oil temp.	Rebuild turbo. Clean oil cooler (if fitted); change engine oil and filter. Monitor oil temp.
	• Turbo rotating assy. out of balance.	Rebuild and balance turbo.
	• Bad operating practices—application of full throttle upon startup and/or hot shutdown.	Rebuild turbo. Train operators. A prelube system may extend turbo life in harsh operating environments.

3. Rotating assembly makes
rubbing contact with case.
Leading causes are:

• Bearing failure (see above).	Rebuild turbo. Determine cause of failure and correct.
• Entry of foreign matter into compressor.	Rebuild turbo. Inspect air cleaner and repair as necessary.
• Excessive turbo speed.	Rebuild turbo. Turbo rpm is a function of exhaust gas temperature (engine load). Check air filter for restrictions, bar over engine to verify that crankshaft rotates freely. If necessary, adjust governor to reduce engine power output.
• Improper assembly.	Repair turbo, review service procedures.

Excessive oil consumption, blue or white exhaust smoke

Symptom	Probable cause	Corrective Actions
1. Condition may be accompanied by oil stains in the inlet and/ or outlet ducting, or, in extreme cases, by oil drips from the turbocharger housing.	1. High restriction in the air inlet.	Clean or repalce filter element.
	2. Worn turbocharger seals, possibly associated with bearing failure and/or wheel imbalance.	Repair turbo.

Abnormal turbocharger lag

Symptom	Probable cause	Corrective Actions
1. Engine power output exhibits a pronounced "flat spot" during cceleration. Fuel system appears to function normally.	1. Carbon buildup on compressor wheel and housing.	Disassemble and clean turbo.

Unusual noise or vibration

Symptom	Probable cause	Corrective Actions
1. "Chuffing" noise, often most pronounced during acceleration.	1. Surging caused by restriction at compressor discharge nozzle.	Disassemble and clean turbo.
2. Knocking or squeal.	1. Rotating elements in contact with housing because of bearing failure or impact damage.	Replace or rebuild turbo.

Table 9-1. Continued

Symptom	Probable cause	Corrective Actions
3. Speed-sensitive vibration.	1. Loose turbocharger mounts.	Tighten.
	2. Rotating elements in contact with housing.	Replace or rebuild turbo.
	3. Severe turbine shaft imbalance.	Rebuild and balance.

When dismantling a turbocharger and related hardware, make a careful tally of all fasteners, lockwashers, and small parts removed. Be absolutely certain that all are accounted for before starting the engine. Immediately shut down the engine if the turbocharger makes unusual noise or vibrates.

Turbocharging (and supercharging generally) put severe stress on lubrication, air inlet, crankcase ventilation, and exhaust systems.

Lubrication system Elevated combustion pressure contaminates the oil with blowby gases and promotes oxidation by raising crankcase oil temperature. That fraction of the oil diverted to the turbocharger can undergo 80°F temperature rise in its passage over the bushings. Change lube oil and filter(s) frequently.

The cost and disposal problems associated with filters make reusable filters attractive for fleet operators. Racor, a division of Parker Hannifin, manufactures a series of liquid filters with washable, stainless-steel elements and a Tattle-Tale light that alerts the operator when the filter needs to be cleaned.

Frequently inspect the turbocharger and its oiling circuitry for evidence of leaks that, if neglected, can draw down the crankcase.

Air inlet system Dust particles, entering through a poorly maintained filter or through leaks in the ducting rapidly erode the compressor wheel. Leaks downstream of the compressor cost engine power and waste fuel.

Crankcase ventilation system This system removes combustion residues and, in the process, subjects the crankcase to a slight vacuum. In normal operation, fresh air enters through the breather filter and crankcase vapors discharge to the atmosphere (prepollution engines) or to the intake side of the turbo compressor. The latter arrangement, known as positive crankcase ventilation (PCV) is rapidly becoming the norm.

Under severe load, blowby gases accumulate faster than they can be vented and escape through the breather filter. These flow reversals, which occur more frequently in turbocharged engines, tend to clog the filter. Restrictions at the filter allow corrosive gases to linger in the crankcase. A partially functional filter can also pressurize the crankcase under the severe blowby conditions that accompany heavy loads. Oil seals and gaskets leak as a consequence.

Under light loads, reduced flow through the breather depressurizes the crankcase. Low crankcase pressures encourage oil to migrate into the turbocharger and collect in the aftercooler on engines with positive crankcase ventilation. Because oil is a fairly good thermal insulator, the efficiency of the aftercooler suffers. Tests of a Detroit Diesel engine, conducted by Diesel Research, Inc., established that oil migration resulting from a 40% efficient breather cost 6% of engine output. The engine in question had 4000 hours on the clock.

Exhaust system Restrictions downstream of the turbine—crimped pipes, abrupt changes in direction, clogged mufflers—reduce turbo efficiency and, in extreme cases, represent enough load to induce boost. However, exhaust leaks between the engine and turbocharger are more typical. The pipe rusts out (especially if wrapped in insulation), fatigues, or cracks under the 1000°F-plus temperature. If OEM parts do not give satisfactory service, you might try calling a Flexonics applications engineer (312-837-1811). The company makes a line of exhaust tubing, fitted with metal bellows to absorb thermal expansion.

Turbocharger inspection

The seven-step inspection procedure outlined here was adapted, with modifications, from material supplied by John Deere.

1. *Turbo housing* Before disconnecting the oil lines, examine external surfaces of the housing for oil leaks, which would almost certainly mean turbo seal failure.
2. *Compressor housing inlet and wheel* Inspect the compressor wheel for erosion and impact damage. Erosion comes about because of dust intrusion; impact damage is prima facie evidence of negligence. Carefully examine the housing ID and compressor blade tips for evidence of rubbing contact, which means bearing failure.
3. *Compressor housing outlet* Check the compressor outlet for dirt, oil, and carbon accumulations. Dirt points to a filtration failure; oil suggests seal failure, although other possibilities exist, such as clogged turbo-oil return line or crankcase breather. Carbon on the compressor wheel might suggest some sort of combustion abnormality, but the phenomenon is also seen on healthy engines. I can only speculate about the cause.
4. *Turbine housing inlet* Inspect the inlet ports for oil, heavy carbon deposits, and erosion. Any of these symptoms suggest an engine malfunction.
5. *Turbine housing outlet and wheel* Examine the blades for impact damage. Look for evidence of rubbing contact between the turbine wheel and the housing, which would indicate bearing failure.
6. *Oil return port* The shaft is visible on most turbochargers from the oil return port. Excessive bluing or coking suggests lubrication failure, quite possibly caused by hot shutdowns.
7. *Bearing play measurements* Experienced mechanics determine bearing condition by feel, but use of a dial indicator gives more reliable results. Note that measurement of radial, or side-to-side, bearing clearance involves moving the shaft from one travel extreme to another, 180° away (Fig. 9-8A). Hold the shaft level during this operation, because a rocking motion would muddy the results. Axial motion is measured as travel between shaft thrust faces (Fig. 9-8B). In very general terms, subject to correction by factory data for the unit in question, we would be comfortable with 0.002 in. radial

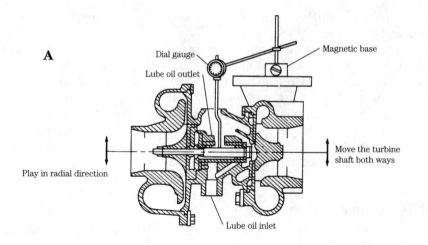

A

Dial gauge

Lube oil outlet

Magnetic base

Move the turbine
shaft both ways

Play in radial direction

Lube oil inlet

B

Magnet base

Turbine wheel chamber

Dial gauge

Move the turbine shaft
in the shaft axial direction

Rotor play in the shaft
axial direction

9-8
A dial indicator is used to
measure radial (A) and axial (B)
shaft play, specified as total
indicator movement. Take off the
radial measurement at a point
near the center of the shaft, with
shaft held level throughout its
range of travel. Yanmar Diesel Engine Co., Ltd.

and 0.003 in. axial play. Note that Schwitzer and small foreign types tend to
be set up tighter.

Overhaul

Only the most general instructions can be provided here, because construction
details, wear limits, and torque specifications vary between make and model. It
should also be remarked that American mechanics do not, as a rule, attempt turbocharger repairs. The defective unit is simply exchanged for another one.

However, there are no mysteries or secret rites associated with turbocharger
work. Armed with the necessary documentation and the one indispensable special
tool—a turbine-shaft holding fixture—any mechanic can replace the bearings and
seals, which is what the usual field overhaul amounts to.

The holding fixture secures the integral turbine wheel and shaft during removal
and installation of the compressor nut. Figure 9-9 illustrates plans for one such fixture, used on several Detroit Diesel applications. Dimensions vary, of course, with
turbocharger make and model.

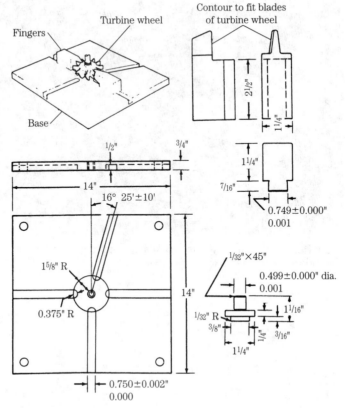

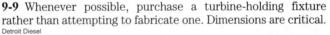

9-9 Whenever possible, purchase a turbine-holding fixture rather than attempting to fabricate one. Dimensions are critical.
Detroit Diesel

Factory-supplied documentation should alert you to any peculiarities of the instrument, such as a left-handed compressor-nut thread or the need to heat the compressor wheel prior for removal. Special precautions include the following:

- Do not use a wire wheel or any other sort of metallic tool on turbo wheels or shaft. Remove carbon deposits with a soft plastic scraper and one of the various solvents sold for this purpose. Light scratches left by wheel contact on the turbo housings can be polished out with an abrasive.
- Do not expose the turbine shaft to bending forces of any magnitude. Pull the shaft straight out of its bearings. Use a double u-joint between the socket and the wrench when removing and torquing the compressor-wheel nut.
- Do send out the rotating assembly for shaft alignment and balancing (Fig. 9-10).
- Do exercise extreme cleanliness during all phases of the operation.
- Do pre-lubricate the bearings with motor oil and pre-fill the bearing housing prior to starting the engine.
- Do make certain that all fasteners and small tools are accounted for and not lurking within the turbocharger or its plumbing.

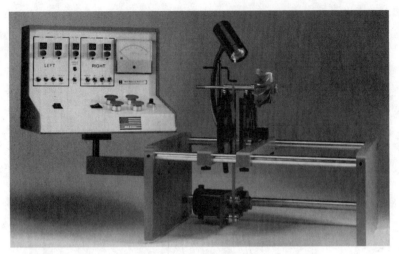

9-10 Heins Balancing Systems makes specialized tools for turbocharger rebuilders, including this precision balancer.

10
CHAPTER

Electrical fundamentals

Most small diesel engines are fitted with an electric starter, battery, and generator. The circuit might include glow plugs for cold starting and electrically operated instruments such as pyrometers and flow meters. The diesel technician should have knowledge of electricity.

This knowledge cannot be gleaned from the hardware. Just looking at an alternator will not tell you much about its workings. The only way to become even remotely competent in electrical work is to have some knowledge of basic theory. This chapter is a brief, almost entirely nonmathematical, discussion of the theory.

Electrons

Atoms are the building blocks of all matter. These atoms are widely distributed; if we enlarged the scale to make atoms the size of pinheads, there would be approximately one atom per cubic yard of nothingness. But as tiny and as few as they are, atoms (or molecules, which are atoms in combination) are responsible for the characteristics of matter. The density of a substance, its chemical stability, thermal and electrical conductivity, color, hardness, and all its other characteristics are fixed by the atomic structure.

The atom is composed of numerous subatomic particles. Using high-energy disintegration techniques, scientists are discovering new particles almost on an annual basis. Some are reverse images of the others; some exist for only a few millionths of a second. But, we are only interested in the relatively gross particles whose behavior has been reasonably well understood for generations.

In broad terms the atom consists of a nucleus and one or more electrons in orbit around it. The nucleus has at least one positively charged *proton* and might have one or more electrically neutral *neutrons*. These particles make up most of the atom's mass. The orbiting *electrons* have a negative charge. All electrons, as far as we know, are identical. All have the same electrical potency. Their orbits are balanced by centripetal force and the pull of the positive charge of the nucleus (Fig. 10-1).

A fundamental electrical law states unlike charges attract, like charges repel. Thus two electrons, both carrying negative charges, repel each other. Attraction between opposites and the repulsion of likes are the forces that drive electrons through circuits.

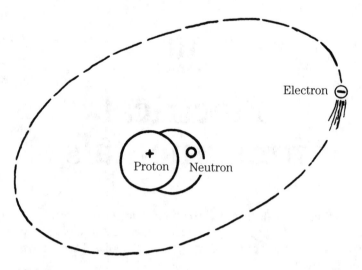

10-1 Representation of an electron orbiting around a nucleus consisting of a proton and a neutron.

Circuits

The term *circuit* comes from a Latin root meaning *circle*. It is descriptive because the path of electrons through a conductor is always in the form of a closed loop, bringing them back where they started.

A battery or generator has two *terminals,* or posts. One terminal is negative. It has a surplus of electrons available. The other terminal has a relative scarcity of electrons and is positive.

A *circuit* is a pathway between negative and positive terminals. The relative ease with which the electrons move along the circuit is determined by several factors. One of the most important is the nature of the conductor. Some materials are better conductors than others. Silver is at the top of the list, followed by copper, aluminum, iron, and lead. These and other conductors stand in contrast to that class of materials known as insulators. Most gases, including air, wood, rubber, and other organic materials, most plastics, mica, and slate are insulators of varying effectiveness.

Another class of materials is known as *semiconductors*. They share the characteristics of both conductors and insulators and under the right conditions can act as either. Silicon and germanium semiconductors are used to convert alternating current to direct current at the alternator and to limit current and voltage outputs. These devices are discussed in more detail later in this chapter.

The ability of materials to pass electrons depends on their atomic structures. Electrons flood into the circuit from the negative pole. They encounter other electrons already in orbit. The newcomers displace the orbiting electrons (like repels like) toward the positive terminal and, in turn, are captured by the positive charge on the nucleus (unlike attracts). This game of musical chairs continues until the circuit is broken or until the voltages on the terminals equalize. Gold, copper, and other conductors give up their captive electrons with a minimum of fuss. Insulators hold their electrons tightly in orbit and resist the flow of current.

You should not have the impression that electrons flow through various substances in a go/no-go manner. All known materials have some reluctance to give up their orbiting electrons. This reluctance is lessened by extreme cold, but never reaches zero. And all insulators "leak" to some degree. A few vagrant electrons will pass through the heaviest, most inert insulation. In most cases the leakage is too small to be significant. But it exists and can be accelerated by moisture saturation and by chemical changes in the insulator.

Circuit characteristics

Circuits, from the simplest to the most convoluted, share these characteristics:
- Electrons move from the negative to the positive terminal. Actually the direction of electron movement makes little difference. In fact, for many years it was thought that electrons moved in the opposite direction—from positive to negative.
- The circuit must be complete and unbroken for electrons to flow. An incomplete circuit is described as *open*. The circuit might be opened deliberately by the action of a switch or a fuse, or it might open of its own accord as in the case of a loose connection or a broken wire. Because circuits must have some rationale besides the movement of electrons from one pole to the other, they have *loads,* which convert electron movement into heat, light, magnetic flux, or some other useful quantity. These working elements of the circuit are also known as *sinks*. They absorb, or sink, energy and are distinguished from the sources (battery and the generator). A circuit in which the load is bypassed is described as *shorted*. Electrons take the easiest, least resistive path to the positive terminal.
- Circuits can feed single or multiple loads. The arrangement of the loads determines the circuit type.

Series & parallel circuits

A *series* circuit has its loads connected one after the other like beads on a string (Fig. 10-2A). There is only one path for the electrons between the negative and positive terminals. In a string of Christmas tree lights, each lamp is rated at 10 V; 11 in series required 110 V. If any lamp failed, the string went dark.

Pure series circuits are rare today. (They can still be found as filament circuits in some alternating current/direct current radios and in high-voltage applications such as airport runway lights.) But switches, rheostats, relays, fuses, and other circuit controls are necessarily in series with the loads they control.

Parallel circuits have the loads arranged like the rungs of a ladder, to provide multiple paths for current (Fig. 10-2). When loads are connected in parallel, we say they are *shunt*. Circuits associated with diesel engines usually consist of parallel loads and series control elements. The major advantage of the parallel arrangement from a serviceman's point of view is that a single load can open without affecting the other loads. A second advantage is that the voltage remains constant throughout the network.

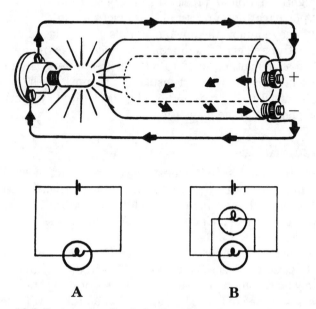

A   **B**

10-2 Basic series circuit (top) and its schematic (A). In B, a second lamp has been added with the first.

Single- & two-wire circuits

However the circuit is arranged—series, parallel, or in some combination of both—it must form a complete path between the positive and negative terminals of the source. This requirement does not mean the conductor must be composed entirely of electrical wire. The engine block, transmission, and mounting frame are not the best conductors from the point of view of their atomic structures. But because of their vast cross-sectional area, these components have almost zero resistance.

In the single-wire system the battery is *grounded,* or as the British say, *earthed* (Fig. 10-3). These terms seem to have originated from the power station practice of using the earth as a return. With the exception of some Lucas and CAV systems, most modem designs have the negative post grounded. The "hot" cable connects the positive post to the individual loads, which are grounded. Electrons flow from the negative terminal, through the loads, and back to the battery via the wiring.

The single-wire approach combines the virtues of simplicity and economy. But it has drawbacks. Perhaps the most consequential is the tendency for the connections to develop high resistances. One would think that a heavy strap bolted to the block

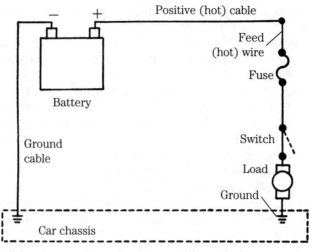

10-3 The principle of grounding.

would offer little resistance. Unfortunately this is not the case. A thin film of oil, rust, or a loose connection is enough to block the flow of current.

Corrosion problems are made more serious by weather exposure and *electrolysis*. Electrolysis is the same phenomenon as occurs during electroplating. When current passes through two dissimilar metals (e.g., copper and cast iron) in the presence of damp air, one of the metals tends to disintegrate. In the process it undergoes chemical changes that make it a very poor conductor.

Another disadvantage of the single-wire system, and the reason it is not used indiscriminately on aircraft and ocean-going vessels, is the danger of short circuits. Contact between an uninsulated hot wire and the ground will shunt the loads down circuit.

The *two-wire system* (one wire to the load, and a second wire from it to the positive terminal) is preferred for critical loads and can be used in conjunction with a grounded system. Electrolytic action is almost nil, and shorts occur only in the unlikely event that two bare wires touch.

Electrical measurements

Voltage is a measure of electrical pressure. In many respects it is analogous to hydraulic pressure. The unit of voltage is the volt, abbreviated *V.* Thus we speak of a 12-V or 24-V system. Prefixes, keyed to the decimal system, expand the terms so we do not need to contend with a series of zeros. The prefix *K* stands for *kilo,* or 1000. A 50-kV powerline delivers 50,000 V. At the other end of the scale, *milli* means ¹⁄₁₀₀₀, and one millivolt (1 mV) is a thousandth of a volt.

The *ampere,* shortened sometimes to *amp* and abbreviated *A* is a measure of the quantity of electrons flowing past a given point in the circuit per second. One ampere represents the flow of $6.25 \times 10_{18}$ electrons per second. Amperage is also referred to as *current intensity* or *quantity.* From the point of view of the loads on the circuit, the amperage is the *draw*. A free-running starter motor might draw 100 A, and three times as much under cranking loads.

Resistance is measured in units named after G. S. Ohm. Ohms are expressed by the last letter of the Greek alphabet, omega (Ω). Thus we might speak of a 200 Ω resistance.

The resistance of a circuit determines the amount of current that flows for a given applied voltage. The resistance depends on the atomic structure of the conductor—how tightly the electrons are held captive in their orbits—and on certain physical characteristics. The broader the cross-sectional area of the conductor, the less opposition to current. And the longer the path formed by the circuit between the poles of the voltage source, the more resistance. Think of these two dimensions in terms of ordinary plumbing. The *resistance* to the flow of a liquid in a pipe is inversely related to its diameter (decreases as diameter increases) and directly related to its length. Resistance in the pipe produces heat, exactly as does resistance in an electrical conductor.

Resistance is generally thought of as the electrical equivalent of friction—a kind of excise tax that we must pay to have electron movement. There are, however, positive uses of resistance. Resistive elements can be deliberately introduced in the circuit to reduce current in order to protect delicate components. The heating effect of resistance is used in soldering guns and irons and in the glow plugs employed as starting aids in diesel engines.

Ohm's law

The fundamental law of simple circuits was expressed by the Frenchman G. S. Ohm in the early 1800s in a paper on the effects of heat on resistance. The law takes three algebraic forms, each based on the following relationship: a potential of 1 V drives 1 A through a resistance of 1 Ω. In the equations the symbol E stands for *electromotive force*, or, as we now say, volts. An I represents *intensity*, or current, and R is resistance in ohms.

The basic relationship is expressed as:

$$I = \frac{E}{R}$$

This form of the equation states that current in amperes equals voltage in volts divided by resistance in ohms. If a circuit with a resistance of 6 Ω is connected across a 12-V source, the current is 2 A (12/6 = 2). Double the voltage (or halve the resistance), and the current doubles. Figure 10-4 shows this linear (straight-line graph) relationship.

Another form of Ohm's law is:

$$R = \frac{E}{I}$$

or resistance equals voltage divided by current. The 12-V potential of our hypothetical circuit delivers 2 A, which means the resistance is 6 Ω. The relationship between amperage and resistance with a constant voltage is illustrated in Fig. 10-5.

Another way of expressing Ohm's law is:

$$E = IR$$

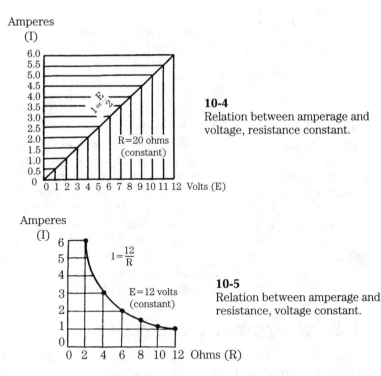

10-4
Relation between amperage and voltage, resistance constant.

10-5
Relation between amperage and resistance, voltage constant.

or voltage equals current multiplied by resistance. Two amperes through 6 Ω requires a potential of 12 V. Double the resistance, and twice the voltage is required to deliver the same amount of current (Fig. 10-6).

Various memory aids have been devised to help students remember Ohm's law. One involves an Indian, an eagle, and a rabbit. The Indian (I) sees the eagle (E) flying over the rabbit (R); this gives the relationship $I = E/R$. The eagle sees both the Indian and the rabbit in the same level, or $E = IR$. And the rabbit sees the eagle over the Indian, or $R = E/I$. A visual aid is shown in Fig. 10-7. The circle is divided into three segments. To determine which form of the equation to use, put your finger on the quantity you want to solve for.

All of this might seem academic, and in truth, few mechanics perform calculations with Ohm's law. It is usually easier to measure all values directly with a meter. Still, it is very important to have an understanding of Ohm's law. It is the best description of simple circuits we have.

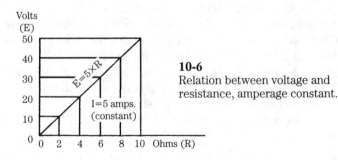

10-6
Relation between voltage and resistance, amperage constant.

10-7
Ohm's law: Cover the unknown
with your finger to determine
which of the three equations to
use.

For example, suppose the wiring is *shorted:* electrons have found a more direct (shorter) path to the positive terminal of the battery. Ohm's law tells us certain facts about the nature of shorts, which are useful in troubleshooting. First, a short increases the current in the affected circuit, because current values respond inversely to resistance. This increase will generate heat in the conductor and might even carbonize the insulation. Voltage readings will be low because the short has almost zero resistance. Now suppose we have a partially open circuit caused by a corroded terminal. The current through the terminal will be reduced, which means that the lights or whatever other load is on the circuit will operate at less-than-peak output. The terminal will be warm to the touch, because current is transformed to heat by the presence of resistance. Voltage readings from the terminal to ground will be high on the source side of the resistance and lower than normal past the bottleneck.

Direct & alternating current

Diesel engines might employ direct or alternating currents. The action of *direct,* or *unidirectional,* current can be visualized with the aid of the top drawing in Fig. 10-8. *Alternating current* is expressed by the opposed arrows in the lower drawing. Flow is from the negative to the positive poles of the voltage source, but the poles exchange identities, causing the current reversal. The positive becomes negative, and the negative becomes positive.

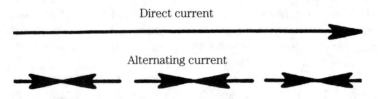

Direct current

Alternating current

10-8 Direct and alternating current.

The graph in Fig. 10-9 represents the rise, fall, and reversal of alternating current. Because the amplitude changes over time, alternating voltage and current values require some qualification. The next drawing illustrates the three values most often used.

Peak-to-peak values refer to the maximum amplitude of the voltage and amperage outputs in both directions. In Fig. 10-10 the peak-to-peak value is 200 V_{p-p} (or 200 A_{p-p}). The half-cycle (alternation) on top of the zero reference line is considered to be positive; below the line alternation is negative.

The *average,* or *mean,* value represents an average of all readings. In Fig. 10-10 it is 63.7 units.

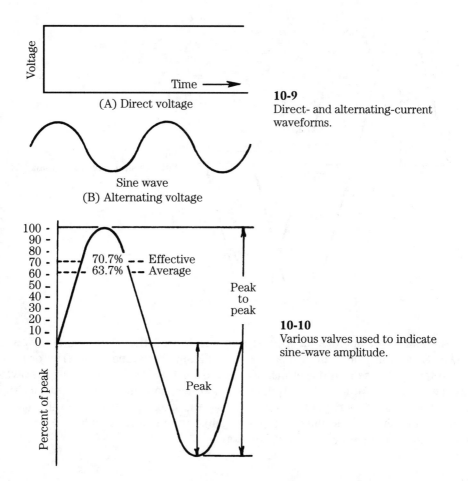

10-9
Direct- and alternating-current waveforms.

10-10
Various valves used to indicate sine-wave amplitude.

The *root mean square* (rms) value is sometimes known as the *effective* value. One rms ampere has the same potential for work as 1-A direct current. Unless otherwise specified, alternating current values are *rms* (effective) values and are directly comparable to direct current in terms of the work they can do. Standard meters are equipped with appropriate scales to give rms readings.

The illustration depicts one complete cycle of alternating current. The number of cycles completed per second is the frequency of the current. One cycle per second is the same as one *hertz* (1 Hz). Domestic household current is generated at 60 Hz. The alternating current generators used with diesel engines are variable-frequency devices because they are driven by the engine crankshaft. At high speed a typical diesel alternating current generator will produce 500-600 Hz.

In addition to these special characteristics, alternating current outputs can be superimposed upon each other (see Fig. 10-11). The horizontal axis (X-axis) represents zero output—it is the crossover point where alternations reverse direction. In this particular alternator 360 degrees of rotation of the armature represents a complete output cycle from maximum positive to maximum negative and back to maximum positive. The X-axis can be calibrated in degrees or units of time (assuming

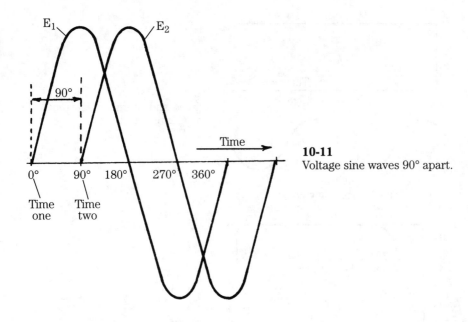

10-11
Voltage sine waves 90° apart.

that the alternator operates at a fixed rpm). Note that there are two sine waves shown. These waves are out of phase; E_1 leads E_2 by 90 degrees of alternator rotation. Alternators used on diesel installations usually generate three output waves 120 degrees apart. Multiphase outputs are smoother than single-phase outputs and give engineers the opportunity to build multiple charging circuits into the alternator. Circuit redundancy gives some assurance against total failure; should one charging loop fail, the two others continue to function.

Magnetism

Electricity and magnetism are distinct phenomena, but related in the sense that one can be converted into the other. Magnetism can be employed to generate electricity, and electricity can be used to produce motion through magnetic attraction and repulsion.

The earth is a giant magnet with magnetic poles located near the geographic poles (Fig. 10-12). Magnetic *fields* extend between the poles over the surface of the earth and through its core. The field consists of *lines of force*, or *flux*. These lines of force have certain characteristics, which, although they do not fully explain the phenomenon, at least allow us to predict its behavior. The lines are said to move from the north to the south magnetic pole just as electrons move from a negative to a positive electrical pole. Magnetic lines of force make closed loops, circling around and through the magnet. The poles are the interface between the internal and external paths of the lines of force. When encountering a foreign body, the lines of force tend to stretch and snap back upon themselves like rubber bands. This characteristic is important in the operation of generators.

South magnetic Pole North geographic Pole

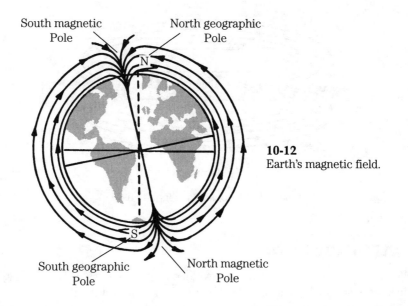

10-12
Earth's magnetic field.

South geographic Pole North magnetic Pole

Lines of force penetrate every known substance, as well as the emptiness of outer space. However, they can be deflected by soft iron. This material attracts and focuses flux in a manner analogous to the action of a lens on light. Lines of force "prefer" to travel through iron, and they digress to take advantage of the *permeability* (magnetic conductivity) of iron (Fig. 10-13).

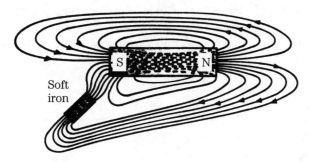

Soft iron

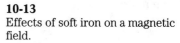

10-13
Effects of soft iron on a magnetic field.

The permeability of iron is exploited in almost all magnetic machines. Generators, motors, coils, and electromagnets all have iron cores to direct the lines of force most efficiently.

When free to pivot, a magnet aligns itself with a magnetic field, as shown in Fig. 10-14. *Unlike magnetic poles attract, like magnetic poles repel.* The south magnetic pole of the small magnet (compass needle) points to the north magnetic pole of the large bar magnet.

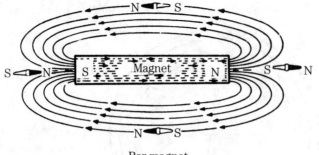

Bar magnet

10-14 Compass deflections in a magnetic field.

Electromagnets

Electromagnets use electricity to produce a magnetic field. Electromagnets are used in starter solenoids, relays, and voltage-and current-limiting devices. The principle upon which these magnets operate was first enunciated by the Danish researcher Hans Christian Oersted. In 1820 he reported a rather puzzling phenomenon: A compass needle, when placed near a conductor, deflected as the circuit was made and broken. Subsequently it was shown that the needle reacted because of the presence of a magnetic field at right angles to the conductor. The field exists as long as current flows, and its intensity is directly proportional to the amperage.

The field around a single strand of conductor is too weak to have practical application. But if the conductor is wound into a coil, the weak fields of the turns reinforce each other (Fig. 10-15). Further reinforcement can be obtained by inserting an iron bar inside the coil to give focus to the lines of force. The strength of the electromagnet depends on the number of turns of the conductor, the length/width ratio of the coil, the current strength, and the permeability of the core material.

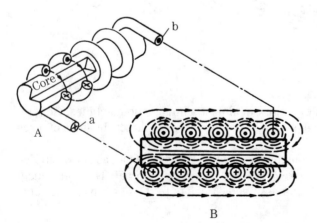

10-15
Electromagnet construction showing how magnetic fields mutually reinforce one another.

Voltage sources

Starting and charging systems employ two voltage sources. The generator is electromagnetic in nature and is the primary source. The battery, which is electro-chemical in nature, is carried primarily for starting. Also, in case of overloads, the battery can send energy into the circuit.

Generator principles

In 1831 Michael Faraday reported that he had induced electricity in a conductor by means of magnetic action. His apparatus consisted of two coils, wound over each other but not electrically connected. When he made and broke the circuit to one coil, momentary bursts of current were induced in the second. Then Faraday wound the coils over a metal bar and observed a large jump in induced voltage. Finally he re-produced the experiment with a permanent magnet. Moving a conductor across a magnetic field produced voltage. The intensity of the voltage depended on the strength of the field, the rapidity of movement, and the angle of movement. His generator was most efficient when the conductor cut the lines of force at right angles.

The rubber band effect mentioned earlier helps one to visualize what happens when current is induced in the conductor. A secondary magnetic field is set up, which resists the movement of the conductor through the primary field. The greater the amount of induced current, the greater the resistance, and the more power required to overcome it.

Alternator

Figure 10-16 illustrates Faraday's apparatus. Note that there must be relative movement between the magnetic field and conductor. Either can be fixed as long as one can move. Note also that the direction of current is determined by the direction of movement. Of course, this shuttle generator is hardly efficient; the magnet or conductor must be accelerated, stopped, and reversed during each cycle.

The next step was to convert Faraday's laboratory model to rotary motion (Fig. 10-17). The output alternates with the position of the armature relative to the fixed magnets. In Fig. 10-17A the windings are parallel to the lines of force and the output is zero. Ninety degrees later the output reaches its highest value because the armature windings are at right angles to the field. At 180 degrees of rotation (*C*) the output is again zero. This position coincides with a polarity shift. For the remainder of the cycle, the movement of the windings relative to the field is reversed.

To determine the direction of current flow from a generator armature, position the left hand so that the thumb points in the direction of current movement and the forefinger points in the direction of magnetic flux (from the north to the south pole). Extend the middle finger 90 degrees from the forefinger; it will point to the direction of current. As convoluted as the lefthand rule for generators sounds, it testifies to the fact that polarity shifts every 180 degrees of armature rotation. This is true of all dynamos, whether the machine is dc generator or ac generator (alternator).

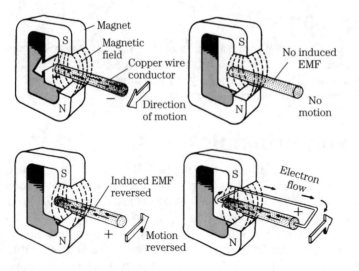

10-16 Voltage produced by magnetism.

The frequency in hertz depends on the rotational speed of the armature and on the number of magnetic poles. Figure 10-18 illustrates a four-pole (two-magnet) alternator, which for a given rpm, has twice the frequency of the single-pole machine in the previous illustration. Normally, alternators supplied for diesel engines have four magnetic poles and are designed to peak at 6000 rpm. Their frequency is given by:

$$f = \frac{P \times \text{rmp}}{120}$$

where f is frequency in hertz and P is the number of poles. Thus, at 6000 rpm, a typical alternator should deliver current at a frequency of 2000 Hz.

The alternators depicted in Figs. 10-17 and 10-18 are simplifications. Actual alternators differ from these drawings in two important ways. First, the fields are not fixed, but rotate as part of an assembly called the *rotor*. Secondly, the fields are not permanent magnets; instead they are electromagnets whose strength is controlled by the voltage regulator. Electrical connection to the rotor is made by a pair of brushes and slip rings.

dc generator

Direct-current generators develop an alternating current that is mechanically rectified by the commutator—brush assembly. The commutator is a split copper ring with the segments insulated from each other. The brushes are carbon bars that are spring-loaded to bear against the commutator. The output is pulsating direct current, as shown in Fig. 10-19. The pulse frequency depends on the number of armature loops, field poles, and rpm. A typical generator has 24–28 armature loops and 4 poles. The fields are electromagnets and are energized by 8–12% of the generator output current. The exact amount of current detailed for this *excitation* depends on the regulator, which in turn, responds to the load and state of charge of the generator. A few surviving generators achieve output regulation by means of a third brush.

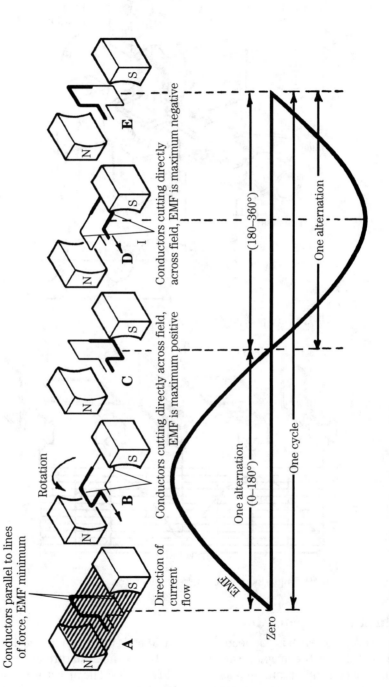

Conductors parallel to lines
of force, EMF minimum

Rotation

Direction of
current
flow

Conductors cutting directly across field,
EMF is maximum positive

Conductors cutting directly
across field, EMF is maximum negative

One alternation
(0–180°)

(180–360°)

One cycle

One alternation

Zero

EMF

10-17 Alternating current generator (or alternator).

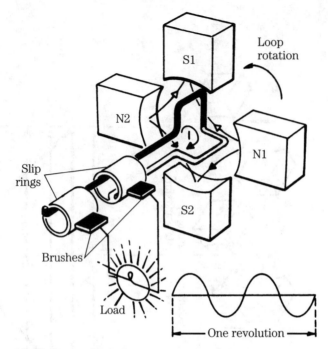

10-18 Four-pole alternator.

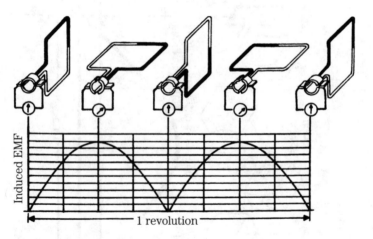

10-19 Single-loop direct current generator.

Third-brush generator

The third brush is connected to the field coils as shown in Fig. 10-20. At low rotational speeds the magnetic lines of force bisect the armature in a uniform manner. But as speed and output increase, the field distorts. The armature generates its own field, because a magnetic field is created at right angles to a conductor when current flows through it. The resulting field distortion, sometimes called *magnetic whirl*,

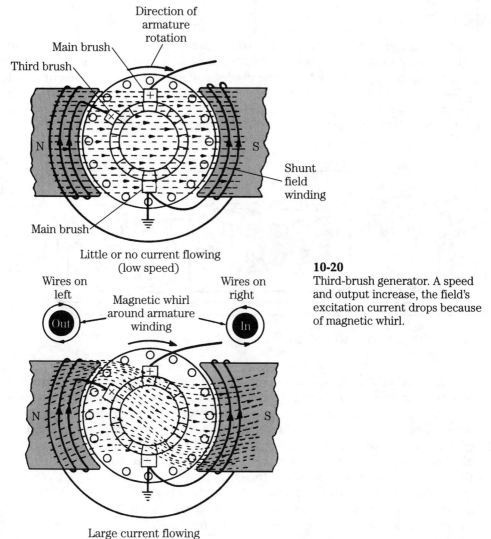

10-20
Third-brush generator. A speed and output increase, the field's excitation current drops because of magnetic whirl.

places the loops feeding the third brush in an area of relative magnetic weakness. Consequently, less current is generated in these loops, and the output to the fields is lessened. The fields become correspondingly weak, and the total generator output as defined by the negative and positive brushes remains stable or declines.

The output depends on the position of the third brush, which can be moved in or out of the distorted field to adjust output for anticipated loads. Moving the brush in the direction of armature rotation increases the output; moving it against the direction of rotation reduces the output. When operated independently from the external circuit, the brushes must be grounded to protect the windings. Many of these generators have a fuse in series with the fields.

dc motors

One of the peculiarities of a direct current generator is that it will "motor" if the brushes are connected to a voltage source. In like manner a direct current motor will "gen" if the armature is rotated by some mechanical means. Some manufacturers of small-engine accessories have taken advantage of this phenomenon to combine both functions in a single housing. One such system is the *Dynastart* system employed on many single-cylinder Hatz engines (shown schematically in Fig. 10-21).

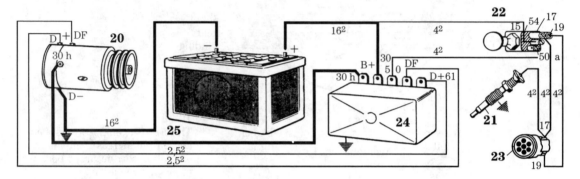

10-21 Combination generator and starter motor. Teledyne Wisconsin Motor

Figure 10-22 illustrates the motor effect. In the top sketch the conductor is assumed to carry an electric current toward you. The magnetic flux around it travels in a clockwise direction. Magnetic lines of force above the conductor are distorted and stretched. Because the lines of force have a strong elastic tendency to shorten, they push against the conductor. Placing a loop of wire in the field (bottom sketch) instead of a single conductor doubles the motor effect. Current goes in the right half of the loop and leaves at the left. The interaction of the fields causes the loop (or, collectively, the armature) to turn counterclockwise. Reversing the direction of the current would cause torque to be developed in a clockwise direction.

Starter motors are normally wired with the field coils in series with the armature (Fig. 10-23). Any additional load added to a series motor will cause more current in the armature and correspondingly more torque. Because this increased current must pass through the series field, there will be a greater flux. Speed changes rapidly with load. When the rotational speed is low, the motor produces its maximum torque. A starter might draw 300 A during cranking, more than twice that figure at stall.

Storage batteries

Storage batteries accumulate electrical energy and release it on demand. The familiar lead-acid battery was invented more than 100 years ago by Gaston Plante. It suffers from poor energy density (watt-hours per pound) and poor power density (watts per pound). The average life is said to be in the neighborhood of 360 complete

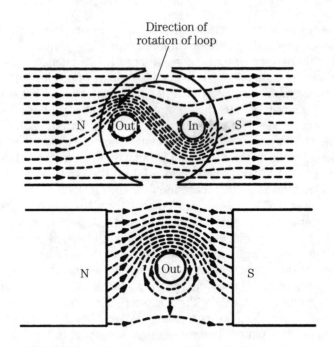

Direction of
rotation of loop

10-22 Motor effect created by current bearing conductor
in a magnetic field.

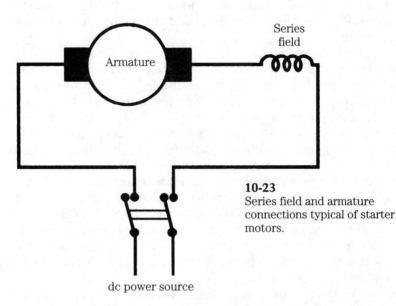

Armature

Series
field

10-23
Series field and armature
connections typical of starter
motors.

dc power source

charge-discharge cycles. During charging, the lead-acid battery shows an efficiency
of about 75%; that is, only three-quarters of the input can be retrieved.

Yet it remains the only practical alternative for automotive, marine, and most
stationary engine applications. Sodium-sulfur, zinc-air, lithium-halide, and lithium-

chlorine batteries all have superior performance, but are impractical by reason of cost and, in some cases, the need for complex support systems.

The lead-acid battery consists of a number of cells (hence the name *battery*) connected in series. Each fully charged cell is capable of producing 2.2 V. The number of cells fixes the output; 12-V batteries have six cells; 24-V batteries have 12. The cells are enclosed in individual compartments in a rubberoid or (currently) high-impact plastic case. The compartments are sealed from each other and, with the exception of Delco and other "zero maintenance" types, open to the atmosphere. The lower walls of the individual compartments extend below the plates to form a sediment trap. Filler plugs are located on the cover and can be combined with wells or other visual indicators to monitor the electrolyte level.

The cells consist of a series of lead plates (Fig. 10-24) connected by internal straps. Until recently the straps were routed over the top of the case, making convenient test points for the technician. Unfortunately, these straps leaked current and were responsible for the high self-discharge rates of these batteries.

The plates are divided into positive and negative groups and separated by means of plastic or fiberglass sheeting. Some very large batteries, which are built almost entirely by hand, continue to use fir or Port Orford cedar separators. A few batteries intended for vehicular service feature a loosely woven fiberglass padding between the separators and positive plates. The padding gives support to the lead filling and reduces damage caused by vibration and shock.

Both sets of plates are made of lead. The positive plates consist of a lead gridwork that has been filled with lead oxide paste. The grid is stiffened with a trace of antimony. Negative plates are cast in sponge lead. The plates and separators are immersed in a solution of sulfuric acid and distilled water. The standard proportion is 32% acid by weight.

The level of the electrolyte drops in use because of evaporation and hydrogen loss. (Sealed batteries have vapor condensation traps molded into the roof of the cells.) It must be periodically replenished with distilled water. Nearly all storage batteries are shelved dry, and filled upon sale. Once the plates are wetted, electrical energy is stored in the form of chemical bonds.

When a cell is fully charged the negative plates consist of pure sponge lead (Pb in chemical notation), and the positive plates are lead dioxide (PbO_2, sometimes called lead peroxide). The electrolyte consists of water (H_2O) and sulfuric acid (H_2SO_4). The fully charged condition corresponds to drawing A in Fig. 10-25. During discharge both sponge lead and lead dioxide become lead sulfate ($PbSO_4$). The percentage of water in the electrolyte increases because the SO_4 radical splits off from the sulfuric acid to combine the plates. If it were possible to discharge a lead-acid battery completely, the electrolyte would be safe to drink. In practice, batteries cannot be completely discharged in the field. Even those that have sat in junkyards for years still have some charge.

During the charge cycle the reaction reverses. Lead sulfate is transformed back into lead and acid. However, some small quantity of lead sulfate remains in its crystalline form and resists breakdown. After many charge-discharge cycles, the residual sulfate permanently reduces the battery's output capability. The battery then is said to be *sulfated*.

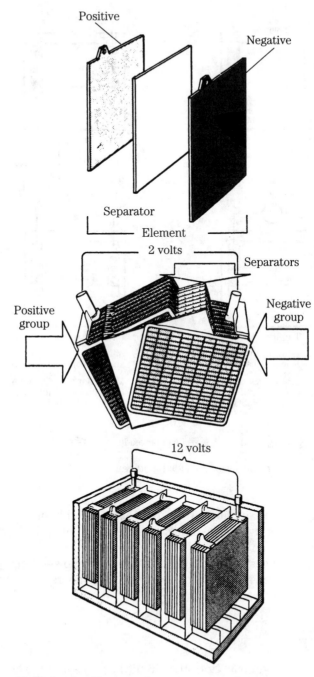

10-24 Lead-acid storage battery construction.

The rate of self-discharge is variable and depends on ambient air temperature, the cleanliness of the outside surfaces of the case, and humidity. In general it averages 1% a day. Batteries that have discharged below 70% of full charge might sulfate.

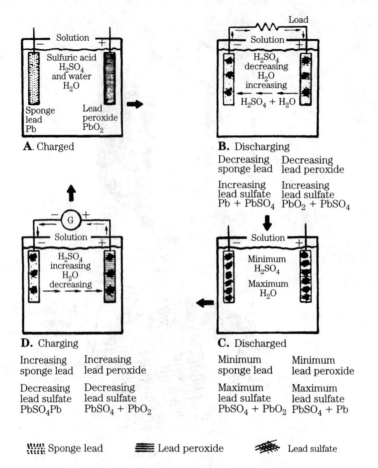

10-25 Chemical action in a lead-acid cell.

Sulfation becomes a certainty below the 70% mark. In addition to the possibility of damage to the plates, a low charge brings the freezing point of the electrolyte to near 32°F.

Temperature changes have a dramatic effect on the power density. Batteries are warm-blooded creatures and perform best at room temperatures. At 10°F the battery has only half of its rated power.

Switches

Switches neither add energy to the circuit nor take it out. Their function is to give flexibility by making or breaking circuits or by providing alternate current paths. They are classified by the number of *poles* (movable contacts) and *throws* (closed positions). Figure 10-26 illustrates four standard switch types in schematic form.

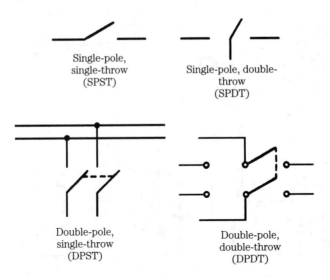

10-26 Schematic symbols of commonly used switches.

A single-pole, single-throw switch is like an ordinary light switch in that it makes and breaks a single circuit with one movement from the rest position. Single-pole, double-throw (SPDT) switches have three terminals, but control a single circuit with each throw. As one is completed, the other is opened. You can visualize an spdt as a pair of spst switches in tandem. Double-pole, double-throw switches control two circuits with a single movement. Think of the dpdt as two dpst switches combined, but working so that when one closes the other opens.

Self-actuating switches

Not all switching functions are done by hand. Some are too critical to trust to the operator's alertness. In general two activating methods are used. The first depends on the effect of heat on a bimetallic strip or disc to open or close contacts. Heat can be generated through electrical resistance or, as is more common, from the engine coolant. The bimetallic element consists of two dissimilar metals bonded back-to-back. Because the coefficient of expansion is different for the metals, the strip or disc will deform inward, in the direction of the least expansive metal. Figure 10-27 illustrates this in a flasher unit. Other switches operate by the pressure of a fluid acting against a diaphragm. The fluid can be air, lube oil, or brake fluid. The most common use of such switches is as oil pressure sensors (Fig. 10-28).

Relays

Relays are electrically operated switches. They consist of an electromagnet, a movable armature, a return spring, and one or more contact sets. Small currents excite the coil, and the resulting magnetic field draws the armature against the coil. This linear movement closes or opens the contacts. The attractiveness of relays is that small currents—just enough to excite the electromagnet—can be used to

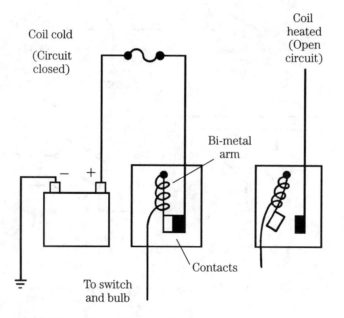

10-27 Bimetallic switch operation.

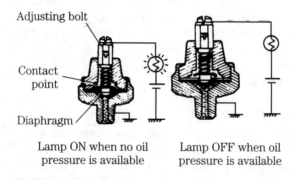

10-28 Diaphragm switch (oil pressure).

switch large currents. You will find relays used to activate charging indicator lamps (Fig. 10-29), some starter motors, horns, remotely controlled air trips, and the like.

Solenoids are functionally similar to relays. The distinction is that the movable armature has some nonelectrical chore, although it might operate switch contacts in the bargain. Many starter motors employ a solenoid to lever the pinion gear into engagement with the flywheel. The next chapter discusses several of these starters.

Circuit protective devices

Fuses and fusible links are designed to vaporize and open the circuit during overloads. Some circuits are routed through central fuse panels. Others can be protected by individual fuses spliced into the circuit at strategic points.

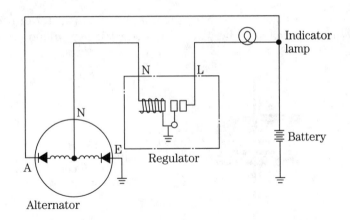

10-29 Relay-tripped charging indicator lamp.

Fusible links are appearing with more frequency as a protection for the battery-regulator circuit (Fig. 10-30). Burnout is signaled by swollen or discolored insulation over the link. New links can be purchased in bulk for soldered joints, or in precut lengths for use with quick-disconnects. Before replacing the link, locate the cause of failure, which will be a massive short in the protected circuit.

On the other hand, a blown fuse is no cause for alarm. Fuses fatigue and become resistive with age. And even the best-regulated charting circuit is subject to voltage and current spikes, which can take out a fuse but otherwise are harmless. Chronic failure means a short or a faulty regulator.

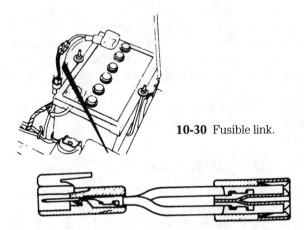

10-30 Fusible link.

Capacitors (condensers)

Two centuries ago it was thought that Leyden jars for storing charges actually condensed electrical "fluid." These glass jars, wrapped with foil on both the inner and outer surfaces, became known as *condensers*. The term lingers in the vocabulary of automotive and marine technicians, although *capacitor* is more descriptive of the devices' capacity for storing electricity.

A capacitor consists of two plates separated by an insulator or *dielectric* (Fig. 10-31). In most applications the plates and dielectric are wound on each other to save space.

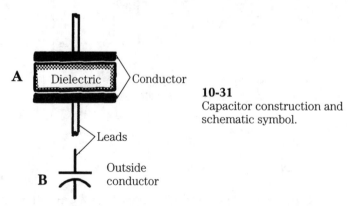

10-31
Capacitor construction and schematic symbol.

A capacitor is a kind of storage tank for electrons. In function it is similar to an accumulator in hydraulic circuits. Electrons are attracted to the negative plate by the close proximity of the positive plate. Storing the electrons requires energy, which is released when the capacitor is allowed to discharge. Shorting the plates together sends free electrons out of the negative plate and to the positive side. There is a to-and-fro shuttle of electrons between the plates until the number of electrons on the plates is equalized. The number of oscillations depends on circuit resistance and *reactance* (or the reluctance of the capacitor to charge because of the mutual repulsion of electrons on the negative plate).

Capacitors block direct current, but pass alternating current. The electrons do not physically penetrate the dielectric. The distortions in the orbits of the electrons in the dielectric displace or attract electrons at the plates. The effect on alternating current in a circuit is often as though the dielectric did not exist at least at the higher frequencies.

Capacitors are found at the generator or alternator, where they shunt high-frequency alternating current to ground to muffle radio interference.

Solid-state components

Although the transistor was invented in the Bell Laboratories in 1947, it and its progeny remain a mystery to most mechanics. These components are used in all alternators to rectify alternating current to direct current and can be found in current regulators produced by CAV, Lucas, Bosch, and Delco. In the future they will have a much wider use and possibly will substitute for relays.

Solid-state components are made of materials called semiconductors, which fall midway in the resistance spectrum between conductors and insulators. In general these components remain nonconductive until subjected to a sufficient voltage.

Silicon and germanium are the most-used materials for semiconductors. Semiconductor crystals for electronics are artificial crystals, grown under laboratory conditions. In the pure state these crystals are electrically just mediocre conductors. But if minute quantities of impurities are added in the growth state, the crystals acquire special electrical characteristics that make them useful for electronic devices. The process of adding impurities is known as *doping*.

When silicon or germanium is doped with arsenic or phosphorus, the physical structure remains crystalline. But for each atom of impurity that combines with the intrinsic material, one free electron becomes available. This material has electrons to carry current and is known as *n*-type semiconductor material.

On the other hand the growing crystals might be doped with indium, aluminum, boron, or certain other materials and result in a shortage of electrons in the crystal structure. "Missing" electrons can be thought of as electrical *holes* in the structure of the crystal. These holes accept electrons that blunder by, but the related atom quickly releases the newcomer. When a voltage is applied to the semiconductor, the holes appear to drift through the structure as they are alternately filled and emptied. Crystals with this characteristic are known as *p*-type materials, because the holes are positive charge carriers.

Diode operation

When *p*-type material is mated with *n*-type material, we have a *diode*. This device has the unique ability to pass current in one direction and block it in the other. When the diode is not under electrical stress, *p*-type holes in the *p*-material and electrons in the *n*-material complement each other at the interface of the two materials. This interface, or junction, is electrically neutral, and presents a barrier to the charge carriers. When we apply negative voltage to the *n*-type material, electrons are forced out of it. At the same time, holes in the *p*-type material migrate toward the direction of current flow. This convergence of charge carriers results in a steady current in the circuit connected to the diode. Applying a negative voltage to the *n*-type terminal so the diode will conduct is called *forward biasing*.

If we reverse the polarity—that is, apply positive voltage to the *n*-type material—electrons are attracted out of it. At the same time, holes move away from the junction on the *p* side. Because charge carriers are thus made to avoid the junction, no current will pass. The diode will behave as an insulator until the reverse-bias voltage reaches a high enough level to destroy the crystal structure.

Diodes are used to convert alternating current from the generator to pulsating direct current. An alternator typically has three pairs of diodes mounted in heat sinks (heat absorbers).

Diodes do have some resistance to forward current and so must be protected from heat by means of shields and sinks. In a typical alternator the three pairs of diodes that convert alternating current to pulsating direct current are pressed into the aluminum frame. Heat generated in operation passes out to the whole generator.

Solid-state characteristics

As useful as solid-state components are, they nevertheless are subject to certain limitations. The failure mode is absolute. Either the device works or it doesn't. And in the event of failure, no amount of circuit juggling or tinkering will restore operation. The component must be replaced. Failure might be spontaneous—the result of manufacturing error compounded by the harsh environment under the hood—or it might be the result of faulty service procedures. Spontaneous failure usually occurs during the warranty period. Service-caused failures tend to increase with time and mileage as the opportunities for human error increase.

These factors are lethal to diodes and transistors:

High inverse voltage resulting from wrong polarity Jumper cables connected backwards or a battery installed wrong will scramble the crystalline structure of these components.

Vibration and mechanical shock Diodes must be installed in their heat sinks with the proper tools. A 2 × 4 block is not adequate.

Short circuits Solid-state devices produce heat in normal operation. A transistor, for example, causes a 0.3 V–0.7 V drop across the collector-emitter terminals. This drop, multiplied by the current, is the wattage consumed. Excessive current will overheat the device.

Soldering The standard practice is for alternator diodes to be soldered to the stator leads. Too much heat at the connection can destroy the diodes.

11
CHAPTER

Starting & generating systems

The diesel engines we are concerned with are almost invariably fitted with electric starter motors. A number of engines, used to power construction machinery and other vehicles that are expected to stand idle in the weather, employ a gasoline engine rather than an electric motor as a starter. A motor demands a battery of generous capacity and a generator to match.

Starting a cold engine can be somewhat frustrating, particularly if the engine is small. The surface/volume ratio of the combustion space increases disproportionately as engine capacity is reduced. The heat generated by compression tends to dissipate through the cylinder and head metal. In addition, cold clearances might be such that much of the compressed air escapes past the piston rings. Other difficulties include the effect of cold on lube and fuel oil viscosity. The spray pattern coarsens, and the drag of heavy oil between the moving parts increases.

Starting has more or less distinct phases. Initial or breakaway torque requirements are high because the rotating parts have settled to the bottom of their journals and are only marginally lubricated. The next phase occurs during the first few revolutions of the crankshaft. Depending on ambient temperature, piston clearances, lube oil stability, and the like, the first few revolutions of the crankshaft are free of heavy compressive loads. But cold oil is being pumped to the journals, which collects and wedges between the bearings and the shafts. As the shafts continue to rotate, the oil is heated by friction and thins, progressively reducing drag. At the same time, cranking speed increases and compressive loads become significant. The engine accelerates to *firing speed*. The duration from breakaway to firing speed depends on the capacity of the starter and battery, the mechanical condition of the engine, lube oil viscosity, ambient air temperature, the inertia of the flywheel, and the number of cylinders. A single-cylinder engine is at a disadvantage because it cannot benefit from the expansion of other cylinders. Torque demands are characterized by sharp peaks.

Starting aids

It is customary to include a cold-starting position at the rack. This position provides extra fuel to the nozzles and makes combustion correspondingly more likely.

Lube oil and water immersion heaters are available that can be mounted permanently on the engine. Lube oil heaters are preferred and can be purchased from most engine builders. Good results can be had by heating the oil from an external heater mounted below the sump. Use an approved type to minimize the fire hazard. Alternatively, one can drain the oil upon shutdown and heat it before starting. The same can be done with the coolant, although temperatures in both cases should be kept well below the boiling temperature of water to prevent distortion and possible thermal cracking.

If extensive cold weather operation is intended or if the engine will be stopped and started frequently, it is wise to add one or more additional batteries wired in parallel. Negative-to-negative and positive-to-positive connections do not alter the output voltage, but add the individual battery capacities.

Once chilled beyond the cloud point, diesel fuel enters the gelling stage. Flow through the system is restricted, filter efficiency suffers, and starting becomes problematic. Racor is probably the best known manufacturer of fuel heaters, which are available in a variety of styles. Several combine electric resistance elements with a filter, to heat the fuel at the point of maximum restriction. Another type incorporates a resistance wire in a flexible fuel line.

Makers of indirect injection engines generally fit glow plugs as a starting aid (Fig. 11-1). These engines would be extremely difficult to start without some method of heating the air in the prechamber. A low-resistance filament (0.25 to 1.5 ohms, cold) draws heavy current to generate 1500°F at the plug tip. Early types used exposed filaments, which sometimes broke off and became trapped between the piston and chamber roof with catastrophic effects on the piston and (when made of aluminum) the head. Later variants contain the filament inside of a ceramic cover, which eliminates the problem. However, ceramic glow plugs are quite vulnerable to damage when removed from the engine and must be handled with extreme care.

Heating body   Terminal

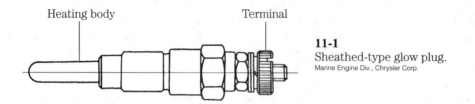

11-1
Sheathed-type glow plug.
Marine Engine Div., Chrysler Corp.

In all cases, glow plugs are wired in parallel and controlled by a large power relay. Test filament continuity with an ohmmeter.

Primitive glow-plug systems are energized by a switch, sometimes associated with a timer, and nearly always in conjunction with a telltale light. The more sophisticated systems used in contemporary automobiles automatically initiate glow-plug operation during cranking and, once the engine starts, gradually phase out power.

Two types of circuits are encountered, both built around a solid-state module with an internal clock. The pulsed system opens the glow-plug power circuit for progressively longer intervals as the engine heats and the timer counts down. In the Ford/Navistar version of this circuit, glow-plug resistance varies with tip temperature, so that the plugs themselves function as heat sensors. Note that these low-resistance devices self-destruct within seconds of exposure to steady-state battery voltage. Pulsed glow plugs can be tested with a low-voltage ohmmeter and plug operation can be observed by connecting a test lamp between the power lead and the glow-plug terminal. Normally, if the circuit pulses, it can be considered okay; when in doubt, consult factory literature for the particular engine model.

Most manufacturers take a less ambitious approach, and limit glow-plug voltage by switching a resistor into the feed circuit. During cold starts, a relay closes to direct full battery voltage to the glow plugs; as the engine heats (a condition usually sensed at the cylinder-head water jacket), the first relay opens and a second relay closes to switch in a large power resistor. Power is switched off when the module times out.

Starting fluid can be used in the absence of intake air heaters. In the old days a mechanic poured a spoonful of ether on a burlap rag and placed it over the air intake. This method is not the safest nor the most consistent; too little fluid will not start the engine, and too much can cause severe detonation or an intake header explosion. Aerosol cans are available for injection directly into the air intake. Use as directed in a well-ventilated place.

More sophisticated methods include pumps and metering valves in conjunction with pressurized containers of starting fluid. Figure 11-2 illustrates a typical metering valve. The valve is tripped only once during each starting attempt, to forestall explosion. Caterpillar engines are sometimes fitted with a one-shot starting device consisting of a holder and needle. A capsule of fluid is inserted in the device and the needle pierces it, releasing the fluid.

The starter motor should not be operated for more than a few seconds at a time. Manufacturers have different recommendations on the duration of cranking, but none suggests that the starter button be depressed for more than 30 seconds. Allow a minute or more between bouts for cooling and battery recovery.

Wiring

Figure 11-3 illustrates a typical charging/starting system in quasi-realistic style. The next drawing (Fig. 11-4) is a true schematic of the same system, encoded in a way that conveys the maximum amount of information per square inch.

Neither of these drawings is to scale and the routing of wires has been simplified. The technician needs to know where wires terminate, not what particular routes they take to get there. When routing does become a factor, as in the case electronic engine control circuitry, the manufacturer should provide the necessary drawings.

Color coding

As is customary, the schematic in Fig. 11-4 indicates the color of the insulation and the size of the various conductors. Color coding has not been completely standardized, but most manufacturers agree that black should represent ground. This

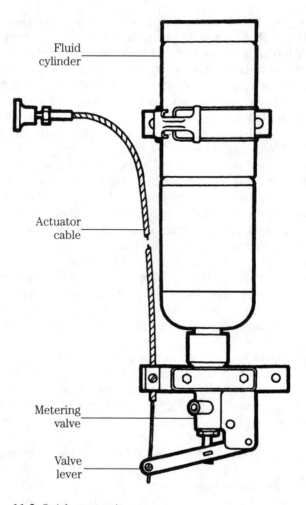

Fluid cylinder

Actuator cable

Metering valve

Valve lever

11-2 Quick-start unit. Detroit Diesel

does not mean that colors change in a purely arbitrary fashion. In the example schematic, red denotes the positive side of the battery (the "hot" wire); white/blue is associated with the temperature switch; orange with the tachometer, and so on.

Two-color wires consist of a base, or primary, color and a tracer. The base color is always first in the nomenclature. Thus, white/blue means a wire with a blue stripe.

Wiring repairs

The material that follows applies to simple dc starting, charging, and instrumentation circuits. Cutting or splicing wiring harnesses used with electronic engine management systems can do strange things to the computer.

The first consideration when selecting replacement wire is its current-carrying capacity, or gauge. In the context of diesel engines, two more or less interchange-

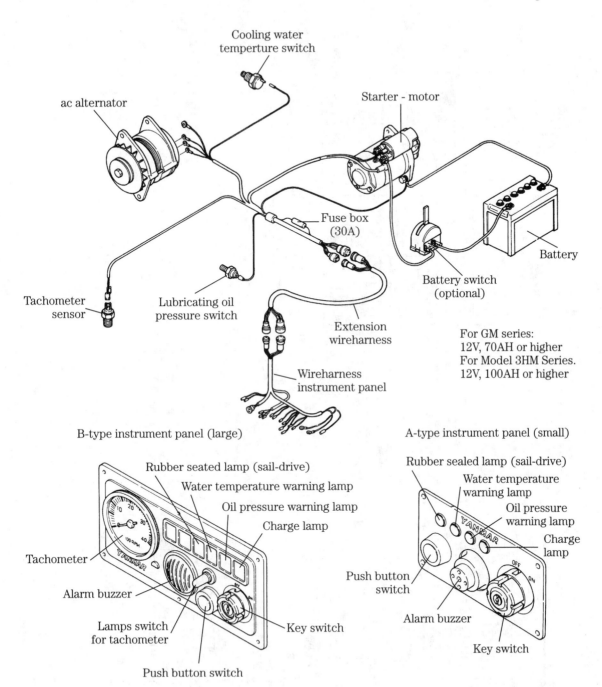

Cooling water
temperture switch

ac alternator

Starter - motor

Battery

Fuse box
(30A)

Battery switch
(optional)

Tachometer
sensor

Lubricating oil
pressure switch

Extension
wireharness

For GM series:
12V, 70AH or higher
For Model 3HM Series.
12V, 100AH or higher

Wireharness
instrument panel

B-type instrument panel (large)

A-type instrument panel (small)

Rubber seated lamp (sail-drive)

Water temperature warning lamp

Oil pressure warning lamp

Charge lamp

Tachometer

Alarm buzzer

Lamps switch
for tachometer

Key switch

Push button switch

Rubber sealed lamp (sail-drive)

Water temperature
warning lamp

Oil pressure
warning lamp

Charge
lamp

Push button
switch

Alarm buzzer

Key switch

11-3 Wiring layout for a Yanmar Marine application.

able standards apply: the SAE (Society of Automotive Engineers) and JIS (Japanese Industrial Standard).

For most wire, the smaller the SAE gauge number, the greater the cross-sectional area of the conductor. Thus, #10 wire will carry more current than #12.

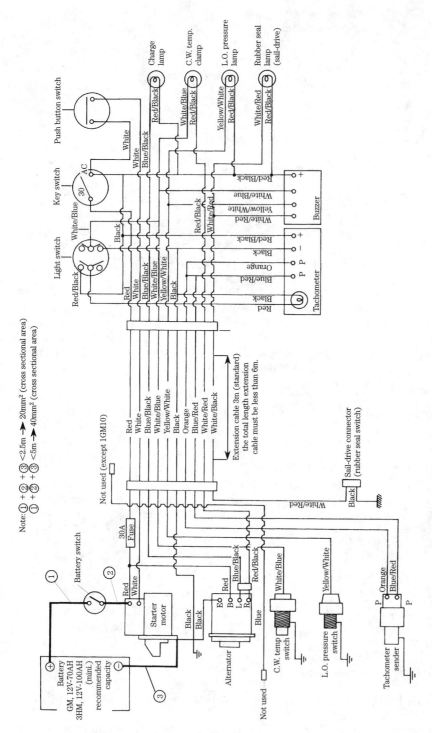

11-4 A detailed schematic for the wiring layout shown previously.

The schema reverses when we get into heavy cable: 4/0 is half again as large as 2.0 and has a correspondingly greater current capacity. Note that one must take these current values on faith.

The JIS standard eliminates much confusion by designating wire by type of construction and cross-sectional area. Thus, JIS AV5 translates as automotive-type (stranded) wire with a nominal conductor area of 5mm². The Japanese derive current-carrying capacity from conductor temperature, which cannot exceed 60°C (140° F). Ambient temperature and wire type affect the rating, as shown in back Table 4-2.

Table 11-1. SAE/metric wire sizes

SAE wire size	Sectional area		Outer dia.		Conductor resistance	Allowable continuous load
	Circular mils	mm²	in.	mm	(Ω/km)	current (A)
20	1,094	0.5629	0.040	1.0	32.5	7
18	1,568	0.8846	0.050	1.2	20.5	9
16	2,340	1.287	0.060	1.5	14.1	12
14	3,777	2.091	0.075	1.9	8.67	16
12	5,947	3.297	0.090	2.4	5.50	22
10	9,443	5.228	0.115	3.0	3.47	29
8	15,105	7.952	0.160	3.7	2.28	39
6	24,353	13.36	0.210	4.8	1.36	88
4	38,430	20.61	0.275	6.0	0.871	115
2	63,119	35.19	0.335	8.0	0.510	160
1	80,010	42.73	0.375	8.6	0.420	180
0	100,965	54.29	0.420	9.8	0.331	210
2/0	126,822	63.84	0.475	10.4	0.281	230
3/0	163,170	84.96	0.535	12.0	0.211	280
4/0	207,740	109.1	0.595	13.6	0.165	340

Source: Marine Engine Div., Chrysler Corp.

Vinyl-insulated, stranded copper wire is standard for engine applications. Teflon insulation tolerates higher temperatures than vinyl and has better abrasion resistance. But Teflon costs more and releases toxic gases when burned. In no case should you use Teflon-insulated wire in closed spaces.

All connections should be made with terminal lugs. Solder-type lugs (Fig. 11-5A) can provide mechanically strong, low-resistance joints and are infinitely preferable to the crimp-on terminals shown in Fig. 11-5B. Insulate with shrink tubing. When shrink tubing is impractical (as when insulating a Y-joint), use a good grade of vinyl electrician's tape. The 3-M brand costs three times more than the imported variety and is worth every penny.

Figure 11-6 illustrates how stranded wire is butt spliced. Cut back the insulation ¾ in. or so, and splay the strands apart. Push the wires together, so that the strands interleave, and twist. Apply a small amount of solder to the top of the joint, heating from below.

Table 11-2. Allowable amperage and voltage drop (JIS AV wire).

Ambient temp.	30°C		40°C		50°C	
Allowable amperage/ Voltage drop	A	mv/m	A	mv/m	A	mv/m
Nominal section (mm²)						
0.5	11	414	9	338	6	226
0.85	15	356	12	284	8	190
1.25	19	310	15	245	10	163
2	25	251	20	201	14	140
3	34	216	28	178	19	121
5	46	185	37	149	26	104
8	60	158	49	129	34	90
15	82	129	66	104	46	72
20	109	110	86	87	61	62
30	152	90	124	73	87	51
40	170	83	139	68	98	48
50	195	75	159	61	113	43
60	215	70	175	57	124	40
85	254	62	207	51	146	36
100	294	56	240	46	170	33

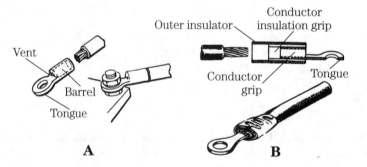

11-5 (A) Solder-type terminal lug. (B) Crimp-on terminal lug.

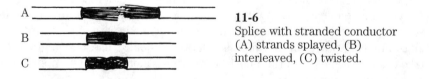

11-6
Splice with stranded conductor
(A) strands splayed, (B)
interleaved, (C) twisted.

Soldering

Most mechanics believe they know how to solder, but few have received any training in the art. Applied correctly, solder makes a molecular bond with the base metal. Scrape conductors bright to remove all traces of oxidation.

Use a good grade of low-temperature, rosin-core solder, such as Kessler "blue," which consists of 60% tin and 40% lead. A 250-W gun should be adequate for all but the heaviest wiring.

The tip of a nonplated soldering iron should be dressed with a file down to virgin copper and tinned, or coated with molten solder. Silver solder is preferable for tinning because it melts at a higher temperature than lead-based solder and so protects the tip from corrosion. Retighten the tip periodically.

The following rules were developed from experience and from a series of experiments conducted by the military:

- Use a minimum amount of solder.
- Wrapping terminal lugs and splicing ends with multiple turns of wire does not add to the mechanical strength of the joint and increases the heat requirement. Wrap only to hold the joint while soldering.
- Heat the connection, not the solder. When the parts to be joined are hot enough, solder will flow into the joint.
- Do not move the parts until the solder has hardened. Movement while the solder is still plastic will produce a highly resistive "cold" joint.
- Use only enough heat to melt the solder. Excessive heat can damage nearby components and can crystallize the solder.
- Allow the joint to air cool. Dousing a joint with water to cool it weakens the bond.

Starter circuits

Before assuming that the motor is at fault, check the battery and cables. The temperature-corrected hydrometer reading should be at least 1.240, and no cell should vary from the average of the others by more than 0.05 point. See that the battery terminal connections are tight and free of corrosion.

Excessive or chronic starter failure might point to a problem that is outside the starter itself. It could be caused by an engine that is out of tune and that consequently requires long cranking intervals.

Starter circuit tests

There are several methods that you can use to check the starting-circuit resistance. One method is to open all the connections, scrape bright, and retighten. Another method requires a low-reading ohmmeter of the type sold by Sun Electric and other suppliers for the automotive trades. But most mechanics prefer to test by voltage drop.

Connect a voltmeter as shown in Fig. 11-7. The meter shunts the positive, or hot battery post and the starter motor. With the meter set on a scale above battery voltage, crank. Full battery voltage means an open in the circuit.

If the starter functions at all, the reading will be only a fraction of this. Expand the scale accordingly. A perfect circuit will give a zero voltage drop because all current goes to the battery. In practice some small reading will be obtained. The exact figure depends on the current draw of the starter and varies between engine and starter motor types. As a general rule, subject to modification by experience, a 0.5-V

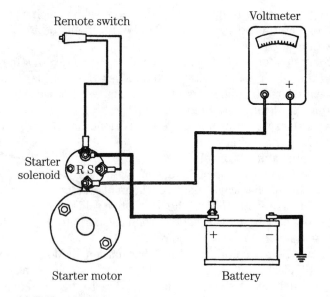

11-7 Hot side voltage drop test.

drop is normal. Much more than this means: (1) resistance in the cable, (2) resistance in the connections (you can localize this by repeating the test at each connection point), or (3) resistance in the solenoid.

Figure 11-8 shows the connections for the ground-side check. A poor ground, and consequent high voltage on the meter, can occur at the terminals, the cable, or between the starter motor and engine block. If the latter is the case, remove the motor and clean any grease or paint from the mounting flange.

Starter motors

Starter motors are series-wound; i.e., they are wound so that current enters the field coils and goes to the armature through the insulated brushes. Because a series-wound motor is characterized by high no-load rpm, some manufacturers employ limiting coils in shunt with the fields. The effect is to govern the free-running rpm and prolong starter life should the starter be energized without engaging the flywheel.

The exploded view in Fig. 11-9 illustrates the major components of a typical starter motor. The frame (No. 1) has several functions. It locates the armature and fields, absorbs torque reaction, and forms part of the magnetic circuit.

The field coils (No. 2) are mounted on the pole shoes (No. 3) and generate a magnetic field, which reacts with the field generated in the armature to produce torque. The pole pieces are secured to the frame by screws.

The armature (No. 4) consists of a steel form and a series of windings, which terminate at the commutator bars. The shaft is integral with it and splined to accept the starter clutch.

The end plates (5 and 6) locate the armature by means of bronze bushings. The commutator end plate doubles as a mounting fixture for the brushes, while the power takeoff side segregates the starter motor from the clutch.

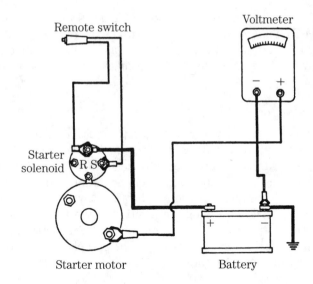

11-8 Ground circuit test.

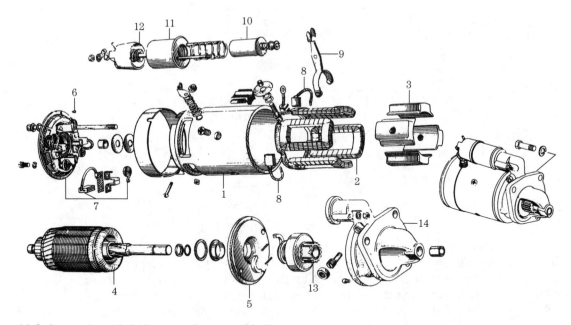

11-9 Starter in exploded view and as assembled. Lehman Manufacturing Co., Inc.

The insulated (hot) brushes (No. 7) provide a current path from the field coils through the commutator and armature windings to the grounded brushes (No. 8).

Engagement of this particular starter is done by means of a yoke (No. 9), which is pivoted by the solenoid plunger (No. 10) in response to current flowing through the solenoid windings (No. 11). Movement of the plunger also trips a relay (No. 12) and energizes the motor. The pinion gear (No. 13) meshes with the ring gear on the rim of the flywheel. The pinion gear is integral with an overrunning clutch.

The starter drive housing supports the power takeoff end of the shaft and provides an accurately machined surface for mounting the starter motor to the ending block or bellhousing.

Brushes

Before any serious work can be done, the starter must be removed from the engine, degreased, and placed on a clean bench. Disconnect one or both cables at the battery to prevent sparking; disconnect the cable to the solenoid and the other leads that might be present (noting their position for assembly later); and remove the starter from the flywheel housing. Starters are mounted with a pair of cap screws or studs.

Remove the brush cover, observing the position of the screw or snap, because wrong assembly can short the main cable or solenoid wire (Fig. 11-10). Hitachi starters do not have an inspection band as such. The end plate must be removed for access to the brushes and commutator.

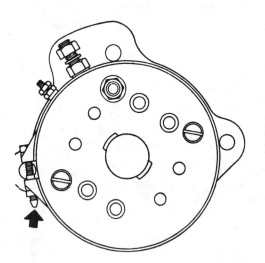

11-10
CAV CA-45C starter; band location in inspection. GM Bedford Diesel

Brushes are sacrificial items and should be replaced when worn to half their original length. The rate of wear should be calculated so that the wear limit will not be reached between inspection periods. Clean the brush holders and commutator with a preparation intended for use on electrical machinery. If old brushes are used, lightly file the flanks at the contact points with the holders to help prevent sticking. New brushes are contoured to match the commutator, but should be fitted by hand. Wrap a length of sandpaper around the commutator—do not use emery cloth—and turn in the normal direction of rotation. Remove the paper and blow out the dust.

Try to move the holders by hand. Most are riveted to the end plate and can become loose, upsetting the brush-commutator relationship. With an ohmmeter, test the insulation on the hot-side brush holders (Fig. 11-11). There should be no continuity between the insulated brush holders and the end plate. Brush spring tension is an important and often overlooked factor in starter performance. To measure it, you

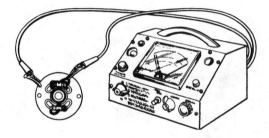

11-11
Testing brush holder insulation.
Tecumseh Products Co.

will need an accurate gauge such as one supplied by Sun Electric. Specifications vary between makes and models, but the spring tension measured at the free (brush) end of the spring should be at least 1½ lb. Some specifications call for 4 lb.

The commutator bars should be examined for arcing, scores, and obvious eccentricity. Some discoloration is normal. If more serious faults are not apparent, buff the bars with a strip of 000 sandpaper.

Armature

Further disassembly requires that the armature shaft be withdrawn from the clutch mechanism. Some starters employ a snap ring at the power takeoff end of the shaft to define the outer limit of pinion movement. Others use a stop nut (Fig. 11-12, No. 49). The majority of armatures can be withdrawn with the pinion and overrunning clutch in place. The disengagement point is at the yoke (Fig. 11-9, No. 9) and sleeve on the clutch body. Remove the screws holding the solenoid housing to the frame and withdraw the yoke pivot pin. The pin might have a threaded fastener with an eccentric journal, as in the case of Ford designs. The eccentric allows for wear compensation. Or it might be a simple cylinder, secured by a flanged head on one side and a cotter pin or snap ring on the other. When the pin is removed there will be enough slack in the mechanism to disengage the yoke from the clutch sleeve, and the armature can be withdrawn. Observe the position of shims—usually located between the commutator and end plate—and, on CAV starters, the spring-loaded ball. This mechanism is shown in Fig. 11-12 as 18 and 34. It allows a degree of end float so that the armature can recoil if the pinion and flywheel ring gear do not mesh on initial contact.

The armature should be placed in a jig and checked for trueness because a bent armature will cause erratic operation and might, in the course of long use, destroy the flywheel ring gear. Figure 11-13 illustrates an armature chucked between lathe centers. The allowable deflection at the center bearing is 0.002 in., or 0.004 in. on the gauge. With the proper fixtures and skill with an arbor press, an armature shaft can be straightened, although it will not be as strong as it was originally and will be prone to bend again. The wiser course is to purchase a new armature.

Make the same check on the commutator. Allowable out-of-roundness is 0.012–0.016 in., or less, depending on the rate of wear and the intervals between inspection periods. Commutators that have lost their trueness or have become pitted should be turned on a lathe. The cutting tool must be racked more for copper than for steel. Do not allow the copper to smear into the slots between the segments. Chamfer the end

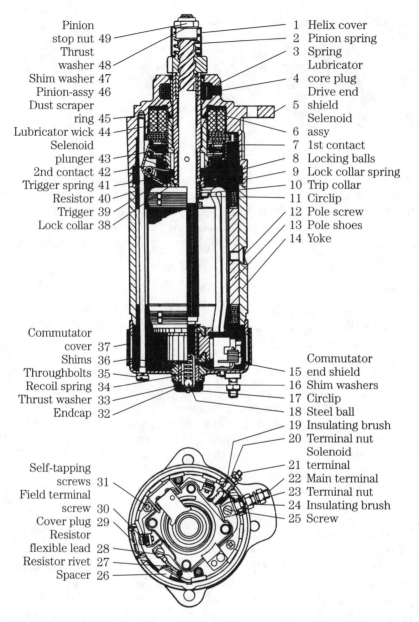

Pinion stop nut 49
Thrust washer 48
Shim washer 47
Pinion-assy 46
Dust scraper ring 45
Lubricator wick 44
Selenoid plunger 43
2nd contact 42
Trigger spring 41
Resistor 40
Trigger 39
Lock collar 38

1 Helix cover
2 Pinion spring
3 Spring
4 Lubricator core plug
5 Drive end shield
6 Selenoid assy
7 1st contact
8 Locking balls
9 Lock collar spring
10 Trip collar
11 Circlip
12 Pole screw
13 Pole shoes
14 Yoke

Commutator cover 37
Shims 36
Throughbolts 35
Recoil spring 34
Thrust washer 33
Endcap 32

15 Commutator end shield
16 Shim washers
17 Circlip
18 Steel ball
19 Insulating brush
20 Terminal nut
21 Solenoid terminal
22 Main terminal
23 Terminal nut
24 Insulating brush
25 Screw

Self-tapping screws 31
Field terminal screw 30
Cover plug 29
Resistor flexible lead 28
Resistor rivet 27
Spacer 26

11-12 Typical CAV starter. GM Bedford Diesel.

of the commutator slightly. Small imperfections can be removed by chucking the commutator in a drill press and turning against a single-cut file.

It is necessary that the insulation (called, somewhat anachronistically, *mica*) be buried below the segment edges; otherwise, the brushes will come into contact with the insulation as the copper segments wear. Undercutting should be limited to 0.015 in. or so. Tools are available for this purpose, but an acceptable job can be done with

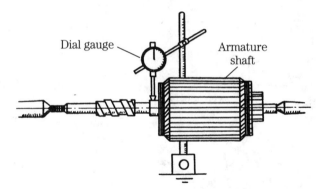

11-13 Checking armature shaft defection. Marine Engine Div., Chrysler Corp.

a hacksaw blade (flattened to fit the groove) and a triangular file for the final bevel cut (Fig. 11-14).

Inspect the armature for evidence of overheating. Extended cranking periods, dragging bearings, chronically low battery charge, or an under-capacity starter will cause the solder to melt at the commutator-armature connections. Solder will be splattered over the inside of the frame. Repairs might be possible because these connections are accessible. Continued overheating will cause the insulation to flake and powder. Discoloration is normal, but the insulation should remain resilient.

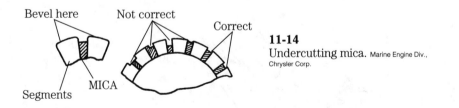

11-14 Undercutting mica. Marine Engine Div., Chrysler Corp.

Armature insulation separates the nonferrous parts (notably the commutator segments) from the ferrous parts (the laminated iron segments extending to the outer diameter of the armature and shaft). Make three tests with the aid of a 120-V continuity lamp or a *megger* (meter for very high resistances). An ordinary ohmmeter is useless to measure the high resistances involved. In no case should the lamp light up or resistance be less than 1 megohm.

Warning: Exercise extreme care when using 120-V test equipment.

- Test between adjacent commutator segments.
- Test between individual commutator segments and the armature form (Fig. 11-15).
- Test between armature or commutator segments and the shaft.

It is possible for the armature windings to short, thus robbing starter torque. Place the armature on a growler and rotate it slowly while holding a hacksaw blade over it as shown in Fig. 11-16. The blade will be strongly attracted to the armature segments because of the magnetic field introduced in the windings by the growler.

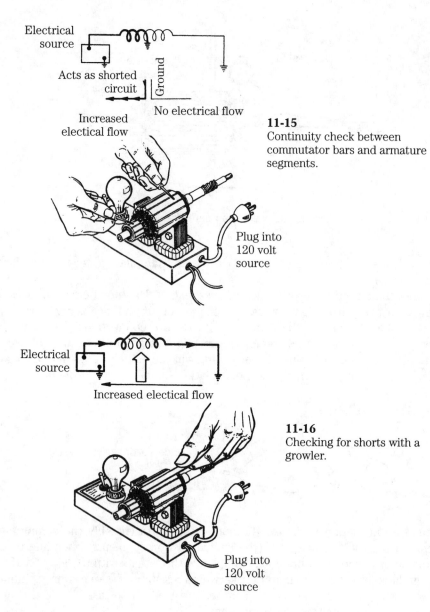

Electrical
source

Acts as shorted
circuit

Ground

No electical flow

Increased
electical flow

11-15
Continuity check between
commutator bars and armature
segments.

Plug into
120 volt
source

Electrical
source

Increased electical flow

11-16
Checking for shorts with a
growler.

Plug into
120 volt
source

But if the blade does a Mexican hat dance over a segment, you can be sure that the
associated winding is shorted.

Field coils

After armatures, the next most likely source of trouble is the field coils. Field
hookups vary. The majority are connected in a simple series circuit, although you
will encounter starters with *split fields*, each pair feeding off its own insulated
brush. Field resistance values are not as a rule supplied in shop manuals, although a
persistent mechanic can obtain this and other valuable test data by writing the
starter manufacturer. Be sure to include the starter model and serial number.

In the absence of a resistance test, which would detect intracoil shorts, the only tests possible are to check field continuity (an ordinary ohmmeter will do) and to check for shorts between the windings and frame. Connect a lamp or megger to the fields and touch the other probe to the frame, as illustrated in Fig. 11-17. Individual fields can be isolated by snipping their leads.

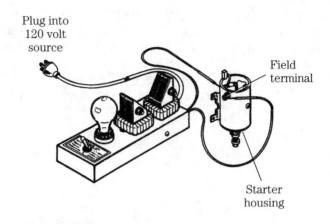

11-17 Checking for grounded fields. Tecumseh Products Co.

The fields are supported by the pole shoes, which in turn, are secured to the frame by screws. The screws more often than not will be found to have rusted to the frame. A bit of persuasion will be needed, in the form of penetrating oil and elbow grease. Support the frame on a bench fixture and, with a heavy hammer, strike the screwdriver exactly as if you were driving a spike. If this does not work, remove the screw with a cape chisel. However, you might (depending on the starter make and your supply of junk parts) have difficulty in matching the screw thread and head fillet.

Coat the screw heads with Loctite before installation and torque securely. Be sure that the pole shoe and coil clears the armature and that the leads are tucked out of the way.

Bearings

The great majority of starters employ sintered bronze bushings. In time these bushings wear and must be replaced to ensure proper teeth mesh at the flywheel and prevent armature drag. In extreme cases the bushings can wear down to their bosses so that the shaft rides on the aluminum or steel end plates. The old bushings are pressed out and new ones pressed in. Tools are available to make this job easier, particularly at the blind boss on the commutator end plate.

Without these tools, the bushing can be removed by carefully ridging and collapsing it inward, or by means of hydraulic pressure. Obtain a rod that matches shaft diameter. Pack the bushing with grease and hammer the rod into the boss. Because the grease cannot easily escape between the rod and bushing, it will lift the bushing up and out. Because of the blind boss, new bushings are not reamed.

Final tests

The tests described thus far have been *static* tests. If a starter fails a static test, it will not perform properly, but passing does not guarantee the starter is faultless. The only sure way to test a starter—or, for that matter, any electrical machine—is to measure its performance under known conditions, and against the manufacturer's specifications or a known-good motor.

No-load performance is checked by mounting the starter in a vise and monitoring voltage, current, and rpm. Figure 11-18 shows the layout. The voltage is the reference for the test and is held to 12 V. Current drain and rpm are of course variable, depending on the resistance of the windings, their configuration, and whether or not speed-limiting coils are provided. You can expect speeds of 4000–7000 rpm and draws of 60–100 A. In this test we are looking for low rpm and excessive current consumption.

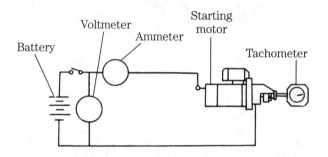

11-18 No-load starter test. Marine Engine Div., Chrysler Corp.

The locked-rotor and stall test requires a scale to accept the pinion gear (Fig. 11-19). It must be made quickly, before the insulation melts. Mount the motor securely and lock the pinion. Typically the voltage will drop to half the normal value. Draw might approach or, with the larger starters, exceed 100 A.

Solenoids

Almost all diesel starters are engaged by means of a solenoid mounted on the frame. At the same time the solenoid plunger moves the pinion, it closes a pair of contacts to complete the circuit to the starter motor. In other words the component consists of a solenoid or *linear motor* and a relay (Fig. 11-20). Some circuits feature a second remotely mounted relay, as shown in Figs. 11-20 and 11-21. The circuit depicted in the earlier illustration was designed for marine use.

The starter switch might be 30 ft. or more from the engine. To cut wiring losses a second relay is installed, which energizes the piggyback solenoid. The additional relay (2St in Fig. 11-21) also gives overspeed protection should the switch remain closed after the engine fires. It functions in conjunction with a direct current generator. Voltage on the relay windings is the difference between generator output and battery terminal voltage. When the engine is cranking, generator output is function-

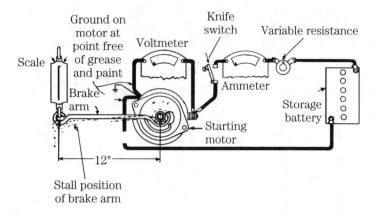

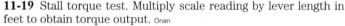

11-19 Stall torque test. Multiply scale reading by lever length in feet to obtain torque output. Onan

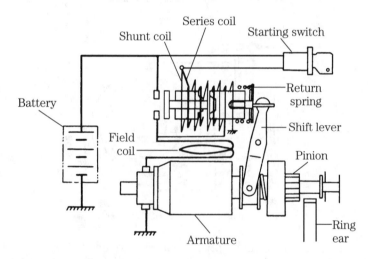

11-20 Solenoid internal wiring diagram.

ally zero. The relay closes and completes the circuit to the solenoid. When the engine comes up to speed, generator output bucks battery voltage and the relay opens, automatically disengaging the starter. As admirable as such a device is, it should not be used continually, because some starter overspeeding will still occur, with detrimental effects to the bearings.

Should the generator circuit open, the starter will be inoperative because the return path for the overspeed relay is through the generator. In an emergency the engine can be started by bridging the overspeed terminals. Of course, the transmission or other loads must be disengaged and, in a vehicle, the handbrake must be engaged.

In addition, you will notice that the solenoid depicted in Fig. 11-21 has two sets of contacts. One set closes first and allows a trickle of current to flow through the resistor (represented by the wavy line above the contact arm) to the motor. The armature

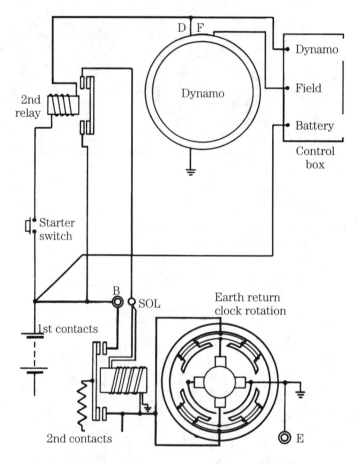

11-21 Overspeed relay and solenoid wiring diagram. GM Bedford Diesel

barely turns during the engagement phase. But once engaged a trigger is released and the second set of contacts closes, shunting the resistor and applying full battery current to the motor. This circuit, developed by CAV, represents a real improvement over the brutal spin-and-hit action of solenoid-operated and inertia clutches and should result in longer life for all components, including the flywheel ring gear.

Solenoids and relays can best be tested by bridging the large contacts. If the starter works, you know that the component has failed. Relays are sealed units and are not repairable. But most solenoids are at least amenable to inspection. Repairs to the series or shunt windings (illustrated in Figs. 11-20 and 11-21) are out of the question unless the circuit has opened at the leads. Contacts can be burnished, and some designs have provision for reversing the copper switch element.

Starter drives

Most diesel motors feature positive engagement drives energized by the solenoid. The solenoid is usually mounted piggyback on the frame and the pinion moved by means of a pivoted yoke (Fig. 11-20).

Regardless of mechanical differences between types, all starter drives have these functions:

- The pinion must be moved laterally on the shaft to engage the flywheel.
- The pinion must be allowed to disengage when flywheel rpm exceeds pinion rpm.
- The pinion must be retracted clear of the flywheel when the starter switch is opened.

Figure 11-22 illustrates a typical drive assembly. The pinion moves on a helical thread. Engagement is facilitated by a bevel on the pinion and the ring gear teeth of the flywheel. Extreme wear on either or both profiles will lock the pinion.

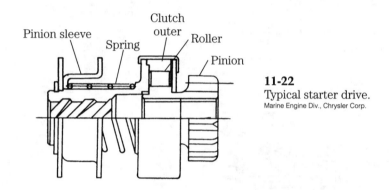

11-22
Typical starter drive.
Marine Engine Div., Chrysler Corp.

The clutch shown employs ramps and rollers. During the motor drive phase the rollers are wedged into the ramps (refer to Fig. 11-23). When the engine catches, the rollers are freed and the clutch overruns. Other clutches employ balls or, in a few cases, ratchets. The spring retracts the drive when the solenoid is deactivated.

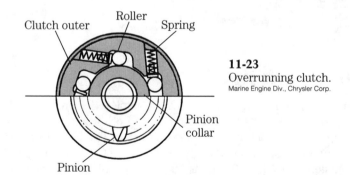

11-23
Overrunning clutch.
Marine Engine Div., Chrysler Corp.

Inspect the pinion teeth for excessive wear and chipping. Some battering is normal and does not affect starter operation. The clutch mechanism should be disassembled (if possible), cleaned, and inspected. Inspect the moving parts for wear or deformation, with particular attention to the ratchet teeth and the ramps. Lubricate with Aero Shell 6B or the equivalent. Sealed drives should be wiped with a solvent-wetted rag. Do not allow solvent to enter the mechanism, because it will dilute the lubricant and cause premature failure. Test the clutch for engagement in one direction of pinion rotation and for disengagement in the other.

Adjustment of the *pinion throw* is important to ensure complete and full mesh at the flywheel. Throw is measured between the pinion and the stop ring as shown in Fig. 11-24. Adjustment is by adding or subtracting shims at the solenoid housing (Fig. 11-25), moving the solenoid mounting bolts in their elongated slots, or turning the yoke pin eccentric.

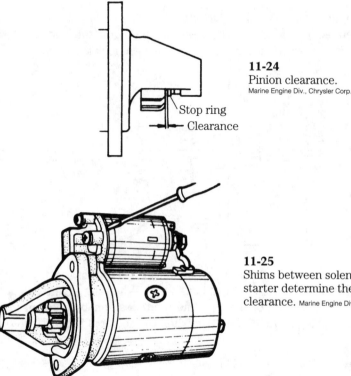

11-24
Pinion clearance.
Marine Engine Div., Chrysler Corp.

Stop ring
Clearance

11-25
Shims between solenoid body and starter determine the pinion clearance. Marine Engine Div., Chrysler Corp.

Charging systems

The charging system restores the energy depleted from the battery during cranking and provides power to operate lights and other accessories. It consists of two major components: an alternator and a regulator. The circuit can be monitored by an ammeter or a lamp and is usually fused to protect the generator windings.

Alternators generate a high frequency 3-phase ac voltage, which is rectified (i.e., converted to direct current) by internal diodes. Wrong polarity will destroy the diodes and can damage the wiring harness. Observe the polarity when installing a new battery or when using jumper cables. Before connecting a charger to the battery, disconnect the cables. Should the engine be started with the charger in the circuit the regulator might be damaged. Isolate the charging system before any arc welding is done. Do not disconnect the battery or any other wiring while the alternator is turning. And, finally, do not attempt to polarize an alternator. The exercise is fruitless and can destroy diodes.

Initial tests

The charge light should be on with the switch on and the engine stopped. Failure to light indicates an open connection in the bulb itself or in the associated wiring. Most charging-lamp circuits operate by a relay under the voltage regulator cover. Lucas systems employ a separate relay that responds to heat. The easiest way to check either type is to insert a 0–100-A ammeter in series with the charging circuit. If the meter shows current and the relay does not close, one can safely assume that it has failed and should be replaced. The Lucas relay can be tested as shown in Fig. 11-26. You will need a voltage divider and 2.2-W lamp. Connect clip A to the 12-V terminal. The lamp should come on. Leaving the 12-V connection in place, connect clip B to the 6-V tap. The bulb should burn for 5 seconds or so and go out. Move B to the 12-V post and hold for no more than 10 seconds. Then move it to the 2-V (single cell) tap. The bulb should come on within 5 seconds. These units do not have computerlike precision, and some variation can be expected between them. But the test results should roughly correlate with the test procedure. Do not attempt to repair a suspect relay.

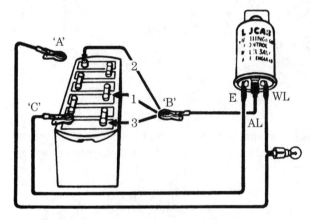

11-26
Testing Lucas charging lamp relay. To distinguish these relays from turn-signal flashers, Lucas has coded them green.
GM Bedford Diesel

Test the alternator output against the meter on the engine or by inserting a test meter in series between the B terminal and battery. Voltage is monitored with a meter in parallel with the charging circuit. Discharge the battery by switching on the lights and other accessories. Connect a rheostat or carbon pile across the battery for a controlled discharge. (Without this tool you will be reduced to guessing about alternator condition.) With the load set at zero, start the engine and operate at approximately midthrottle. Apply the load until the alternator produces its full rated output. If necessary open the throttle wider. An output 2–6 A below rating often means an open diode. Ten amps or so below rating usually means a shorted diode. The alternator might give further evidence of a diode failure by whining like a wounded banshee.

The voltage should be 18–20 V above the nominal battery voltage under normal service conditions. It might be higher by virtue of automatic temperature compensation in cold weather.

Assuming that the output is below specs, the next step is to isolate the alternator from the regulator. Disconnect the field (F or FD) terminal from the regulator and ground it to the block. Load the circuit with a carbon pile to limit the voltage output. Run the engine at idle. In this test we have dispensed with the regulator and are protecting the alternator windings with carbon pile. No appreciable output difference between this and the previous tests means that the regulator is doing its job. A large difference would indicate that the regulator was defective.

Late-production alternators often have integrated regulators built into the slip ring end of the unit. Most have a provision for segregating alternator output from the regulator so that "raw" outputs can be measured. The Delcotron features a shorting tab. A screwdriver is inserted into an access hole in the back of the housing (Fig. 11-27); contact between the housing and the tab shorts the fields.

Note: The tab is within ¾ in. of the casting. Do not insert a screwdriver more than 1 in. into the casting.

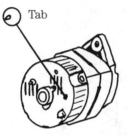

11-27
Location of Delcotron shorting tab.

Bench testing

Disconnect the battery and remove the alternator from the engine at the pivot and belt-tensioning bracket. Three typical alternators are shown in exploded view in Figs. 11-28 through 11-30.

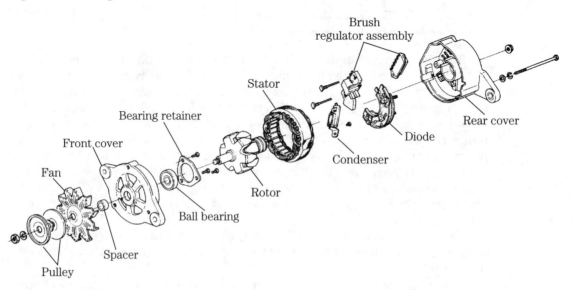

11-28 US generic pattern alternator.

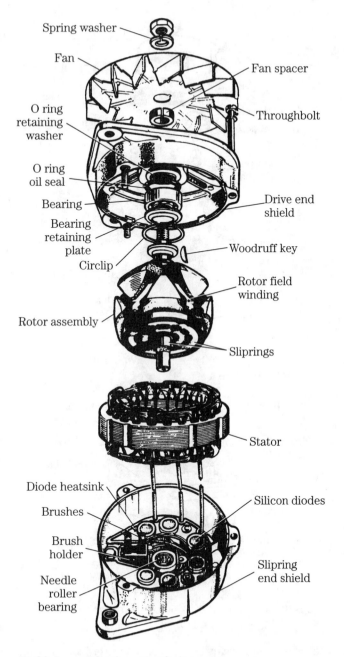

Spring washer

Fan

Fan spacer

O ring
retaining
washer

Throughbolt

O ring
oil seal

Bearing

Drive end
shield

Bearing
retaining
plate

Circlip

Woodruff key

Rotor field
winding

Rotor assembly

Sliprings

Stator

Diode heatsink

Brushes

Silicon diodes

Brush
holder

Needle
roller
bearing

Slipring
end shield

11-29 Lucas 10-AC or 11-AC alternator. GM Bedford Diesel

Remove the drive pulley. A special tool might be needed on some of the automotive derivations. Hold the fan with a screwdriver and turn the fan nut counterclockwise. Tap the sheave and fan off the shaft with a mallet. Remove the throughbolts and separate the end shields.

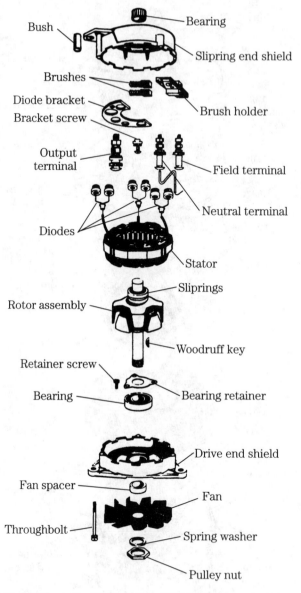

Bush — Bearing — Sliprings end shield

Brushes — Brush holder

Diode bracket —

Bracket screw —

Output terminal — Field terminal

Neutral terminal

Diodes — Stator

Sliprings

Rotor assembly —

Woodruff key

Retainer screw —

Bearing — Bearing retainer

Drive end shield

Fan spacer — Fan

Throughbolt — Spring washer

Pulley nut

11-30 Prestolite CAB-1235 or CAB-1245 alternator.

GM Bedford Diesel

Inspect the brushes for wear. Some manufacturers thoughtfully provide a wear limit line on the brushes (Fig. 11-31). Clean the holders with Freon or some other non-petroleum-based solvent and check the brushes for ease of movement. File lightly if they appear to bind. The slip rings should be miked for wear and eccentricity. Ten to twelve thousandths should be considered the limit (Fig. 11-32). Slip rings are usually, but not always, integral with the rotor. Removable rings are chiseled off and new ones pressed into place. Fixed rings can be restored to concentricity with light machining.

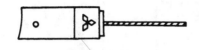

11-31
Brush showing wear limit line.
Chrysler Corp.

Wear limit line

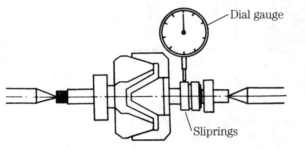

Dial gauge

11-32
Determining slipring
concentricity. Marine Engine Div.,
Chrysler Corp.

Sliprings

Determine the condition of the rotor insulation with a 120-V test lamp (Fig. 11-33). The slip rings and their associated windings should be insulated from the shaft and pole pieces. If you have access to an accurate, low-range ohmmeter, test for continuity between slip rings. The resistance might lead one to suspect a partial open; less could mean an intracoil short.

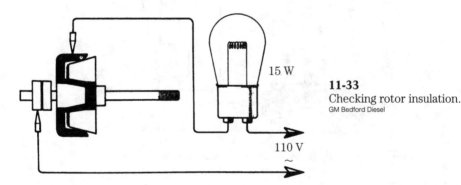

15 W

11-33
Checking rotor insulation.
GM Bedford Diesel

110 V
~

The stator consists of three distinct and independent windings whose outputs are 120 degrees apart. It is possible for one winding to fail without noticeably affecting the others. Peak alternator output will, of course, be reduced by one third. Disconnect the three leads going to the stator windings. Many European machines have these leads soldered, while American designs generally have terminal lugs. When unsoldering, be extremely careful not to overheat the diodes. Exposure to more than 300°F will upset their crystalline structure. Test each winding for resistance. Connect a low-range ohmmeter between the natural lead and each of three winding leads as shown in Fig. 11-34. Resistance will be quite low—on the order of 5 or 6 ohms—and becomes critical when one group of windings gives a different reading than the others.

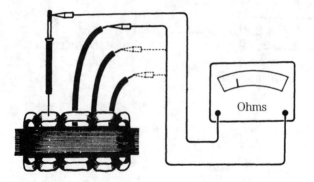

11-34 Comparison test between stator windings.
GM Bedford Diesel

Test the stator insulation with a 120-V lamp connected as shown in Fig. 11-35. There should be no continuity between the laminations and windings.

The next step is to check the diodes (Fig. 11-36). You might already have had some evidence of diode trouble in the form of alternator whine or blackened varnish on the stator coils. The diodes must be tested with an ohmmeter or a test lamp of the same voltage as generator output.

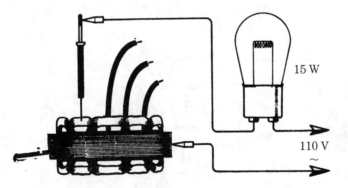

15 W

110 V
∼

11-35 Comparison test between stator windings. GM Bedford Diesel

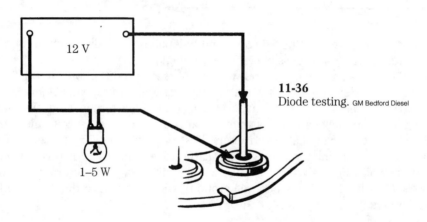

12 V

11-36
Diode testing. GM Bedford Diesel

1–5 W

Test each diode by connecting the test leads and then reversing their polarity. The lamp should light in one polarity and go out in the other. Failure to light at all means an open diode; continuous burning means the diode has shorted. In either case it must be replaced. You might use a low-voltage ohmmeter in lieu of a lamp. Expect high (but not infinite) resistance with one connection, and low (but not zero) resistance when the two leads are reversed.

To simply service and limit the need for special tools, some manufacturers package mounting brackets with their diodes. The bracket is a heat sink and must be in intimate contact with the diode case. Other manufacturers take the more traditional approach and supply individual diodes, which must be pressed (not hammered) into their sinks. K-D Tools makes a complete line of diode removal and installation aids, including heat sink supports and diode arbors of various diameters. Figure 11-37 shows a typical installation with an unsupported heat sink. Other designs might require support.

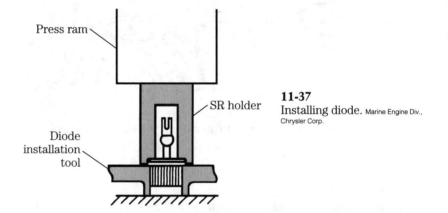

11-37
Installing diode. Marine Engine Div., Chrysler Corp.

Soldering the connections is very critical. Should the internal temperature reach 300°F the diode will be ruined. Use a 150 W or smaller iron and place a thermal shunt between the soldered joint and the diode (Fig. 11-38). The shunt might be in the form of a pair of needle-nosed pliers or copper alligator clips. In some instances there might not be room to shunt the heat load between the diode and joint (Fig. 11-38B). It is only necessary to twist the leads enough to hold them while the solder is liquid. Work quickly and use only enough solder to flow between the leads. More solder merely increases the thermal load and increases the chances that the diode will be ruined.

Alternator bearings are sealed needle and ball types. They are not to be disturbed unless noisy or rough. Then bearings are pressed off and new ones installed with the numbered end toward the arbor.

Voltage regulation

Most alternator-based charging circuits employ voltage regulation (as opposed to voltage and current regulation). The regulator can be external to the alternator or integral with it. External regulators can be mechanical or solid state.

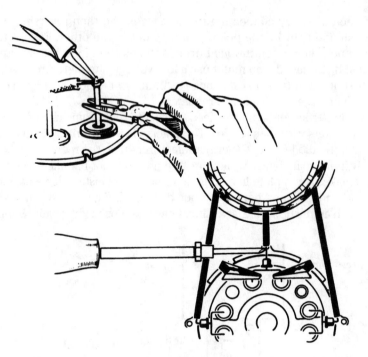

11-38 Two forms of heatsinks to protect the diode when soldering.
GM Bedford Diesel

External regulators

Figure 11-39 is a schematic of a typical mechanical voltage regulator. The voltage-sensing winding is in parallel with the output and drives NC (normally closed) contacts. As the generator comes up to speed, voltage increases until the winding develops a strong enough field to open the contacts. Rotor output then passes through dropping resistor RF.

Depending on the make and model, voltage adjustment is accomplished by bending the stationary contact, moving the hinges in elongated mounting holes, or by screw (Fig. 11-40). In theory, the correct point gap should correspond with an output voltage of approximately 15 V at 68°F or 28 V for 24-V systems. In practice, better results are had by measuring alternator output voltage at the battery terminals. Assuming that specifications are available, core and yoke gap adjustments also can be made.

Clean oxidized contacts with a riffle file or a diamond-faced abrasive strip. Do not use sandpaper or emery cloth. Inspect the dropping resistor (often found on the underside of the unit), springs, and contact tips for evidence of overheating. Check the regulator ground connection.

Before discarding a defective regulator, attempt to discover why it failed. Burnt points or heat-discolored springs mean high resistance in the charging circuit or a bad regulator ground.

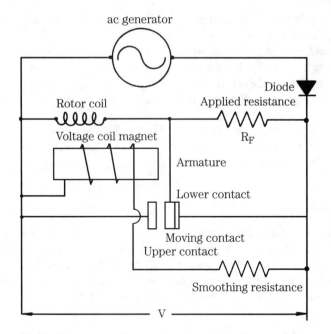

11-39 Voltage regulator in alternator circuit.
Marine Engine Div., Chrysler Corp.

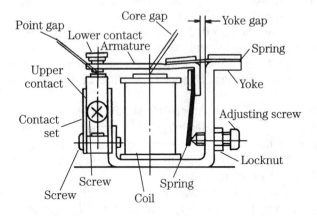

11-40 A typical relay and its adjustment points.
Marine Engine Div., Chrysler Corp.

Solid-state regulators

Transistorized regulators are capable of exceedingly fine regulation, partially because there are no moving parts. Durability is exceptional. On the other hand any internal malfunction generally means that the unit must be replaced. As a rule, no repairs are possible. Failure can occur because of manufacturing error (this usually shows up in the first few hours of operation and is covered by warranty), high current draws, and voltage spikes.

The mechanic must be particularly alert when working with transistorized circuits. The cautions that apply to alternator diodes apply with more force to regulators if only because regulators are more expensive. Do not introduce stray voltages, cross connections, reverse battery polarity, or open connections while the engine is running.

The regulator might be integral with the alternator or might be contained in a separate box. In general, no adjustment is possible; however, the Lucas 4TR has a voltage adjustment on its bottom, hidden under a dab of sealant.

Batteries

The battery has three functions: provide energy for the starting motor; stabilize voltages in the charging system; and, for limited periods, provide energy for the accessories in the event of charging-circuit failure. Because it is in a constant state of chemical activity and is affected by temperature changes, aging, humidity, and current demands, the battery requires more attention than any other component in the electrical system.

Battery ratings

Starting a diesel engine puts a heavy drain on the battery, especially in cold climates. One should purchase the best quality and the largest capacity practical. The physical size of the battery is coded by its *group number*. The group number has only an indirect bearing on electrical capacity but does ensure that replacements will fit the original brackets. In some instances a larger capacity battery might require going to another group number. Expect to modify the bracket and possibly to replace one or more cables.

The traditional measure of a battery's ability to do work is its *ampere-hour* (A-hr.) *capacity*. The battery is discharged at a constant rate for 20 hr. so that the potential of each cell drops to 1.75 V. A battery that will deliver 6 A over the 20-hr. period is rated at 120 A-hr. (6 A × 20 hr.). You will find this rating stamped on replacement batteries or in the specifications.

Like all rating systems, the ampere-hour rating is best thought of as a yardstick for comparison between batteries. It has absolute validity only in terms of the original test. For example, a 120-A-hr. battery will not deliver 120 A for 1 hr., nor will it deliver 1200 A for 6 minutes.

Cranking-power tests are more meaningful because they take into account the power loss that lead-acid batteries suffer in cold weather. At room temperature the battery develops its best power; power output falls off dramatically around 0°F. At the same time, the engine becomes progressively more difficult to crank and more reluctant to start. Several cranking-power tests are in use.

Zero cranking power is a hybrid measurement expressed in volts and minutes. The battery is chilled to 0°F; depending on battery size, a 150- or 300-A load is applied. After 5 seconds the voltage is read for the first part of the rating. Discharge continues until the terminal voltage drops to 5 V. The time in minutes between full charge and effective exhaustion is the second digit in the rating. The higher these two numbers are for batteries in the same load class, the better.

The *cold cranking performance* rating is determined by lowering the battery temperature to 0°F (or, in some instances, 20°F) and discharging for 30 seconds at such a rate that the voltage drops below 1.2 V per cell. This is the most accepted of all cold weather ratings and has become standard in specification sheets.

Battery tests

As the battery discharges, some of the sulfuric acid in the electrolyte decomposes into water. The strength of the electrolyte in the individual cells is a reliable index of the state of charge. There are several ways to determine acidity, but long ago technicians fixed on the measurement of specific gravity as the simplest and most reliable.

The instrument used is called a *hydrometer*. It consists of a rubber bulb, a barrel, and a float with a graduated tang. The graduations are in terms of specific gravity. Water is assigned a specific gravity of 1. Pure sulfuric acid is 1.83 times heavier than water and thus has a specific gravity of 1.83. The height of the float tang above the liquid level is a function of fluid density, or specific gravity. The battery is said to be fully charged when the specific gravity is between 1.250 and 1.280.

An accurate hydrometer test takes some doing. The battery should be tested prior to starting and after the engine has run on its normal cycle. For example, if the engine is shut down overnight, the test should be made in the morning, before the first start. Water should be added several operating days before the test to ensure good mixing. Otherwise the readings can be deceptively low.

Use a hydrometer reserved for battery testing. Specifically, do not use one that has been used as an antifreeze tester. Trace quantities of ethylene glycol will shorten the battery's life.

Place the hydrometer tip above the plates. Contact with them can distort the plates enough to short the cell. Draw in a generous supply of electrolyte and hold the hydrometer vertically. You might have to tap the side of the barrel with your fingernail to jar the float loose. Holding the hydrometer at eye level, take a reading across the fluid level. Do not be misled by the meniscus (concave surface, Fig. 11-41) of the fluid.

American hydrometers are calibrated to be accurate at 80°F. For each 10°F above 80°F, add 4 points (0.004) to the reading; conversely, for each 10°F below the standard, subtract 4 points. The standard temperature for European and Japanese hydrometers is 20°C, or 68°F. For each 10°C increase add 7 points (0.007); subtract a like amount for each 10°C decrease. The more elaborate hydrometers have a built-in thermometer and correction scale.

All cells should read within 50 points (0.050) of each other. Greater variation is a sign of abnormality and might be grounds for discarding the battery. The relationship between specific gravity and state of charge is shown in Fig. 11-42.

The hydrometer test is important, but by no means definitive. The state of charge is only indirectly related to the actual output of the battery. Chemically the

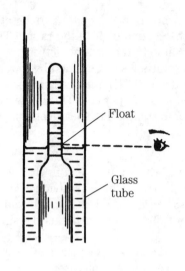

11-41
Reading hydrometer.
Marine Engine Div., Chrysler Corp.

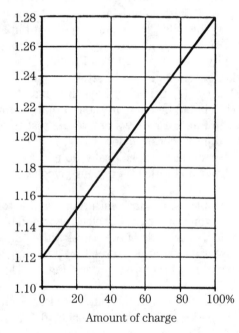

11-42
Relationship between state of charge and specific gravity.
Marine Engine Div., Chrysler Corp.

battery might have full potential, but unless this potential passes through the straps and terminals, it is of little use.

Perhaps the single most reliable test is to load the battery with a rheostat or carbon pile while monitoring the terminal voltage. The battery should be brought up to full charge before the test. The current draw should be adjusted to equal three times the ampere-hour rating. Thus, a 120 A-hr. battery would be discharged at a rate of 360 A. Continue the test for 15 seconds and observe the terminal voltage. At no time should the voltage drop below 9.5 V.

In this test, sometimes called the *battery capacity* test, we used voltage as the telltale. But without a load, terminal voltage is meaningless. The voltage remains almost constant from full charge to exhaustion.

Battery maintenance

The first order of business is to keep the electrolyte level above the plates and well into the reserve space below the filler cap recesses. Use distilled water. Tap water might be harmful, particularly if it has iron in it.

Inspect the case for cracks, acid seepage, and (in the old asphalt-based cases) softening. Periodically remove the cable clamps and scrape them and the battery terminals. Look closely at the bond between the cables and clamp. The best and most reliable cables have forged clamps, solder-dipped for conductivity. Replace spring clip and other clever designs with standard bolt-up clamps sweated to the cable ends. After scraping and tightening, coat the terminals and clamps with grease to provide some protection from oxidation.

The battery case should be wiped clean with a damp rag. Dirt, spilled battery acid, and water are conductive and promote self-discharge. Accumulated deposits can be cleaned and neutralized with a solution of baking soda, water, and detergent. Do not allow any of the solution to enter the cells, where it would dilute the electrolyte. Rinse with clear water and wipe dry.

Charging

Any type of battery charger can be used—selenium rectifier, tungar rectifier, or, reaching way back, mercury arc rectifier. Current and voltage should be monitored and there should be a provision for control. When charging multiple batteries from a single output, connect the batteries in series as shown in Fig. 11-43.

Batteries give off hydrogen gas, particularly as they approach full charge. When mixed with oxygen, hydrogen is explosive. Observe these safety precautions:

- Remove all filler caps (to prevent pressure rise should the caps be clogged).
- Charge in a well ventilated place remove from open flames or heat.
- Connect the charger leads before turning the machine on. Switch the machine off before disconnecting the leads.

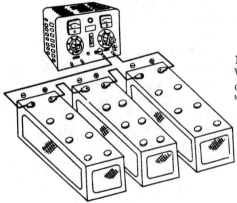

11-43
When charging multiple batteries, connect in series.
Marine Engine Div., Chrysler Corp.

In no case should the electrolyte temperature be allowed to exceed 115°F. If your charger does not have a thermostatic control, keep track of the temperature with an ordinary thermometer.

Batteries can be charged by any of three methods. *Constant-current* charging is by far the most popular. The charging current is limited to one-tenth of the ampere-hour rating of the battery. Thus a 120 A-hr. battery would be charged at 12 A. Specific gravity and no-load terminal voltage are checked at 30-minute intervals. The battery might be said to be fully charged when both values peak (specific gravity 1.127–1.129, voltage 15–16.2 V) and hold constant for three cranking intervals.

A *quick charge*, also known as a booster or hotshot, can bring a battery back to life in a few minutes. The procedure is not recommended in any situation short of an emergency, because the high-power boost will raise the electrolyte temperature and might cause the plates to buckle. Disconnect the battery cables to isolate the generator or alternator if such a charge is given the battery while it is in place.

A constant-voltage charge can be thought of as a compromise between the hotshot and the leisurely constant-current charge. The idea is to apply a charge by keeping charger voltage 2.2–2.4 V higher than terminal voltage. Initially the rate of charge is quite high; it tapers off as the battery approaches capacity.

Battery hookups

One of the most frequently undertaken field modifications is the use of additional batteries. The additional capacity makes starts easier in extremely cold weather and adds reliability to the system.

To increase capacity add one or more batteries in parallel—negative post connected to negative post, positive post to positive post. The voltage will not be affected, but the capacity will be the sum of all parallel batteries.

Connecting in series—negative to positive, positive to negative—adds voltage without changing capacity. Two 6-V batteries can be connected in series to make a physically large 12-V battery

Cable size is critical because the length and cross-sectional area determine resistance. Engineers at International Harvester have developed the following recommendations, which can be applied to most small, high-speed diesel engines:

- Use cables with integral terminal lugs.
- Use only rosin or other noncorrosive-flux solder.
- Terminal lugs must be stacked squarely on the terminals. Haphazard stacking of lugs should be avoided.
- Where the frame is used as a ground return, it must be measured and this distance added to the cable length to determine the total length of the system. Each point of connection with the frame must be scraped clean and tinned with solder. There should be no point of resistance in the frame such as a riveted joint. Such joints should be bridged with a heavy copper strap.
- Pay particular attention to engine-frame grounds. If the engine is mounted in rubber it is, of course, electrically isolated from the frame.
- Check the resistance of the total circuit by the voltage drop method or by means of Ohm's law ($R = E/I$).

- Use this table as an approximate guide to cable size for standard-duty cranking motors.

System voltage	Maximum resistance	Cable size & length
12 V	0.0012 ohm	Less than 105 in., No. 0 105 to 132 in., No. 00 132 to 162 in., No. 000 162 to 212 in., No. 0000 in parallel.
24 V	0.0020 ohm	Less than 188 in., No. 0 188 to 237 in., No. 00 237 to 300 in., No. 0000 300 to 380 in., No. 0000, or two No. 0 in parallel.

- Suggested battery capacity varies with system voltage (the higher the voltage, the less capacity needed for any given application), engine displacement, compression ratio, ambient air temperature, and degree of exposure. Generalizations are difficult to make, but typically a 300 CID engine with a 12-V standard-duty starter requires a 700-A battery for winter operation. This figure is based on SAE J-5371 specifications and refers to the 30-second output of a chilled battery. At 0°F capacity should be increased to 900 A. The International Harvester 414 CID engine requires 1150, or 1400 A at 0°F. Battery capacity needs roughly parallel engine displacement figures, with some flattening of the curve for the larger and easier-to-start units. In extremely cold weather—below -10°F—capacity should be increased by 50% or the batteries heated.
- RV and boat owners often install a second battery to support accessory loads. Arranging matters so that there is always power available for starting requires special hardware; otherwise, the state of charge of either battery is the average of the two. Figure 11-44 illustrates this proposition.

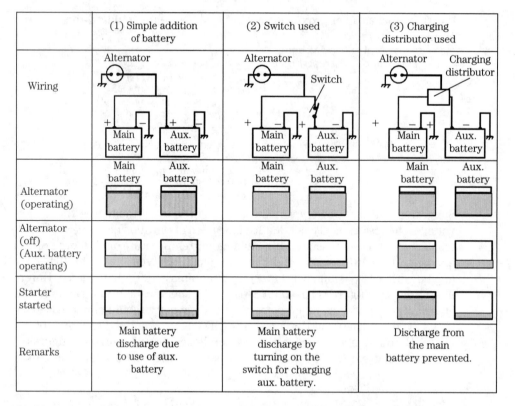

	(1) Simple addition of battery	(2) Switch used	(3) Charging distributor used
Wiring	Alternator Main battery + – Aux. battery + –	Alternator Switch Main battery + – Aux. battery + –	Alternator Charging distributor + Main battery – + Aux. battery –
Alternator (operating)	Main battery Aux. battery	Main battery Aux. battery	Main battery Aux. battery
Alternator (off) (Aux. battery operating)			
Starter started			
Remarks	Main battery discharge due to use of aux. battery	Main battery discharge by turning on the switch for charging aux. battery.	Discharge from the main battery prevented.

11-44 Hard wiring batteries in parallel divides the load equally between both. Installing a manual switch in the B+ line confines the load to the auxiliary battery *until the switch is closed for charging*. When this happens, the main battery promptly discharges into the auxiliary. The best solution is to purchase a charging distributor. To avoid problems, furnish the vendor with a complete schematic and alternator characteristics.

12
CHAPTER

Cooling systems

No engine converts all, or even most, of the heat latent in the fuel to useful work. The diesel is the best of them and still wastes about 60% of the heat input. Some of this waste goes out the exhaust, and the rest is absorbed in the cooling system either directly or indirectly by means of the lube oil. Without some form of cooling the engine would quickly self destruct. The gas temperature in the combustion chamber can exceed 5000°F under normal operation, and gets even hotter when the engine is lugged.

Small diesel engines are air- or liquid-cooled. Air-cooled engines are generally limited to under 40 hp, although Hatz makes a four-cylinder, 245 CID model developing 80 hp. The only customer for larger air-cooled engines is the military.

Air cooling

The major advantage of air cooling is the simplicity and light weight of such a system. There are no pumps, radiators, hoses, oil coolers, and the like to fail and add pounds of dead weight. On the other hand, air-cooled engines are noisy—the same fins that radiate heat also radiate sound. And air-cooled engines do not have the precise temperature regulation that we associate with liquid cooling. Air-cooled engines reach operating temperature quickly, which is all to the good; however, their temperature tends to fluctuate with load and rpm. The flywheel fan pumps more air than is needed at high speeds, and less than is needed when the engine bogs. Of course, this objection does not apply when the load is fixed or when it varies well within the governor's capabilities.

Figure 12-1 shows a typical air-cooled engine. The flywheel has impeller blades cast into its rim. Air moves through the guard and is flung outwards against the shroud and around the cylinders. You might run into an application with reversed air flow. Some generator sets employ the flywheel fan in suction so that the generator gets the cool breeze first.

Air cooling is not entirely maintenance-free. The fins should be brushed and blown clean periodically, and the shrouds must be kept in good repair. Cap screws

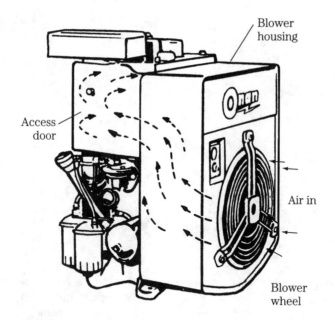

12-1 Typical air flow pattern. <small>Onan</small>

that have vibrated loose can be doctored with a dab of Permatex Lock-Nut or the equivalent. The lube oil level is critical in any engine, and especially in these air pumpers. The lube is a heat sink, and the level might drop more quickly than you expect if you are not accustomed to working with air-cooled engines. Because the engines run hotter than the liquid-cooled types, the cold fit of the parts is a trifle loose. Consequently, oil consumption might be slightly higher, although in theory, everything should be back to normal when operating temperature is reached. For improved temperature control a thermostatically controlled shutter is often employed in this type of system to modulate the air flow (Fig. 12-2). The shutter is opened further by the actuating mechanism in Fig. 12-3 as more cooling is called for.

The shutter-opening temperature is not adjustable, but to ensure complete opening, the power element plunger must contact the shutter roll pin at room temperature. The power element screws should be loosened and the element slid in its mounts until positive contact is made. Failure to open is a serious matter and can quickly put you in the market for a new engine. Check the shutters for freedom of movement and check the Vernatherm element. Checking the element can be done by heating it. At about 120°F the plunger should start to move out of its housing; at 140°F or so the plunger should be extended to its maximum, which is about 0.200 in.

Failure to close completely will rob the engine of power and waste fuel. If the nylon bushings that support the shutter pivots are worn or binding, replace them. Remove the shutters and withdraw the pivot shaft. Install new bearings from the inside of the housing. The large surface serves as a thrust bearing and should be on the inside. End-thrust clearance should be no more than $\frac{1}{32}$ in. One can assume that the actuating rod is in adjustment, because it was set at the factory, and that it will not come out of adjustment unless the housing is warped or otherwise damaged.

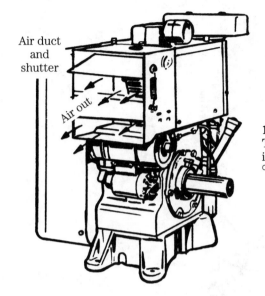

Air duct
and
shutter

Air out

12-2
Thermostatic shutter for
improved temperature control.
Onan

As reliable as this Onan system is, it is not perfect. As a safeguard you should install a thermostatic switch, available from Onan, in series with the solenoid cutoff. The switch is normally closed and opens at 250°F, cutting off the fuel supply. It closes again at about 195°F. To test the thermostatic switch, heat it while monitoring continuity with an ohmmeter across the terminals.

Liquid cooling

Most diesel engines are liquid-cooled. The additional weight and potential for trouble is not an overriding concern in most applications and is compensated for by good control over internal temperatures, relative silence, and ease of manufacture.

The major parts of a liquid-cooled system are the radiator, which dumps heat collected by the coolant into the atmosphere (in a sense, all engines are air-cooled); the circulation pump; the thermostat, which traps some of the coolant in the head for quicker warm-ups; the water jacket surrounding the upper engine; and assorted hoses. Figure 12-4 shows a typical layout as used in a vehicle. Stationary and larger vehicular engines might include such refinements as oil and transmission coolers. Marine engines forego the radiator and circulate raw water or, sometimes, fresh water through a heat exchanger.

Coolant circuits

The basic circuit is shown in Fig. 12-5. With the thermostat closed, water flow is limited to that which can squeeze through a small port in the thermostat body. The engine heats quickly, but without any dead water areas, which could boil. At a predetermined temperature the thermostat opens and circulation is unimpeded.

Air is the enemy of cooling systems and must be kept out of the coolant supply. The usual way to do this is to provide overflow tanks (as in Fig. 12-4) and internal

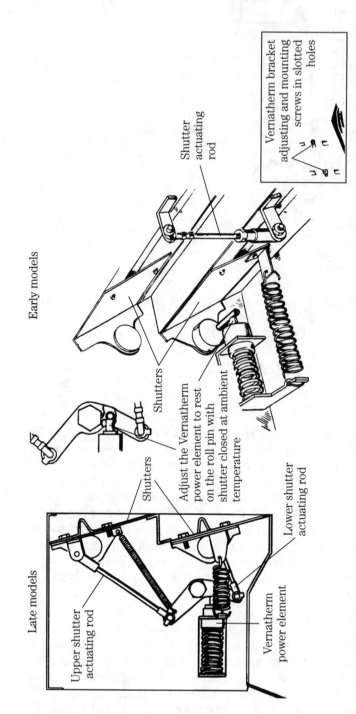

Early models

Shutter actuating rod

Vernatherm bracket adjusting and mounting screws in slotted holes

Late models

Upper shutter actuating rod

Shutters

Adjust the Vernatherm power element to rest on the roll pin with shutter closed at ambient temperature

Shutters

Lower shutter actuating rod

Vernatherm power element

12-3 Shutter operation and adjustment points. Onan

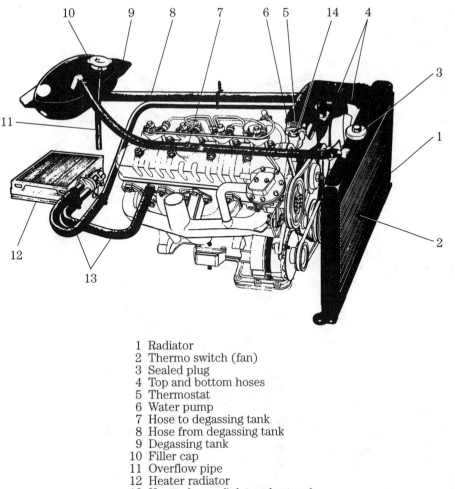

1 Radiator
2 Thermo switch (fan)
3 Sealed plug
4 Top and bottom hoses
5 Thermostat
6 Water pump
7 Hose to degassing tank
8 Hose from degassing tank
9 Degassing tank
10 Filler cap
11 Overflow pipe
12 Heater radiator
13 Heater hoses (inlet and return)
14 Thermistor

12-4 Cooling-system diagram. Peugeot

baffles. The problem is more serious with diesel engines than with gasoline types because diesels tend, as they get older, to leak compression into the coolant. Other sources of air are the pump seals and splash entrapment. Besides reducing the efficiency of the system (air is 3500 times less efficient than water in terms of heat removal), air can damage the pump impeller and cause the coolant to erupt out of the overflow. In the system in Fig. 12-6 the thermostat is shunted by a deaeration line. The line picks up coolant at the highest point in the engine, and one hopes, air as well. When the thermostat opens, coolant still flows through the line because of a built-in restriction in the thermostat body.

The full deaeration system is an elaboration of the one just discussed. As you can see from Fig. 12-7, the thermostat blocks the radiator and directs the coolant back

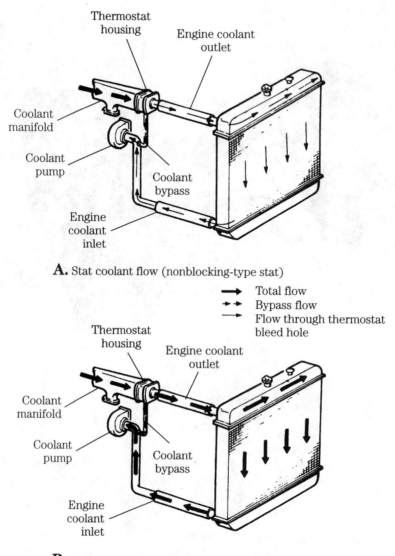

Thermostat housing

Engine coolant outlet

Coolant manifold

Coolant pump

Coolant bypass

Engine coolant inlet

A. Stat coolant flow (nonblocking-type stat)

→ Total flow
➤➤ Bypass flow
→ Flow through thermostat bleed hole

Thermostat housing

Engine coolant outlet

Coolant manifold

Coolant pump

Coolant bypass

Engine coolant inlet

B. Open-stat coolant flow (nonblocking-type stat)

12-5 Basic diesel system. International Harvester

to the engine. The block and head have the benefit of full circulation at all times. The deaeration line picks up air and separates it from the water in the header tank. When the thermostat opens, the deaeration line works in conjunction with a standpipe. This system has advantages, especially in cold climates. The engine warms up faster and, because the radiator cuts in and out of the circuit, the coolant can be at much higher temperatures than the radiator would allow. On the other hand, coolant in the radiator is at near ambient temperature, which means that it can freeze even while the engine is running. Antifreeze is mandatory.

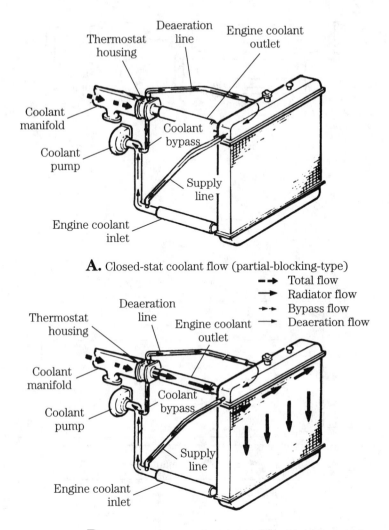

Thermostat housing
Deaeration line
Engine coolant outlet
Coolant manifold
Coolant bypass
Coolant pump
Supply line
Engine coolant inlet

A. Closed-stat coolant flow (partial-blocking-type)

- -→ Total flow
→ Radiator flow
- -→ Bypass flow
—→ Deaeration flow

Thermostat housing
Deaeration line
Engine coolant outlet
Coolant manifold
Coolant bypass
Coolant pump
Supply line
Engine coolant inlet

B. Open-stat coolant flow (partial-blocking-type stat)

12-6 Self-purging system. International Harvester

Marine applications might employ a raw water system as illustrated in Fig. 12-8. The scoop (shown at A) should be of standard marine design, with generous-sized fittings, and barred to give some protection to the raw water pump and heat exchanger. In addition you might wish to install a strainer (D). The seacock (B) should be the gate valve type that opens fully and should have a minimum orifice size of 1 in. NPT (National Pipe Thread) for engines of 200 hp and under. If hosing is used it should be reinforced to withstand the suction of the pump.

Figure 12-9 shows the layout for a fresh water system, which is, in most ways, preferable to using sea water directly as a coolant. The heat exchanger is by far the most critical component and is generally fitted with removable caps to facilitate

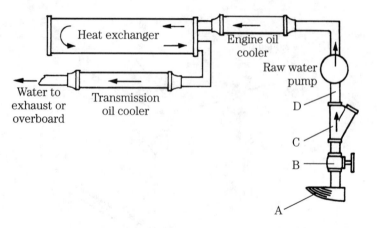

12-7 Raw water system. Ford Diesel

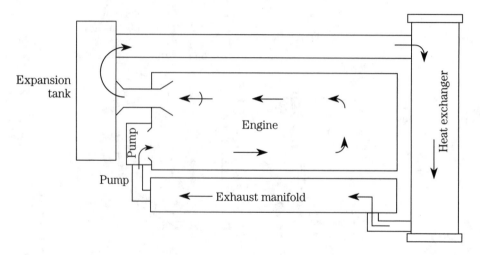

12-8 Fresh water system. Lehman Ford Diesel

cleaning. Periodically rod the tubes with a wooden dowel. Many exchangers have an additional refinement in the shape of a zinc "pencil." The zinc is sacrificial and is attacked before the metal of the exchanger. Replace as needed.

If you opt for a dry manifold you might—depending on engine size, hull shape, and boat use—be able to dispense with the raw water pump. Consult with your dealer.

Perhaps the best test of a closed system is to bleed a known amount of water relative to the total system capacity and observe the action in the header tank under normal loads. The amount of coolant the system can lose without loss of flow or entrapment of air is known as the *drawdown rating*. Figure 12-10 shows acceptable drawdown capacities for systems of up to 360 quarts capacity.

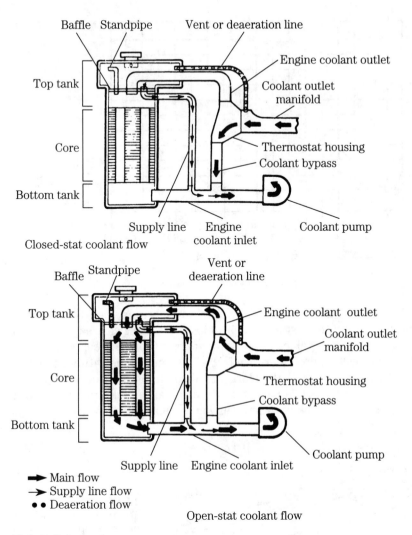

Baffle Standpipe   Vent or deaeration line

Engine coolant outlet

Coolant outlet manifold

Top tank

Core

Thermostat housing

Coolant bypass

Bottom tank

Supply line   Engine coolant inlet

Coolant pump

Closed-stat coolant flow

Baffle Standpipe   Vent or deaeration line

Top tank

Engine coolant outlet

Coolant outlet manifold

Core

Thermostat housing

Coolant bypass

Bottom tank

Coolant pump

Supply line   Engine coolant inlet

➡ Main flow
→ Supply line flow
•• Deaeration flow

Open-stat coolant flow

12-9 Full deaeration system. International Harvester

Radiators

Radiator core designs should be tailored to the application. The canted-tube Z-core is one of the most popular because the lay of the tubes generates air turbulence that "scrubs" the heat away. The straightforward wood core is used in dusty areas—in logging, demolition, rock-quarrying, and underbrush-clearing operations—and is characterized by wide air passages and tubes stacked in parallel. Most cores are made of copper, although aluminum has been tried, and steel is sometimes used in desert and other abrasive environments.

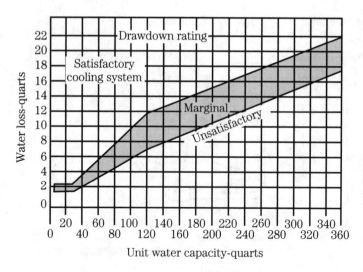

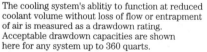

The cooling system's ablitiy to function at reduced coolant volume without loss of flow or entrapment of air is measured as a drawdown rating. Acceptable drawdown capacities are shown here for any system up to 360 quarts.

12-10 Drawdown capacities. International Harvester

The radiator fins should be blown free of debris as needed. Lift trucks used to handle cotton or jute must have their radiator passages cleared every 30 minutes or so. Over-the-highway vehicles can go for years without this service.

Coolant level is another critical factor. It should come within an inch of the bottom of the filter neck or to the mark inscribed on the filter or overflow tank. Sometimes radiators will fool you; there might be an air bubble trapped below the baffle (Fig. 12-11). Double-check with the engine running and the thermostat open.

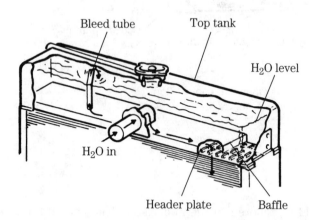

12-11

Top-tank construction. The baffle helps to separate gases and coolant by spreading the outlet flow from the engine over the core with very little turbulence.
International Harvester

The radiator should be reverse-flushed as needed (Fig. 12-12). Drain the system and disconnect both hoses. Connect the flushing gun to the lower hose and to an air source of 100 psi. Turn on the water. When the radiator is full, inject air in short bursts to clear the tubes. Between bursts allow the radiator to fill. Continue until the water flows freely or until the radiator passes a flow test. This test is a measure of the amount of water that falls through the radiator in one minute and, of course, de-

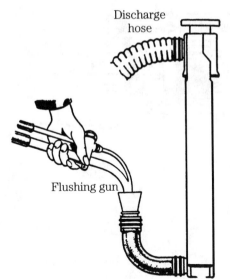

Discharge hose

Flushing gun

12-12
Reverse-flushing the radiator. ɢᴍ
Bedford Diesel

pends on the manufacturer's specifications. Radiators can be cleaned with commercial preparations (some of which are not compatible with antifreeze) or they can be dismantled and boiled. In extreme cases the header tank must be unsoldered and the tubes rodded.

Leaks are usually quite obvious and tend to appear around the joints of the header tank and tubes. Elusive leaks can be pinpointed by placing the radiator in a tub of water and blocking the outlet and applying no more than 15 psi of air to the inlet. Less test pressure is advisable if the radiator is old and fragile.

Most shops prefer to farm out radiator work, having had unfortunate experiences with their own radiator work. But unless the leak is buried within the tube columns or unless other repairs are needed, such as replacement of fins or removal of the header tank, the average mechanic can cope with radiators. The trick is in the soldering.

Use 60/40 straight bar or acid-core solder. Mix a few ounces of flux. The traditional formula is muriatic (dilute hydrochloric) acid and zinc powder. Clean the joint and reclean it. With a pencil torch, heat the area to be mended. Do not employ too much heat and be careful not to heat adjacent tubes or joints. Apply the solder. If it bubbles and skates, the surface is still not clean. It should sink into the metal, leaving a mirrorlike glaze on the surface, which dulls as it hardens. Allow the solder to cool at its own rate and test the radiator in the water tank.

Pressure caps Standard practice is to pressurize the cooling system at a few psi above atmospheric pressure. The additional pressure raises the boiling point of the coolant and allows hotter operation. Internal combustion engines operate best at some temperature approaching 200°F. Another advantage of pressurization is that it delays boiling at high altitudes.

Figure 12-13 illustrates a typical pressure cap. Note that it is somewhat complex, with a seal on the radiator filler tube flange and two valves. The main valve is the pressure relief and opens at 3–7 psi, depending on application. The vacuum valve is located inside the spring and equalizes pressures after shutdown.

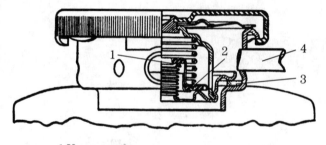

1 Vacuum valve
2 Pressure valve
3 Filler cap scal
4 Overflow pipe

12-13 Cutaway view of pressure cap (installed).
GM Bedford Diesel

A failed pressure cap can cause loss of coolant from boiling and, consequently, overheating of the engine. Caps should be inspected visually and, if the occasion demands it, tested. AC and other suppliers furnish pressure testers that can be used to test the cap and cooling system.

I emphasize that pressure caps are designed to be opened in two stages. The first stage bleeds down the pressure and directs the hot vapor down, away from the mechanic's hand. The second stage releases the cap. Open one too soon and you can get burned.

Radiator hoses Hoses are a necessary evil because in most applications, the engine is free to rock on its mounts, while the radiator is fixed. Eventually hoses will have to be replaced because they fail by the combined action of flexing and heat. Inspect the outside covering for cracks and bulges, especially at the mounting points. Squeeze the hoses to determine if they are still resilient. It is not unusual for the lower hose to collapse from pump suction as the engine runs. Heater hoses should be valved at the engine, and not at the heater. Otherwise stagnant water collects in them during the summer months.

Many patented hose clamps are in use, but most mechanics prefer the stainless steel type with a worm gear that engages serrations on the ribbon. Spring clamps are generally found on late-model equipment and are adequate.

Thermostats

The thermostat is located in a removable housing at the fan end of the engine. Single or double thermostats can be fitted with opening temperatures of from about 150°F to 205°F. The "colder" thermostats were intended to be used with ethyl or methyl alcohol antifreeze. Alcohol boils at 178°F.

The illustration in Fig. 12-14 shows a pill or pellet type, opened and closed. When closed, a small amount of coolant flows through the seepage port to the radiator, but most is recirculated through the bypass valve (No. 8 in the drawing) to the pump and back through the engine. As the engine warms, the wax in the pellet goes into the liquid state and expands. It forces out the pin attached to the frame (No. 4).

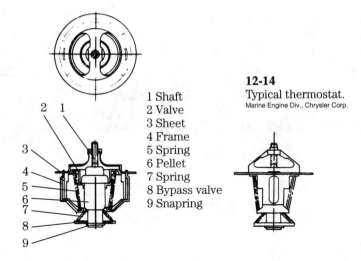

1 Shaft
2 Valve
3 Sheet
4 Frame
5 Spring
6 Pellet
7 Spring
8 Bypass valve
9 Snapring

12-14
Typical thermostat.
Marine Engine Div., Chrysler Corp.

This action pries open the valve (2) against the spring (7) and clears the coolant passage to the radiator. The action is progressive. In the partially open condition the coolant flow splits between the bypass valve and radiator. At full open all of the coolant is directed to the radiator.

Check the thermostat for signs of physical damage. Frames tend to break, and the metal bellows used on many (rather than telescoping tubes shown in Fig. 12-14) can develop cracks at the seams.

To check the action, heat the thermostat in water, supporting it as shown in Fig. 12-15. The temperature rating, usually found stamped on the frame, refers to the temperature at which the thermostat just cracks open. The full-open temperature can be found in the engine builder's specs and should be tested. A partially open thermostat will run you around in circles. High-temperature stats can be tested in a solution of ethylene glycol and water. A 50/50 mix has a boiling point of 227°F.

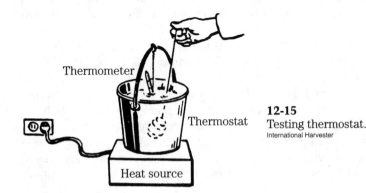

Thermometer

Thermostat

Heat source

12-15
Testing thermostat.
International Harvester

Water pumps

Raw water pumps are rugged devices designed to be mounted at some point remote from the engine. The Jabsco unit in Fig. 12-16 is one of the most popular.

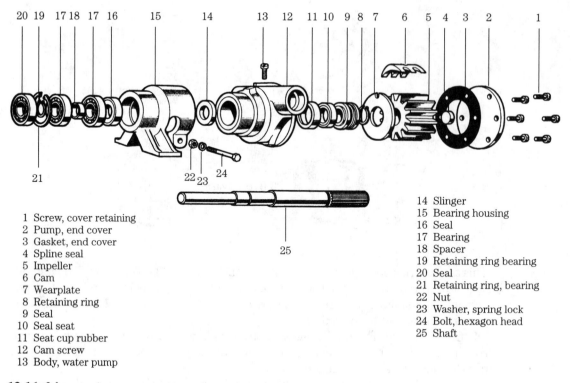

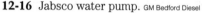

1 Screw, cover retaining
2 Pump, end cover
3 Gasket, end cover
4 Spline seal
5 Impeller
6 Cam
7 Wearplate
8 Retaining ring
9 Seal
10 Seal seat
11 Seat cup rubber
12 Cam screw
13 Body, water pump

14 Slinger
15 Bearing housing
16 Seal
17 Bearing
18 Spacer
19 Retaining ring bearing
20 Seal
21 Retaining ring, bearing
22 Nut
23 Washer, spring lock
24 Bolt, hexagon head
25 Shaft

12-16   Jabsco water pump. GM Bedford Diesel

It has prelubricated bearings and a symmetrical impeller, which means it can be driven in either direction. Intake and discharge ports are at the top of the pump body and must be swapped to match changes in rotation. The pump is lubricated by water and should not be run dry for longer than it takes to prime. This model has no drain cock. In freezing temperatures the end cover screws are loosened to drain.

To disassemble, remove the cover retaining screws, end cover (No. 2), and impeller (5). If the pump has any time on it, the impeller will be stuck. Pull it off the shaft with pliers on two blades. Remove the cam locking screw (13) and the cam (6). The wear plate (7) is now accessible. Replace it as a routine precautionary matter. Support the pump body in a vise and loosen the pinch bolt (24), which secures the pump body to the bearing block. Remove seals and slinger (14) from the pump body. The inner bearing seal in the bearing housing is pried loose first; then the outer seal is driven out from the impeller side (after the shaft and bearings have been pressed out). The last part to remove is the spring clip (19), which secures the bearings to the shaft.

Replace all seals and any other parts that show wear or corrosion damage. Assemble in the reverse order of disassembly. The cam screw (13) is as critical as any other single fastener in the system. Should it vibrate loose, the cam will drop down into the casting and destroy or damage the pump. Secure it with Loctite or the equivalent.

The engine pump (Fig. 12-17) is integral to the cooling system and is present in all applications. Some designs employ the *thermosiphon* principle—the tendency of hot water to rise and displace cold water—but no modern engine depends entirely on this. The pump has been refined to stark simplicity over the years and consists of a pump body, centrifugal impeller, shaft, bearing, and seal pack. No routine maintenance is required (other than keeping the system clean). The pump should be opened if one or more of the following symptoms is present:

- Perceptible friction as the shaft is turned by hand.
- Leaks around the shaft bearing.
- Leaks from the vent hole on the pump body. Some moisture is normal and represents seepage past the seal. But droplets or a continuous stream discharged here point to seal and possibly bearing failure.

To disassemble, remove the fan and detach the pump body from the engine block. Fan hub, impeller, and bearing removal require a 5-ton arbor press. Rebuild kits are available with shaft, impeller, seals, and gaskets (Fig. 12-18).

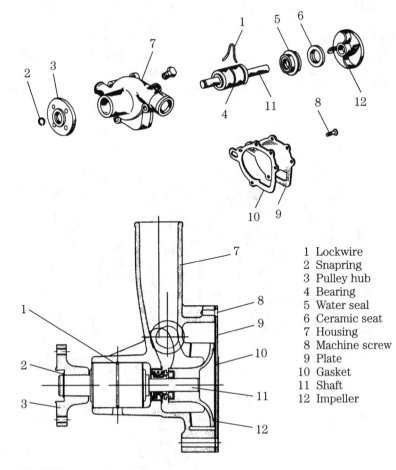

 1 Lockwire
 2 Snapring
 3 Pulley hub
 4 Bearing
 5 Water seal
 6 Ceramic seat
 7 Housing
 8 Machine screw
 9 Plate
10 Gasket
11 Shaft
12 Impeller

12-17 Pump in two views. Marine Engine Div., Chrysler Corp.

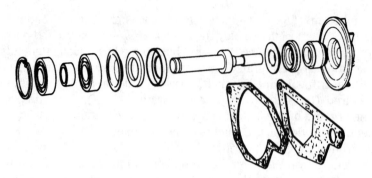

12-18 Pump rebuild kit. Perkins Engines, Inc.

Engine block

The cutaway drawing in Fig. 12-19 illustrates the typical flow pattern through the block and head. Coolant flow is concentrated around the injectors, valve seats, swirl chambers, and valve guides. Although not shown here, a number of engines employ baffles (sometimes called directors) placed in the head casting to direct water over these critical parts. The modern tendency is to keep the water inlet and outlet temperatures within a narrow range to reduce thermal stress and to allow coolant velocity. Low-velocity systems tended to leave a layer of steam clinging to the hot side of the jacket.

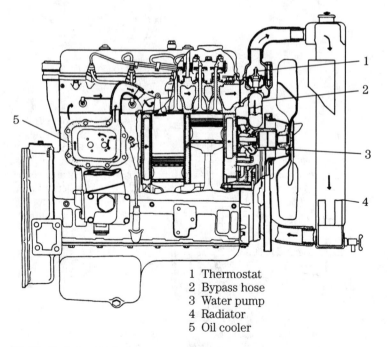

1 Thermostat
2 Bypass hose
3 Water pump
4 Radiator
5 Oil cooler

12-19 Coolant circulation. Marine Engine Div., Chrysler Corp.

The block should be inspected for telltale rust stains at the gasketed joints, and especially around the freeze plugs. These plugs deteriorate from the inside out and should be replaced every few years. If expansion plugs are used, coat the mating surface with 3M weather adhesive cement or a comparable product and drive the plug into place with a mallet. Threaded plugs give better security and should be used whenever possible.

The major difficulty with the water jacket is rust and scale formation. According to General Motors, $\frac{1}{16}$ in. of scale adhering to a 1-in. cast-iron wall is the equivalent in terms of insulation to $4\frac{1}{4}$ in. of cast iron. Loose scale can be removed with a flushing gun as shown in Fig. 12-20. Remove the thermostat and plug the heater hoses. Operate the gun at 100 psi, allowing the jacket to fill with water between air bursts.

Chemicals can be used to dissolve scale and sludge, but do not expect miracles, especially if the water in your locality is hard. The really effective chemicals are corrosive and can damage the radiator and heater core. All such preparations should be neutralized by following to the letter the instructions on the container. The best way to clean a block is to have it boiled. Automotive machinists have the necessary equipment. The parts to be boiled must be stripped of all nonferrous metal, including the camshaft bearings.

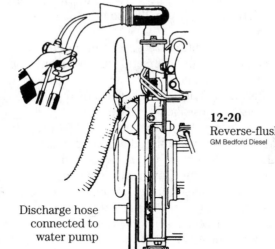

12-20
Reverse-flushing the jacket.
GM Bedford Diesel

Discharge hose
connected to
water pump

Coolant

Ideally no diesel engine should operate without a water filter and conditioner. If the engine is already in service, drain and flush the coolant. It might be necessary to change the filter element at close intervals during the first few hundred hours of operation if the jacket has rusted or silted. Each time the filter is changed, buff the metal element on a wire brush to expose new metal to the coolant. Replace the element as needed. Filters that employ sacrificial elements must be securely grounded to the engine block or mounting-frame members.

Lenrock nonchromate filters give good year-round protection and can tolerate hard water better than most (avoid filters with magnesium parts). But, even with the best filter and conditioner, common sense tells you that the water supply must be carefully evaluated. In general, water should have a total hardness of under 170 ppm (parts per million), and fewer than 100 ppm of sulfates and 40 ppm of chlorides. Dissolved solids should be kept below 340 ppm.

If these contaminants are present, distill or demineralize the supply. The water can be softened by addition of the suitable chemicals if the individual contaminant levels are below those listed and if the total hardness is above 170 ppm.

Inhibitor compounds are available from several manufacturers. They are used singly in bulk form; premixed with antifreeze; or in concert with a filter conditioner. Obtain detailed information from your dealer before using any of these products, because they might not be compatible.

For example, sodium chromate and potassium dichromate are not compatible with ethylene glycol antifreeze. On contact with permanent antifreeze these compounds produce a green slime that will cut the rate of heat transfer enough to cause overheating. It can be removed by flushing, followed by a dose of scaler.

Nonchromate inhibitors—a class that includes nitrates, nitrides, and borates—is recommended because of the convenience. These can be used with permanent-type antifreeze solutions or with water. Soluble oil is now obsolete as a corrosion inhibitor. At concentrations of more than 1% by volume, heat transfer becomes problematic. GM engineers have found that a 2½% solution increases fire deck temperature by as much as 15%.

Alcohol should not be used as antifreeze except in an emergency. It boils at lower temperature than operating temperature and must constantly be replenished.

Ethylene glycol is permanent and generally contains rust inhibitors. These inhibitors should be replenished at 500 hr. with a nonchromate inhibitor such as Nalcool 2000. Do not use antifreeze with sealant.

In general, mix antifreeze with softened water in a 50/50 proportion. This will give freeze protection to −40°F and will raise the boiling point to approximately 227°F. At between 62% and 75% the engine will be protected to almost −70°F, with a commensurate rise in the boiling point. But greater concentrations will lower the freezing point. Under no condition should the glycol mix be richer than 67/40.

The major drawback associated with ethylene glycol is its reaction to oil. Should a leak develop at the head gasket or oil cooler, shut down immediately. Assuming that the engine is in running condition, you might be able to save a complete teardown by flushing with Butyl Cellosolve. This procedure was developed for Detroit Diesel two-cycles and should not necessarily be employed in other engines, which might have higher bearing loadings. Query your dealer or factory rep. The procedure is as follows:

1. Drain the sump.
2. Change the lube oil filter element and clean the housing.
3. Mix two parts Butyl Cellosolve with one part SAE 10 oil.
4. Run the engine at 1000–1200 rpm for 1 hour. Keep a sharp eye on the oil pressure gauge because lubrication is only marginal.
5. Drain, allowing all the oil to escape, and refill with SAE 10 oil.
6. Run for a quarter-hour at the same rpm as before.

7. Drain, replace the filter, and fill the sump up to the mark with standard-grade oil.

8. Run for 30 minutes, shut down, and restart. If the starter drags, you can be fairly sure that there is still ethylene glycol in the system. Flush again, repeating each step of the process.

Glycol ether (methoxypropanol) has been suggested as an alternative to ethylene glycol. However, it is reputed to attack rubber parts, such as the Viton rubber O-rings in Cummings engines, as well as GM head gaskets.

Fans

Most stationary engines have simple, single-speed fans; Magnetic- or viscous-drive fans are used on automotive applications. The reason for this is to save fuel. At more than 35 mph or so, the fan is not needed and absorbs several horsepower, which could be better used to turn the crankshaft. In general these fans are not repairable, although magnetic coils are sometimes available.

Fixed-speed fans can, depending on the design, be installed backwards. Note the lay of the blades before removal. Most applications use the fan in suction, although on industrial trucks and the like, the fan often blows through the radiator.

Cooling efficiency can be increased by modifications to the fan. These modifications should not be undertaken in lieu of repairs to the radiator or other components.

In order of cost progression these modifications are:

1. Move the fan closer to the radiator by means of spacers on the hub. Leave at least ½-in. clearance between the radiator and fan.

2. Employ a venturi-type fan ring. Many industrial engines merely have a box around the fan or a shield at the header tank.

3. Change the fan to a more efficient version. Some consultation with your supplier will be necessary to determine the optimum pitch and number of blades.

4. Change the drive ratio to speed the fan. Contact your dealer before you make this modification because the fan is mounted on the water-pump shaft.

Drive belts should be tightened routinely to specifications. Some give is necessary to protect the bearings and prolong belt life. Examine the belt for oil damage, wear on the flanks, and heat cracks on its internal surface. Any of these faults means that the belt should be replaced. Paired belts are replaced together. Otherwise, sag will develop in the used belt, causing it to slip and wear the pulley flange. International Harvester engineers have measured wear rates under these conditions and report that the ratio is 1:5, with the belt taking the brunt of it. The two drawings in Fig. 12-21 compare a new belt with one that should be discarded. Note that the worn belt has fallen into the bottom of the groove. Any torque it transmits will be at this point, and not on the groove sides.

Cooling-system accessories

Turbocharged engines can be fitted with an intercooler between the turbine outlet and engine intake. The purpose of the intercooler is to reduce charge temperature and thus increase its density for more power. Maintenance is straightforward,

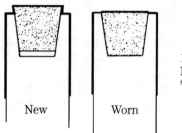

12-21
New and worn V-belts. International Harvester

calling for no special techniques. The tubes are accessible after disassembly and can be cleaned with a brush. If leaks are suspected the unit can be checked by immersing it in water and slightly pressurizing the tube assembly.

Figure 12-22 shows a combined heat exchanger and oil cooler of the type widely used in marine service. Three elements are combined into a single assembly: a header tank for the fresh water system, a fresh water tube cluster, and an oil tube cluster.

To disassemble, remove the fresh water lines and the rod capnut (No. 8). The end covers are now free. Support the oil cooler (16) and the spacing ring (15) because the throughbolt is all that holds these parts to the main casting. Clean the tube stack chemically or with a ⅛-in. brass rod inserted against the direction of water flow. The rod should be dulled on the end, without sharp edges, and must be slowly worked into the tubes. Pressure-test the exchange circuits at 30 psi and the oil cooler at 100 psi. Plugs will have to be fabricated to hold the pressure. The cooler section should be immersed in simmering water to open any holes prior to the test.

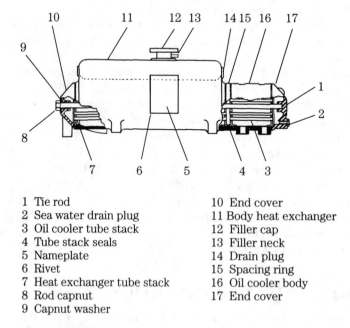

1	Tie rod	10	End cover
2	Sea water drain plug	11	Body heat exchanger
3	Oil cooler tube stack	12	Filler cap
4	Tube stack seals	13	Filler neck
5	Nameplate	14	Drain plug
6	Rivet	15	Spacing ring
7	Heat exchanger tube stack	16	Oil cooler body
8	Rod capnut	17	End cover
9	Capnut washer		

12-22 Combined heat exchanger and oil cooler. GM Bedford Diesel

13
CHAPTER

The new technology

Not very long ago, compression ignition had the characteristics of a mature technology. One could almost say that design had become a ritualized activity, focused on incremental improvements in manufacturing techniques and product performance. Imagination collided with experience and, most of the time, lost. The engineer mentioned in an earlier chapter, who warned against changing the head-bolt spacing on successful designs, was speaking for the majority. The $40 million one company was reported to have spent on cylinder head development disappeared without a ripple, at least so far as the customers could tell.

Mature technologies make the job of the technician more a matter of experience and memory than of learning. What learning there was came in small chunks, filtered by experience. Those old engines were also relatively easy to troubleshoot, because components failed in obvious ways that could be detected by sight or feel.

All that has changed now. Prodded by government regulation and global competition, the diesel engine has mutated into a part-mechanical, part-electronic hybrid, controlled through the mediation of one or more computers. The strategic position occupied by these controls gives them a kind of veto power, so that a failure in the electronics can shut down the engine or seriously compromises its performance.

The new technology also erects an electronic wall between the engine and technicians who operate in the old way, i.e., in terms of purely mechanical models. It remains possible to make a living in this manner; the electronics revolution has not yet impacted smaller diesels and the effect on larger units is, for now, limited to fuel systems. But the days of the traditional wrench twister are numbered. According to *Spectrum*, the journal of the Institute of Electrical and Electronics Engineering, mechanics must "now interact with a computer" and "learn to process information symbolically."

There is still time to come to terms the electronics revolution. But the window of opportunity is closing; by the end of this century, computers will be as integral to engines as pistons & camshafts.

This chapter is intended to familiarize you with some of the basic principles of the new technology. It does not replace the factory documentation supplied with electronically controlled engines. That would require a whole library. But the information presented here should make the subject more accessible and reasonable.

The first order of business is to explain why the computer revolution came just in time to save the diesel engine.

Diesel emissions

Figure 13-1 illustrates the spray pattern produced by a single-nozzle orifice in a high-turbulence direct-injection (DI) chamber. The pattern elongates in response to air flow, with the smaller droplets concentrated on the leading (lower) edge of the pattern. Larger and heavier droplets remain clustered about the core.

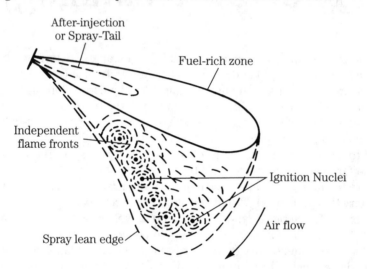

13-1 This drawing illustrates the fuel pattern created by a single-orifice injector in a highly turbulent chamber. Ignition begins as nuclei in the fuel-lean, oxygen-rich zone near the leading (or lower) edge of the pattern. Flame fronts move out these nuclei toward the core of the pattern, where there is relatively little oxygen and a high concentration of fuel.

Ignition begins as a series of small bursts at the interface between the fuel spray and cylinder air, where there is a surplus of oxygen. The bursts combine into a flame fronts that progressively move into the fuel-soaked core of the pattern. Every normal combustion event in a diesel engine begins under oxygen-rich conditions and concludes under oxygen-lean conditions. This variability in fuel/air ratios is a special burden of the diesel engine. (Fuel/air ratios tend to be more uniform in SI combustion chambers, although there is some charge stratification.)

In addition, diesel engines—especially those used to power vehicles—operate under a fairly wide range of loads and speeds. Air turbulence, duration of the expansion stroke, and cylinder temperature vary with the operating mode.

Figures 13-2 through 13-4 plot the three types of emissions considered most critical against rpm and torque for an uncontrolled heavy-duty engine. These three-dimensional graphs, or maps, depict the "natural state" of diesel engines.

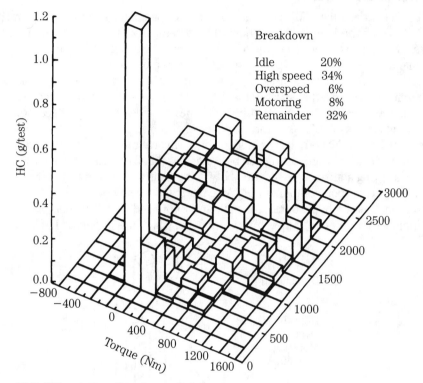

13-2 HC emissions for a heavy-duty engine.

Hydrocarbons

Unburned fuel and lube oil appear in the exhaust as hydrocarbons, or HC. California regulators call exhaust-borne hydrocarbons *volatile organic compounds* (VOC), alternative fuels enthusiasts prefer to speak of *nonmethane hydrocarbons* (NMHC), and some federal agencies use the term *reactive organic compounds*. But, whatever the terminology, it is all HC, which is both a carcinogen and a precursor of photochemical smog.

Hydrocarbons survive their passage through the cylinder when the mixture is either too lean or too rich to burn. Excessively lean mixtures are caused by fuel droplets that break free of spray plume (Fig. 13-1) and diffuse throughout the combustion chamber. The resulting fuel mixture does not support combustion and the vaporized fuel goes out the exhaust. This phenomenon often occurs under light loads and at low speeds, which would explain the HC spike at idle shown in Fig. 13-2.

HC emissions are also generated when the flame is quenched by too rapid infusion of air or by contact with the relatively cool cylinder walls. Data plotted in Fig. 13-2 suggest that the engine in question might be burdened with excessive turbulence that "blows out" the flame at high rpm, where 34% of total HC is released.

Particulate matter

Particulate matter (PM), gives diesel engines a bad name among the general public (Fig. 13-3). In the high concentrations that accompany diesel acceleration and cold starts, PM can be seen as black smoke, felt as a vague stinging sensation, and smelled as the characteristic diesel stink. People instinctively move away from PM and their instincts are correct. These emissions, consisting for the most part of tiny soot spheres, are carcinogenic and lethal, in ways that have not yet been fully understood. The American Lung Association believes that 50,000 Americans die each year because of exposure to high levels of PM, taken for the purpose of the study as particles of less than 10 microns diameter. Nobody has identified the biological mechanism involved, nor does the lung association try to distinguish between diesel PM, dust, and tobacco smoke. But there is a strong statistical connection between inhaling PM and dying early.

The hydrocarbon component of PM, called SOF for soluble organic fraction, consists of combustion byproducts, lube oil, and unburned fuel. This complex chemical mix includes a number of powerful carcinogens, which are implicated in lung cancer.

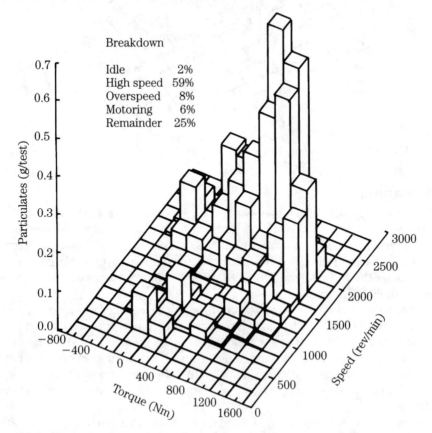

13-3 PM emissions for a heavy-duty engine.

Soot, the SOF carrier, forms in the oxygen-poor region on the trailing edge of the injection plume. Most soot particles oxidize during the expansion stroke, except at full throttle (when the time available for oxidation is diminished) and during the period of turbocharger lag during sudden acceleration. Once the turbocharger reaches speed, the oxygen balance is restored and the exhaust again becomes translucent (although not without some PM).

Oxides of nitrogen

NO_x (a blanket term for No, NO_2, and other oxides of nitrogen) is the joker in the deck. Soot and HC form in oxygen-poor environments, when flame temperatures are low. NO_x is the child of the high temperature, oxygen-rich combustion that occurs on the leading edge of the spray plume. Most soot forms early in the combustion process, when fuel accumulated during ignition lag burns to generate high levels of heat and pressure. Heavy loads, high speeds, or high inlet air temperatures contribute to its formation. In the example engine, nearly 60% of total NO_x output was released at high speed (Fig. 13-4).

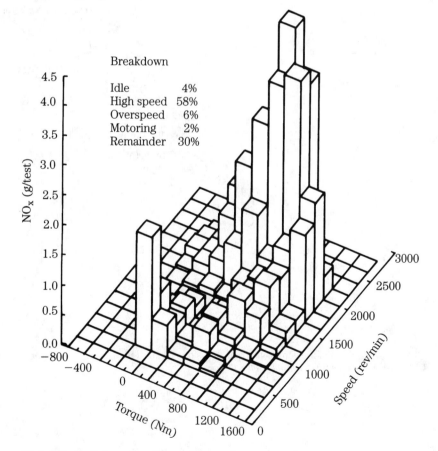

13-4 NO_x emissions for a heavy-duty engine.

NO$_x$ contributes to the production of surface ozone, which is a chemical precursor of acid rain and toxic in its own right.

Carbon monoxide

U.S. federal and California regulations also limit carbon monoxide (CO), but with little direct effect on engine design.

Regulation

Limits on exhaust emissions, imposed by an alphabet soup of government agencies in the U.S. and abroad, drive the new technology (Fig. 13-5). In this country, the major players are the federal Environmental Protection Agency and the California

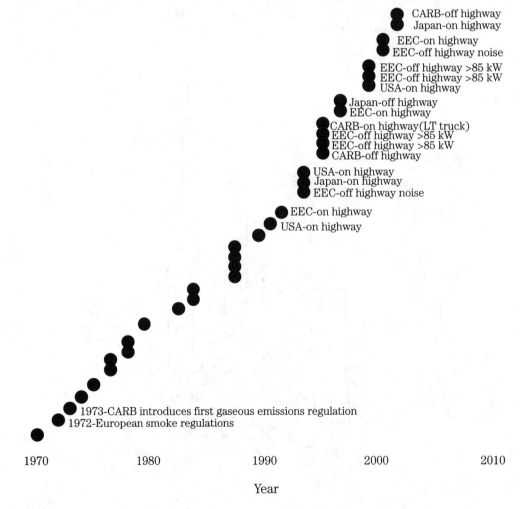

13-5 Worldwide diesel regulations.

Air Resources Board (CARB), which regulates emissions in that state and in a phalanx of northeastern states that follow the California model. Japan and the European Economic Community (EEC) generally follow the broad outlines of U.S. policy. Mexico is expected to follow suit.

At the present time, the only diesel engines not falling under some form of regulation in the U.S. are those used to power off-road equipment (e.g., farm and garden tractors) and certain stationary engines.

Table 13-1 lists the current targets for U.S. coach and heavy-duty truck engine makers. Note that specific technologies are not proscribed; instead, the rule makers (in this case the EPA under authority granted by the Clean Air Act of 1990) establish performance criteria. The EPA does, however, impose a 290,000 zero-maintenance requirement on emissions hardware for heavy-duty engines.

Table 13-1. Heavy duty engine emission standards

Urban bus heavy-duty

(g/bhp-hr measured during EPA heavy-duty engine test)

Model year	NO_x	HC	CO	PM
1990	6.0	1.3	15.5	0.60
1991	5.0	1.3	15.5	0.25
1993	5.0	1.3	15.5	0.10
1994	5.0	1.3	15.5	0.05

Heavy-duty truck

(g/bhp-hr measured during EPA heavy-duty engine test)

Model year	NO_x	HC	CO	PM
1990	6.0	1.3	15.5	0.60
1991	5.0	1.3	15.5	0.25
1994	5.0	1.3	15.5	0.10
1998	4.0	1.3	15.5	0.10

Provisions of the 1990 Clean Air Act also apply to stationary engines. In attainment areas, that is, parts of the country that have legal air quality, 500-hp and larger engines must meet these standards:

HC	1.0 g/bhp/hr
CO	3.0 g/bhp/hr
NO_x	2.0 g/bhp/hr

In the 90 or so non-attainment areas, which include much of the urban landscape of the U.S., stationary diesel emission limits are the same, except that the rules apply to 150 hp and larger engines.

Regional authorities also have their hands in emissions regulations. The most active and powerful of these is the South Coast Air Quality Management District

(SCAQMD) which oversees air quality in Los Angeles and surrounding counties. Stationary engines in the SCAQMD jurisdiction must meet the following standards:

HC	0.6 g/bhp/hr
CO	0.5 g/bhp/hr
NO_x	0.3 g/bhp/hr

Many observers believe that SCAQMD limits will be adopted nationally. As far as engine manufacturers are concerned, they might as well be, because it is exceedingly difficult to design engines for one tiny area of the country.

An unfortunate effect of the bureaucratic approach to emissions control has been to make the control hardware an afterthought, rather than an integral part of engine design. EPA spokesmen have complained about the proliferation of part-time emission controls, which function only when required by test protocols. For example, after about three-quarters throttle, SI auto engines enter the open-loop mode, where most emissions controls are disabled. This is perfectly legal, because existing certification tests do not include an appreciable amount of full throttle operation. The agency had hoped that auto makers would opt a more global approach to the problem of exhaust emissions.

Diesel engine makers follow the example of Detroit and see the emissions problem as a matter of damage control. So far, they have been successful in avoiding radical technologies, such as alternative fuels or PM traps.

Emission control strategies

It is convenient to divide emission controls into three categories: air-system, in-cylinder, and after-treatment.

Air system

The air induction system offers several opportunities to reduce emissions.

Turbocharging The turbocharger is one of the most effective diesel emission controls, particularly when fitted with an aftercooler. Turbocharging strikes at the heart of the problem by reducing fuel consumption (in lb/bhp/hr terms). The vast quantity of air supplied promotes more complete oxidation (thus reducing HC and PM) and exerts a cooling effect that retards NO_x formation.

Most manufacturers use air-to-air aftercoolers, which reduce charge temperature more than would be possible if radiator water were used as the cooling medium. Some Cat designs achieve essentially the same temperature reduction with an air-to-water intercooler, fed by a dedicated water supply.

Onboard microprocessors are frequently used to minimize over-fueling during the transition into turboboost. Formerly this was done less precisely with a vacuum diaphragm that sensed manifold pressure.

Intake pipe tuning The intake manifold is a conduit for a complex interplay of pressure waves, which reflect between the intake valve and the air cleaner. By adjusting the length of the runners, it is possible to arrange matters so that a positive (high pressure) wave arrives just as the valve opens. However, tuned manifolds work only at selected engine speeds and at multiples of those speeds. Even so, the supercharge effect is considerable and, when present, confers the emissions benefits associated with oxygen-rich combustion.

EGR Exhaust gas recirculation is usually considered a stopgap measure, typically used on light trucks and automobiles. EGR reduces flame temperature and, by so doing, retards NO_x formation. Recycling exhaust gas also contaminates the lube oil, increases PM emissions, and compromises fuel efficiency.

Figure 13-6 illustrates the system used on certain Ford products. As configured for this application, EGR remains locked out unless the vehicle is operating at highway speeds, in fifth gear, and with the engine at normal operating temperature. A logic module integrates signals from an engine speed sensor and the transmission to open one or both EGR valves.

PCV Crankcase blowby consists of gases leaked past the piston rings and, on Detroit two-cycles, scavenge air that enters from the air box. Relatively small flows

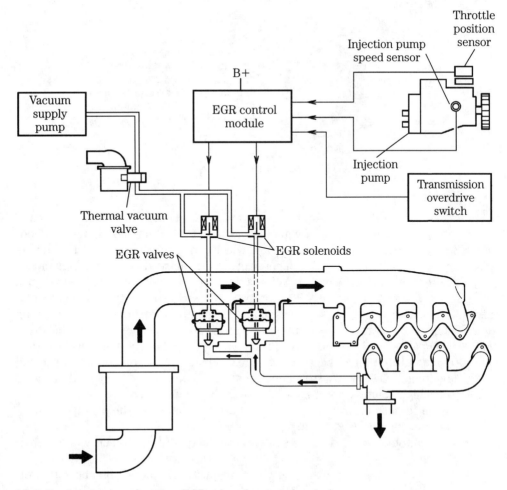

13-6 Ford system employs two EGR valves for staged metering.

are involved, amounting to roughly .25% of exhaust gas volume for a new engine and something less than 2% for an engine midway between overhauls. Blowby was traditionally vented to the atmosphere through a breather. It hardly mattered for an uncontrolled engine.

Blowby gas contains exhaust, metallic particles, and all the pollutants associated with internal combustion. In a controlled engine, blowby can account for a quarter of total allowable HC emissions and, in the case of city buses, half of the allowable PM. These percentages rise with engine wear and increased leakage past the rings.

Consequently, most modern and all emissions-rated engines recycle blowby through the intake manifold, discharging the gas ahead of the turbocharger. Diesel Research, Inc. (P.O. Box 213, Hampton Bays, NY 11946) makes kits to convert American engines with open systems to PCV. The Crankvent filters the incoming crankcase gas, separates the oil, and maintains a slight depression on the crankcase to reduce gasket weep (Fig. 13-7).

13-7
Crankvent filter/separators simplify the conversion from open to closed crankcase ventilation. The system can also be adapted to drain accumulated lube oil from Detroit Diesel air boxes.

In-cylinder

Direct-injection (DI) truck and bus engines have benefited from modifications of the combustion chamber, most of which are aimed at improving air circulation. Soot and HC result from local oxygen deprivation, which can be minimized by increasing the turbulence of the air charge and by opening previously masked areas of the chamber.

Turbulence Inclined valve ports impart "swirl" to the incoming air, which has the effect of stoking the furnace. The intensity of swirl varies with gas velocity and piston speed, so that the designer tries to attain a compromise between good mixing at low speeds and spray-pattern integrity at high speeds. Excessive swirl scatters fuel droplets all over the cylinder, resulting in a lean mixture and high levels of HC.

Crevice volume In a DI chamber, all areas outside the piston bowl are shrouded at tdc and inhibit air flow. These dead spaces can account for 40% of total chamber volume on a high-compression engine. Crevice volume consists of:

- Piston-to-head clearance, or the area between the top of the piston crown and the cylinder head at tdc.
- Valve cutouts, or the notches sometimes milled in the piston crown for valve clearance.
- Top land volume, or the area between the piston and bore, and above the top compression ring.

Serious, even heroic, efforts are underway to reduce crevice volume. For example, the designers of the Navistar 7.9L managed to squeeze down the piston-to-head clearance to 0.025 in., which must have been a formidable undertaking in view of tolerance stackup. Pistons in the same engine come within 0.009 in. of collision with the exhaust valves.

The Deere 7.6L locates its upper ring 6.3 mm below the piston crown to reduce top land crevice volume to 4.5% of chamber volume. Navistar engineers specified a 6.0-mm ring height on the 7.3L T 444E prototype, but had second thoughts and lowered the ring another 3.0 mm on production models. In this case, the tradeoff between greater crevice volume and lower ring temperature seemed appropriate.

Oil consumption New piston ring configurations, better valve seals, and tighter piston-to-bore clearances reduce the amount of oil intrusion into the combustion chamber. As mentioned earlier, lube oil is a major source of HC and is responsible for about 40% of the SOF component in particulates.

Exotic materials Over the longer term, the effort to control exhaust emissions must aim at increasing thermal efficiency. One way to do this is to reduce the amount of heat thrown off to the coolant by insulating the piston and chamber. The true adiabatic engine, which requires no cooling system, has been elusive, but some progress is being made. A good example is the ceramic prechamber used on the Mazda-designed IDI 2L auto engine (Fig. 13-8).

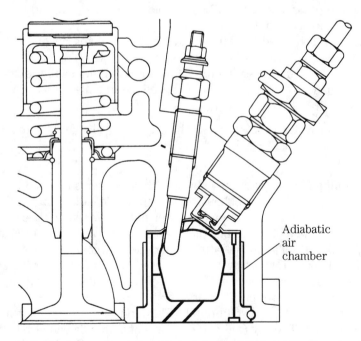

Adiabatic
air
chamber

13-8 Ford ceramic prechamber retains heat that would otherwise be dissipated into the coolant. The term "adiabatic" is a little optimistic, since it suggests zero heat transfer across the ceramic.

After-treatment

Two types of after-treatment receive serious consideration from manufacturers. *PM traps*, sometimes called trap-oxidizers, might be necessary to meet impending urban bus particulate standards. Neither operators nor engine makers look forward to the prospect. Two years ago, Lance Watt, director of engineering for The Flexible Corporation (an Ohio-based coach manufacturer) suggested to me that a reliable PM trap could cost as much as a bus engine.

Experimental PM traps devised for motor vehicles generally use a honeycomb filter, made of ceramic to withstand the high temperatures. Periodically the filter element plugs and must be regenerated like a self-cleaning oven. A catalyst or electric resistance heater supplies the energy. Most of the accumulated PM oxidizes, but some bakes on the filter and must be manually removed. It appears that ash residues from motor oil act as a binder.

Other difficulties surface when regeneration is accomplished "on the fly" with the engine running, as would be desirable for commercial vehicles. Heat input from the heater, exhaust heat (which varies with load and rpm), and the heat of the reaction must be regulated within narrow limits. Too little, and PM does not burn off; too much, and filter melts. All of this can be coped with, but the technology does not come cheap.

Catalytic conversion   Modern SI automobiles are fitted with three-way catalytic converters that simultaneously reduce HC, CO and NO_x. These engines operate at the slightly rich mixtures necessary to equalize CO, which acts as an oxidant, and NO_x production. Unfortunately, the high level of oxygen in diesel exhaust makes such converters impractical. But it is possible to engineer around the problem, at least for stationary, constant-speed plants.

The system developed by Houston Industrial Silencing covers all bases with a muffler, two catalytic converters, and a PM trap (Fig. 13-9). A microprocessor-controlled injection system delivers metered amounts of ammonia to the first-stage converter. Ammonia (NH_3) reacts with NO_x to produce free nitrogen and water:

$$NH_3 + NO_x = N_2 + H_2O$$

A second converter, downstream of the first, causes HC and CO to react with the surplus O_2 in the exhaust to yield carbon dioxide and water.

$$HC + CO + O_2 = CO_2 + H_2O$$

When equipped with platinum catalysts, these systems are said to reduce 90% or more of NO_x present in the exhaust stream and at least 95% of the CO. The vendor has not supplied HC conversion rates which, however, should be comparable to those for CO. Platinum operates in a temperature band of 460 to 540°F, attainable in a well engineered installation operating under constant load.

Base-metal, vanadium catalysts can be used at higher temperatures, for some loss in NO_x conversion efficiency. Either type of catalyst limits ammonia "slip" to two or three parts per million (ppm).

As shown in Fig. 13-9, a solution of 75% water and 25% ammonia is normally used as the oxidant. The dry (anhydrous) form of ammonia can also be injected directly into the exhaust or mixed with steam prior to injection. In any event, the rate of injection must be calibrated for each system by sampling NO_x concentrations at the exhaust outlet.

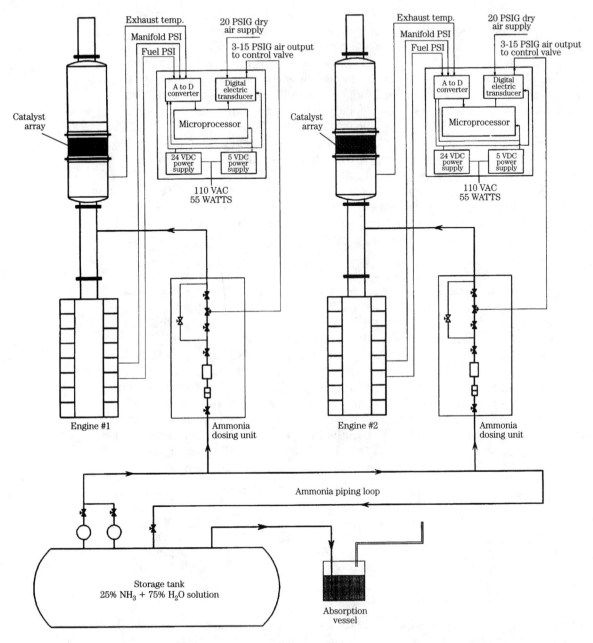

13-9 This custom-engineered system reduces HC, CO, and No$_x$, muffles the engine to hospital standards, and traps PM.

Digital fundamentals

The fuel system is the point of entry for computers and the whole technology associated with digital electronics. Mechanical systems respond too slowly to regulate fuel delivery with the precision necessary for the "green" engine envisioned by government regulators. Several diesel manufacturers already have computer-mediated fuel systems in production and most of the others are known to be working in this direction.

But, once the decision to go ahead with the design is made, something strange happens. The logic of technology (or, maybe, the attraction of bells and whistles) takes over and fairly straightforward fuel management systems metamorphose into global engine management systems. Once you have the computer and sensors in place, it doesn't cost much more to include other functions, such as digital cruise control for highway trucks or electronic weight limiters for cranes and front-loaders. It does, however, make the technology less forgiving by routing all, or nearly all, system functions through the central processing unit.

Technicians need to remember that these systems are built, like Erector sets, from a very finite number of components. Complexity is a surface phenomenon, experienced when you look at a block diagram with 50 inputs and nearly as many outputs. Each function, taken in isolation, is quite simple.

Analog vs. digital

Analog signals fluctuate over time, as shown by the waveform voltage in Fig. 13-10A. The instantaneous value—that is, the value of the signal at any moment in time—conveys information. A little reflection will show that voltage spikes, bad grounds, and loose connections seriously compromise the utility of signal.

Digital signals change intermittently with time, as shown in Fig. 13-10B. Only peak values, corresponding to "1" (on) and "0" (off) register. Because the intermediate values are of no concern, all the computer needs do is to discriminate between binary on and off states, or bits. Depending on the system, a string of 8, 16, or 32 sequential bits make up a word, which conveys information. Circuit noise can destroy part of a word, making it unintelligible, but noise generally does not alter the meaning of complete words in the systematic fashion required to fool the computer.

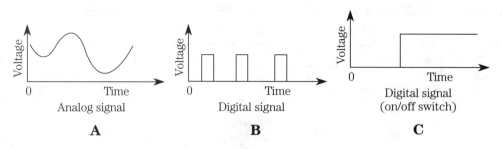

13-10   An analog signal varies continuously with time, as shown in drawing A. The digital signals in B and C have only two values—on (full voltage) and off (zero voltage).

ECM

The electronic control module (ECM) is known variously as the electronic control unit (ECU), computer, microprocessor, or black box. An ECM consists of four major elements, shown in block diagram in Fig. 13-11.

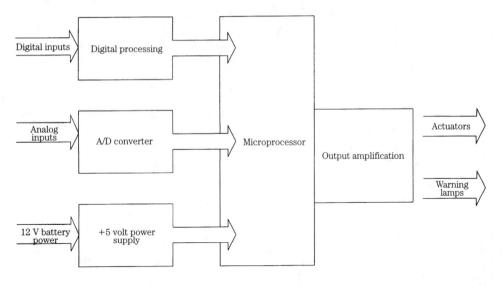

13-11 The basic ECM in block diagram.

Most of what the computer "knows" about engine status comes to it by way of analog sensors. The A/D converter translates analog signals into the digital format required by the microprocessor.

The microprocessor also receives digital inputs originating from other onboard computers, certain diagnostic tools, and from the more sophisticated sensors, such as "drive-by-wire" accelerators. On/off switches also generate digital signals which, in this case, undergo a voltage reduction before entering the microprocessor.

Outputs from the microprocessor are in the form of commands and status reports. The Output Amplification circuit boosts 5-V command signals to the power levels required to drive stepper motors and solenoids. Most system failures occur in the high-amperage output circuit.

The microprocessor in the neural center of the machine, where sensor inputs and operator commands integrate with factory-installed programs to generate the outputs necessary to drive the system. In addition, the microprocessor makes logical decisions, organizes tasks by priority, and exercises memory functions. See Fig. 13-12.

The CPU, or central processing unit, does the computations and decision making. It should be stressed that the computer "brain" does not think, or imagine, or feel. It merely follows instructions encoded in its operating programs.

Engine makers use standard CPUs, most of which were designed for personal computers. For example, early systems used the Intel 8088 chip, which helped to launch the PC revolution, one of the newer management systems use the 32-bit Motorola MC 68020, a chip more often encountered in MacIntosh computers.

ROM (read-only memory) provides the CPU with programs and with reference data of the kind that does not change. In other words, the ROM chip gives the CPU its marching orders and data to steer by. Data is in the form of maps, which plot changes in one parameter against two others. The desired injection timing, for ex-

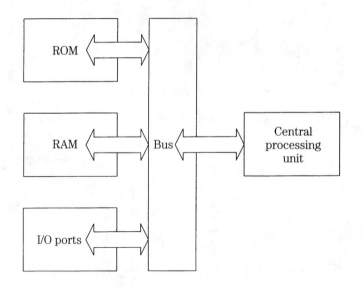

13-12 A microprocessor consists of a central processing unit, two kinds of memory, and input/output ports.

ample, will be plotted against engine load (sensed as manifold pressure) and rpm. The CPU could compute the timing from sensor inputs, but it is faster to look up the answer on a previously calculated map.

ROM memory can neither be erased nor written over, and persists even after system power is switched off. Because programs undergo more or less continuous development, (if only to correct the bugs that appear in service) ROM takes the form of a dedicated chip that plugs into the board for easy replacement.

ROM chips are expensive, because they must be custom-manufactured with the appropriate programs installed. Consequently, most engine makers specify general-purpose PROM (programmable read-only memory) chips and program them in-house, using a machine that "burns in" the desired circuitry. Because the program is physically encrypted, it cannot be altered in the field and remains in place when system power is interrupted.

PROM continues to be the norm for systems built around automotive ECMs; newer, "diesel-dedicated" systems employ an erasable PROM (EPROM) or electrically erasable PROM (EEPROM). As far as the technician is concerned, the "erasability" of these PROM variants is a moot point.

The newest type of ROM, called flash ROM, does impact service. A technician, armed with a PC and the right password, can write over some areas of memory. This feature is useful for fine tuning, but, in no case, can one expect to correct or disguise system malfunctions from the keyboard.

In addition to data and programs kept permanently on file, the microprocessor includes a volatile memory, known as RAM (random-access memory). RAM acts like short-term memory in humans. Data is stored for immediate use and "forgotten" when the power goes down.

Self-diagnosis

Computer-based systems include some sort of self-diagnostic capability. The results of that soul-searching appear as a two-digit code that can be read with the appropriate factory tool or deciphered by counting the number of times the engine trouble light flashes.

Self-diagnostic programs focus on input from the sensors, which are the most vulnerable components of the system. All systems interpret out-of-range values as evidence of sensor or wiring-harness failure. Thus, a manifold pressure reading of, say, 60 psi would generate a trouble code in ROM, illuminate the "Check Engine" light, and, as these systems are generally configured, causes the ECM to substitute a default value for manifold pressure. Like a broken watch, the default is correct only part of the time and engine performance deteriorates to the extent that the failed sensor affects operation. In vehicles this is called the *limp home* mode. Critical parameters, such as low oil pressure or coolant level, can cause the computer to shut down the engine by denying fuel.

Some programs incorporate a refinement, known as a rationality check, to evaluate sensor inputs in the context of the operating environment. For example, a sensor value correlating with 12 psi of manifold turbo boost would pass the out-of-range test. But if that value were generated at idle, a "smart" ECM would flag the sensor on the grounds of irrational behavior.

Solenoids and stepper motors also receive monitoring. The program might sense failure as a function of terminal voltage or as feedback. In the latter case, a telltale reports if the actuator has completed the task it was commanded to perform.

The program might also include provision to exercise sensors and actuators on command from a factory service tool. In this way, the technician can observe the functioning of all electro-mechanical components and, hopefully, replicate the failure registered in computer memory.

Primary factory support takes the form of documentation and special-purpose diagnostic tools, developed in-tandem with system electronics and software. Among the best of these tools is the Caterpillar Electronic Control Analyzer Programmer (ECAP) with its periodically updated and chip-encrypted NEXG4522 Truck Engines Functions Dual Service Program Module. The chip plugs into ECAP to adapt it to various Cat engine families.

ECAP displays engine operating parameters 9.6 seconds before and 3.4 after the appearance of a diagnostic code. This feature, called Engine Snapshot Data, takes some of the mystery out of transient malfunctions. An accessory printer converts inputs and outputs to hardcopy for later analysis. Properly configured, ECAP can read and evaluate electronic injector pulse width—the single most important output variable.

The technician also has other resources, namely, mother wit (which is as good as his or her understanding of the system and its design functions), parts substitution, and better scanners than the factory might supply. Nobody should seriously assume responsibility for computer-driven hardware without access to an oscilloscope.

Applications

Applications of digital electronics occupy a continuum from relatively simple speed-control devices to comprehensive engine management systems.

Governors, large engines

We can begin by discussing some of the add-on electronic governors that are replacing mechanical governors on power generation, rail traction, marine propulsion, and other industrial engines. The primary function of these systems is to regulate engine speed by moving the rack.

As used on large engines, mechanical governors are similar in principle to those discussed in Chapter 6, but have a wider repertoire of functions. These functions include stepped speed control, reduced fuel delivery at high turboboost pressures, and automatic shutdown in the event of loss of oil pressure.

The first step away from the purely mechanical was to relieve the governor of some control functions by using ganged relays. Relay logic, i.e., on/off switches wired to interact with each other in logical patterns, was a crude anticipation of the digital computer. Most relay systems have since been replaced with solid-state programmable logic controllers.

Early electronic governors were analog devices that worked in a manner analogous to mechanical governors. The latter sense engine speed as the force generated by a flyweight against a spring. Analog governors sense speed as a voltage signal, which increases in some fixed proportion to engine rpm. This signal is then compared to a reference voltage, representing the desired speed. A difference between the two voltages energizes a stepper motor to reposition the rack.

The third step in the evolution of speed-control technology was to replace analog governors with digital computers.

Rather than read engine speed as an analog voltage, digital governors count the number of markers (which can be flywheel teeth) that pass a sensing head during a given time frame (Fig. 13-13). The computer then calculates engine speed and compares this number to a reference number representing the desired speed. A discrepancy between the two numbers causes the computer to generate a command signal to an actuator that repositions the rack.

Governors, smaller engines

Most "computerized engine control systems" used on small and mid-sized engines are, in fact, little more than electronic governors with limited self-diagnostic capabilities. Figure 13-14 sketches the system used on the John Deere 7.6L 6076H engine. A switch on the electronic control module or unit (ECU in the drawing) enables the operator to select any of three speed-control programs: true all-speed governing (as would be used for vehicle applications), min-max governing (for power generation), and full power boost. The latter option has a timer associated with it to prevent engine damage. In addition, road-speed limiting and similar options are available.

The ECM uses two methods to control smoke during transients. The first involves the use of sensors to measure manifold air pressure (MAP) and, optionally, manifold

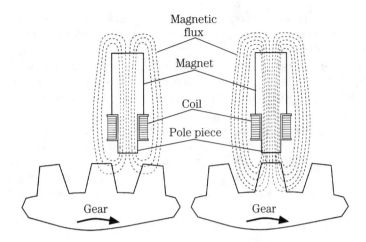

13-13 Engine speed sensors read rpm as magnetically induced voltage fluctuations in a sensor coil. Gear teeth are convenient markers. These circuits may also respond to a magnetic disturbance caused by a tdc reference. The tdc marker can take the form of a modified gear tooth or a permanent magnet fixed at some point near the rim of the gear. Dynalco Controls, a unit of Main Controls Corp.

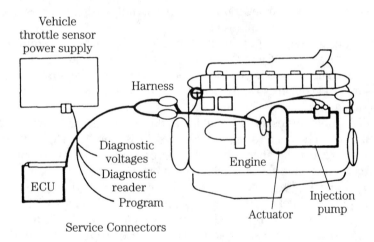

13-14 Electronic throttle control used in certain Deere engines.

air temperature (MAT). If, for example, the vehicle driver floorboards the accelerator at idle, manifold pressure will remain low until the turbo spools up. ECM restrains fuel delivery until the MAP sensor reports that the turbocharger has come up to speed. MAT is useful when operating over a wide range of ambient temperatures.

In addition to the sensor-dependent method described above, the ECM can be programmed to match fuel delivery with torque requirements. Deere engineers say that the programmed method gives better smoke control, because the lag and error associated with sensors are eliminated.

A fuel temperature sensor helps reduce smoke during hot starts. Excess fueling and retarded fuel injection are permitted during cold starts, which are followed by a timed period of fast idle.

The electronic fuel pump uses the same rack-and-plunger mechanism found in mechanically governed inline pumps. The actuator assembly consists of a solenoid to move the rack, a rack-position sensor to provide feedback to ECM, and speed sensor, which operates from a toothed wheel. A sensor measures fuel temperature at the pump inlet and a solenoid valve acts as a kill switch.

Basic ECM inputs are analog throttle, fuel temperature, MAP, MAT, start signal, engine speed, and 12-Vdc power. Outputs include tachometer drive, fault lamp, and actuator current for the fuel-pump solenoid. A socket on the wiring harness can be accessed with a digital voltmeter or special service tool for diagnostic codes.

Engine management systems

Although the line blurs upon close examination, one can distinguish between electronic governors and full-blown engine management systems. The latter provide variable control of injection timing and attempt to monitor and integrate all critical engine functions.

Detroit DDEC ("dee-deck"), or Detroit Diesel Electronic Control, was the first of these systems and, by virtue of its early production lead, remains the most popular. The system is used on several Detroit truck and marine engines, including the new unit-injected Series 55 inline six.

Describing this system puts us deeper into the alphabetical soup. As shown in Fig. 13-15, the electronic control module (ECM) integrates a top dead center ("synch") reference, throttle position input, turboboost pressure, oil temperature and pressure, and coolant level to provide a command pulse to the electric distributor unit (EDU).

The EDU functions as a relay, making and breaking the high-amperage injector-control circuit in response to much weaker signals from the ECM. The distributor module also protects the engine by denying (or, under a program option, limiting) fuel in event of low oil pressure, high oil temperature, or low coolant level.

Caterpillar The way Cat engineers phrase it, three rhyming acronyms sum up the history of diesel fuel system development: MUI, EUI, and HEUI. In 1994, electronic unit injectors (EUI) replaced the mechanical injectors (MUI) formerly used on the 3406 truck engine. Earlier versions of the EUI were used with the 3500 and 3176 engine families. Caterpillar-developed hydraulic electronic unit injectors ("Hewy") run in a competitor's engine.

Figure 13-16 illustrates the current HEUI in cross-section. A roller-tipped rocker-arm follower provides the 28,000 psi injection pressure. The solenoid valve, shown on the left side of the injector, opens the spill/fill port to admit fuel into the injector barrel during the period of injector-plunger retraction. A spring-loaded check valve, located near the injector tip and set to open at 5500 psi, remains closed during the low-pressure fill process.

Rotation of the actuating cam depresses the injector plunger. Injection can be initiated any time after the plunger starts its downward travel. But until the ECM signals the solenoid valve to close the spill/fill port, fuel merely cycles through the EUI.

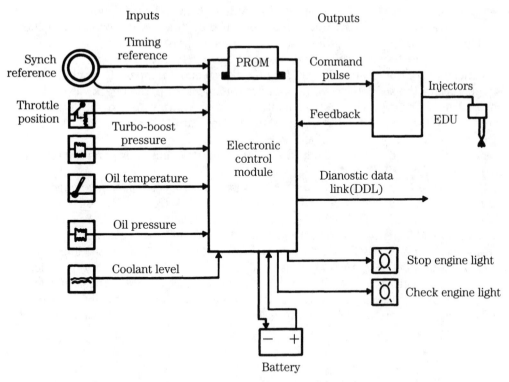

Inputs   Outputs

Synch
reference

Timing
reference

PROM

Command
pulse

Injectors

Throttle
position

Feedback

EDU

Turbo-boost
pressure

Electronic
control
module

Oil temperature

Dianostic data
link(DDL)

Oil pressure

Coolant level

Stop engine light

Check engine light

Battery

13-15 DDEC in block diagram.

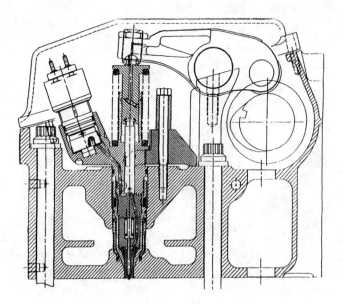

13-16 The HEUI injector in cross-section. A computer-driven solenoid regulates injection timing and fuel quantity.

Upon signal from the electronic control module (ECM), the solenoid valve closes the port, trapping fuel in the injector barrel. Further downward movement of the plunger raises fuel pressure sufficiently to overcome the spring tension against the check valve. Injection into the cylinder commences.

Injection continues until the ECM signals the solenoid to open the spill/fill port. Responding to the sudden loss of pressure at the injector tip, the check valve snaps shut. Injection ceases. The plunger continues on its downward stroke, displacing fuel through the open spill/fill port and into the fuel manifold for return to the tank. This flow of fuel helps cool the injector.

Note that EUI regulates timing (the point of injection initiation) and the rate of fuel delivery (injection duration).

Cat 3116, 3176, 3406E, and 3500 engines use similar electronic control modules, designed and manufactured by TRW's Transportation Electronics Division. Modules incorporate a number of advanced features, such as dual microprocessors and flash memory. PROM chips are no more; armed with a PC and the correct passwords, a technician can program the ECM to reflect operating conditions and engine modifications.

The ECM mounts on the engine (Fig. 13-17), rather than inside of the vehicle cab, as is the usual practice. On-block mounting reduces wiring harness interference, simplifies packaging, and enables the modules to be cooled by fuel flow. A major engineering effort was required to isolate the electronics from the engine bay environment.

13-17 The fuel-cooled electronic control module mounts on the left side of Cat 3406E engines.

Figure 13-18 is a block diagram of the Caterpillar system, which underscores its family resemblance to other engine management systems. Inputs include coolant temperature, MAP, MAT, fuel temperature, and oil pressure. The timing signal is magnetically picked off a camshaft-driven gear that, by virtue of the shape of one of its teeth, also references tdc.

One of the most notable characteristics of the Cat system is the way its sensing capabilities have been extended to include a variety of vehicle functions. A 40-pin connector enables the ECM to communicate directly with electronic transmissions, anti-lock braking systems, and traction control systems. Integrated cruise control, programmed maintenance for both the engine and vehicle, and "drive-by-wire" throttle control are provided.

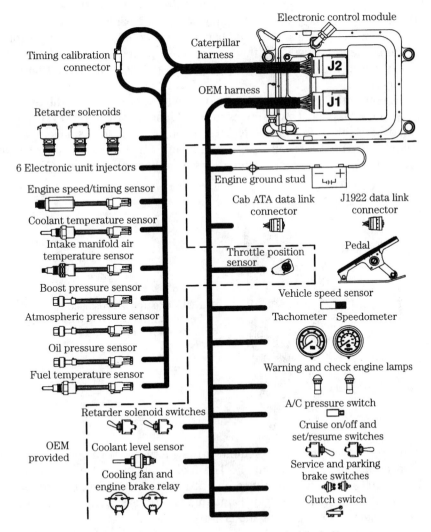

13-18 1994-3406E Cat engine management system in block diagram.

Navistar The electronic control system developed for the new 7.3L T 444E direct-injection engine is built around a variant of the EEC-IV control module, used on Ford automobiles and light trucks since 1982.

Several interesting features are incorporated, including a provision for variable exhaust back pressure. During cold starts, the ECU causes an oil-actuated butterfly valve at the turbocharger outlet to close and partially block the exhaust. The restriction increases the pumping work of the engine and fosters more rapid warm-up. The computer varies the degree of restriction with torque demand, rate of accelerator pedal travel, and other variables to make the device transparent to the operator.

The T 444E is the first and, to date, only commercial application of Caterpillar-designed and fabricated HEUI (hydraulic electronic unit injector) technology. Both Caterpillar and Navistar describe this technology as "revolutionary."

The purpose of the exercise was to substitute hydraulic actuation for the mechanical actuation normally used. Not only would this simplify engine design by eliminating the injector cam and related hardware, it would give global control of all injection parameters. The working fluid, in this case engine lube oil, can be metered by solenoids slaved to the ECM. Timing, injection pressure, and duration become software functions.

HEUI is conceptually simple, as it would have to be to survive in the diesel environment. An internal poppet valve opens to admit lube oil, which reacts against a relatively large piston in the injector body. This piston drives a much smaller piston (for force multiplication) that pressurizes fuel for injection through a conventional multiport nozzle. Fuel is injected at approximately seven times oil pressure. (Fig. 13-19.)

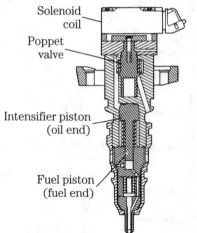

Solenoid coil

Poppet valve

Intensifier piston (oil end)

Fuel piston (fuel end)

13-19
HEUI injector in cross-section. Force multiplication occurs between the oil and fuel ends of the intensifier piston.

The poppet-valve-actuating solenoid draws more electrical power than ECM output processing circuitry can supply. The fuel command pulse goes to an electronic driver unit, where it is amplified and distributed to the appropriate injector. Thus, the latest wear in diesel engines amounts to a transistorized ignition distributor. Rudolf Diesel must be spinning in his grave.

Oil for the injector power circuit is drawn from the main gallery and pressurized by a Rexroth swash-plate pump to between 210 and 1900 psi. The oil is stored in integral manifolds, cast into the cylinder heads. These manifolds function as hydraulic accumulators, storing oil under pressure for starting and dampening the pressure waves created when the injector poppets open and snap closed. A relief valve, under control of the ECU, maintains oil pressure (and its derivative (fuel injection pressure) at the desired levels.

In addition to meeting emissions requirements for 1994 and beyond, flexible injection control enabled Navistar engineers to adjust torque curves to driver expectations. As configured for large trucks, the T 444E builds torque early and hangs on late, in the classic diesel manner. The pickup-truck version of the engine exhibits the peaky torque characteristics expected by drivers more accustomed to gasoline engines.

Index

About the author

Paul Dempsey is the senior consultant for Engineering Management Consultants, a Houston-based consulting firm, and the author of more than 20 technical books. He is a member of Instrument Society of America, International Maintenance Institute, American Petroleum Institute, Automotive Engine Rebuilders Association, International Association of Drilling Contractors, and a standing committee member for the Offfshore Marine Services Association.

Dempsey is the author of more than 20 TAB/McGraw-Hill books, two of which sold more than 100,000 copies. Dempsey's broad familiarily with mechanical devices, combined with his skill and knowledge of writing technical manuscripts, has prompted him to write more than 100 magazine and journal articles dealing with teaching techniques, petroleum-related subjects, and maintenance management.

Other Bestsellers of Related Interest

Small Gas Engine Repair—2nd Edition
Paul Dempsey
With *Small Gas Engine Repair*, do-it-yourselfers can fix any small gas-powered engine on the spot and save hundreds of dollars in technician's fees. The book is also a great source of troubleshooting and preventative maintenance techniques. It contains new sections on Japanese engines, and new starter and ignition systems. Enhanced illustrations and lots of new material are added to this second edition bestseller.
ISBN 0-07-016342-1 $11.95 Paperback

How to Repair Briggs & Stratton Engines—3rd Edition
Paul Dempsey
Dempsey delivers an updated, practical, step-by-step guide to repairing Briggs & Stratton engines. This edition includes the latest information on both new and old engines, as well as a new section on engine components and types. There is comprehensive information on torque specifications, running clearances, emissions control, and part numbers of redesigned and updated components that can be retrofitted to older engines.
ISBN 0-07-016347-2 $13.95 Paperback

Internal Combustion Engines
V. Ganesan
Engineers in a wide range of industries can turn to this expert guide for a detailed treatment of the principles and applications of internal combustion engines. The book covers modern-day environmental and fuel economy standards as well as the latest methods of measurement and testing. This essential reference weaves recent research developments into discussions of processes such as combustion, flame propagation, engine heat transfer, and scavenging and engine emissions - using SI units throughout.
ISBN 0-07-462122-X $65.00 Hardcover

**Marine Diesel Engines: Maintenance, Troubleshooting and Repair
—2nd Edition**

Nigel Calder

Since its publication in 1987, Nigel Calder's *Marine Diesel Engines* has been the most respected and best-selling book ever written to guide the owners of marine diesels through the labyrinth of maintenance and repair. Now completely revised and expanded to cover the larger turbo-charged diesels found aboard cruisers and sportfishermen, the second edition is even more comprehensive and easier to understand, with all-new troubleshooting and maintenance charts, 50 new illustrations, and almost 100 more pages.

ISBN 0-07-009612-0 $24.95 Hardcover